INTRODUCTION TO GENERAL, ORGANIC, AND BIOCHEMISTRY IN THE LABORATORY

SEVENTH EDITION

Morris Hein
Mount San Antonio College

Leo R. Best
Mount San Antonio College

Robert L. Miner
Mount San Antonio College

James M. Ritchey
California State University at Sacramento

BROOKS/COLE
™
THOMSON LEARNING

Australia · Canada · Mexico · Singapore · Spain · United Kingdom · United States

BROOKS/COLE

THOMSON LEARNING

Sponsoring Editor: *Marcus Boggs*
Marketing Manager: *Tom Ziolkowski*
Editorial Assistant: *Emily Levitan*
Production Coordinator: *Stephanie Andersen*
Production Service: *Scratchgravel Publishing Services*

Proofreader: *Kristen Cassereau*
Permissions Editor: *Sue Ewing*
Cover Design: *Vernon T. Boes*
Cover Photo: *Ken Eward / BioGrafx*
Print Buyer: *Kristine Waller*
Printing and Binding: *Webcom Ltd.*

For more information, contact:

BROOKS/COLE PUBLISHING
511 Forest Lodge Road
Pacific Grove, CA 93950 USA
www.brookscole.com
1-800-423-0563 (Thomson Learning Academic Resource Center)

For permission to use material from this work, contact us by
www.thomsonrights.com
fax: 1-800-730-2215
phone: 1-800-730-2214

Printed in Canada

10   9   8   7   6   5   4   3   2   1

ISBN 0-534-38062-X

Contents

STUDY AIDS

EXERCISES

APPENDICES

Preface

This manual is intended for the student who has not had a course in chemistry. The experiments are designed to be challenging but understandable to the student. Experimentation begins with simple laboratory techniques and measurements and progresses to relatively complex procedures. The 42 experiments are graded in difficulty to keep pace with the expanding capability of the student. The number and variety of experiments allow the instructor reasonable flexibility in preparing a laboratory schedule that expands and supports many lecture topics in a one- or two-semester preparatory college chemistry course.

Our major objectives of this flexible laboratory program are to provide experience with (1) hands-on laboratory experimentation, (2) the capabilities and limitations of measurements, (3) a variety of chemical reactions and the equations used to describe them, (4) the collection, analysis, and graphing of data, (5) responsible disposal of chemicals for personal and environmental health, (6) using a computer for graphing of data, (7) drawing valid conclusions from experimental evidence, and (8) support and reinforcement of concepts introduced in the lecture component of the course.

We have tried to establish a balance between descriptive and quantitative experiments. Seven of the experiments are new in this tenth edition of the manual. Ten experiments include unknowns for student analysis, and nine provide opportunities for graphing data.

The format is designed to be helpful and convenient for both student and instructor and includes the following features:

1. A concise discussion of the basic underlying principles for each experiment provides pertinent background material to supplement, not replace, the textbook.

2. Experimental procedures have been extensively tested by many students and provide enough detail for students to work with only general supervision.

3. Report forms for each experiment are cross-referenced to letters and subtitles in the procedure, designed to be completed before leaving the lab session, and relatively easy to grade.

4. The names and formulas of reagents used are listed at the beginning of each experiment.

5. Special safety precautions and waste disposal instructions are indicated when necessary at the point where they are required within the procedure.

6. For the convenience of the instructor and stockroom personnel, the appendices provide (a) an experiment-by-experiment list of special equipment and preparations needed, (b) a list of suggested equipment for student lockers, (c) an experiment-by-experiment list of waste disposal instructions, (d) a list of suggested auxiliary equipment, and (e) a complete list of reagents and details for the preparation of solutions.

7. Six Study Aids provide supplementary material common to several experiments on the important topics of (a) significant figures, (b) chemical formulas and equations, (c) reading and preparing graphs by hand and by computer, (d) use of a scientific calculator, (e) the mole in chemical calculations, and (f) introduction to organic chemistry.

8. The lab manual also contains 26 Exercises, many of which can be used as supplements for a number of experiments. Exercises 25 and 26 (Molecular Models and Isomerism, and Stereoisomerism—Optical Activity) may be used as experiments to give students hands-on experience in these subjects.

We are especially indebted to Dr. Judith N. Peisen and her students at Hagerstown Community College for the development and testing of six of the new experiments. Suggestions and comments from instructors and students are always welcome.

M. Hein
L. R. Best
R. L. Miner
J. M. Ritchey

To the Student

Since your laboratory time is limited, it is important to come to each session prepared by at least one hour of detailed study of the scheduled experiment. This should be considered a standing homework assignment.

Each of the experiments in this manual is composed of four parts:

1. **Materials and Equipment**—a list that includes the formulas of all compounds used.

2. **Discussion**—a brief discussion of the principles underlying the experiment.

3. **Procedure**—detailed directions for performing the experiment with safety precautions clearly noted and disposal procedures for chemical waste provided throughout and identified by a waste icon.

4. **Report for Experiment**—a form for recording data and observations, performing calculations, and answering questions.

Follow the directions in the procedure carefully, and consult your instructor if you have any questions. For convenience, the letters and subtitles in the report form have been set up to correspond with those in the procedure section of each experiment.

As you make your observations and obtain your data, record them on the report form. Try to use your time efficiently; when a reaction or process is occurring that takes considerable time and requires little watching, start working on other parts of the experiment, perform calculations, answer questions on the report form, or clean up your equipment.

Except when your instructor directs otherwise, you should do all the work individually. You may profit by discussing experimental results with your classmates, but in the final analysis you must rely on your own judgment in completing the report form.

⚠ Safety Guidelines

While in the chemistry laboratory, you are responsible not only for your own safety but for the safety of everyone else. *We have included safety precautions in every experiment where needed, and they are highlighted with the icon shown in the title of this section.* Your instructor may modify these instructions and give you more specific directions on safety in your laboratory. If the proper precautions and techniques are used, none of the experiments in this laboratory program are hazardous. But without your reading and following the instructions, without knowledge about handling and disposal of chemicals, and without the use of common sense at all times, accidents can happen. Even when everyone is doing his or her best to comply with the safety guidelines in each experiment, accidents can happen. It is your responsibility to minimize these accidents and know what to do if they happen.

Laboratory Rules and Safety Procedures

1. **Wear protective goggles or glasses** at all times in the laboratory work area. These glasses should wrap around the face so liquids cannot splash into the eye from the side. These goggles are mandated by eye-protection laws and are not optional, even though they may be uncomfortable. Contact lenses increase the risk of problems with eye safety, even when protective goggles are worn. If you wear contact lenses, inform the instructor.

2. **Dress appropriately** for the laboratory. Shoes that do not completely cover the foot are not allowed *(no sandals)*. Long hair should be tied back. Wear a laboratory coat or apron, if available, to protect your clothing.

3. **Keep your benchtop organized as you work.** Put jackets, book bags, and personal belongings away from the work areas. Before you leave, clean your work area and make sure the gas and water are turned off. Clean and return all glassware and equipment to your drawer or the lab bench where you found it.

4. **Keep all stock bottles of solid and liquid reagents in the dispensing area.** Do not bring reagent bottles to your laboratory work area. Use test tubes, beakers, or weigh boats to obtain chemicals from the dispensing areas: (1) the reagent shelf, (2) the balance tables, (3) under the fume hood, and (4) as instructed.

5. **Keep the balance and the area around it clean.** Do not place chemicals directly on the balance pans; place a piece of weighing paper, a weigh boat, or another small container on the pan first, and then weigh your material. Never weigh an object while it is hot.

6. **Check the labels on every reagent bottle carefully.** Many names and formulas appear similar at first glance. Label every beaker, test tube, etc., into which you transfer chemicals. Many labels will contain the National Fire Protection Association (NFPA) diamond label, which provides information about the flammability, reactivity, health effects, and miscellaneous effects for the substance. Each hazard is rated 0 (least hazardous) to 4 (most hazardous). For example, the NFPA label for potassium chromate used in Experiment 1 is shown below.

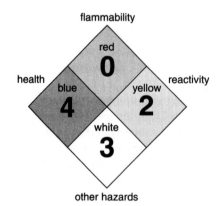

More specific information (the reason for potassium chromate being rated an extreme health hazard, for example) about all known substances is available in the form of Material Safety Data Sheets (MSDS), which many institutions keep on file for chemicals stored and used in their laboratories. MSDSs are usually provided with chemicals by the supplier when they are purchased and are easily obtained from many website sources.

7. **Never return unused chemicals to the reagent bottles.** This is a source of possible contamination of the entire stock bottle. Dispose of unused chemicals exactly as instructed in the waste disposal instructions for that substance, identified by [WASTE DISPOSE OF PROPERLY] throughout each experiment.

8. **Disposal of wastes must follow state and federal guidelines.** Do not put anything into the trash or sink without thinking first. We have tried to anticipate every disposal decision in the procedure and marked the procedure with the waste icon. The following guidelines are the foundation of waste disposal decisions:

   a. Broken glass is put into a clearly marked special container.

   b. Organic solvents are never poured into the sink. They are usually flammable and often immiscible in water. Instead, they are poured into a specially marked container ("waste organic solvents") provided when needed.

   c. Solutions containing cations and anions considered toxic by the EPA are never poured into the sink. They are poured into specially marked containers ("waste heavy metal," etc.) provided when needed. The name of all ions disposed of into a specific bottle must be listed on the label.

   d. Solutions poured in the sink should be washed down with plenty of water.

   e. Some solid chemicals must also be disposed of in specially labeled containers. If you are not sure what to do, ask the instructor.

9. **Avoid contaminating stock solutions.** Do not insert medicine droppers or pipets into reagent bottles containing liquids. Instead, pour a little of the liquid into a small beaker or test tube. If the bottle is fitted with a special pipet that is stored with the bottle, this may not be necessary.

10. **Avoid all direct contact with chemicals.**

    a. Wash your hands anytime you get chemicals on them and at the end of the laboratory session.

    b. If you spill something, clean it up immediately before it dries or gets on your papers or skin.

    c. **Never** pipet by mouth.

    d. **Never** eat, drink, or smoke in the laboratory.

    e. Do not look down into the open end of test tube in which a reaction is being conducted, and do not point the open end of a test tube at someone else.

    f. Inhale odors and chemicals with great caution. Waft vapors toward your nose. The fume hood will be used for all irritating and toxic vapors.

11. **Working with glass requires special precautions:**

    a. Do not heat graduated cylinders, burets, pipets, or bottles with a burner flame.

    b. Do not hold a test tube or beaker in your hand during a chemical reaction.

c. Do not touch glass that has been near a flame or hot plate. Hot glass looks the same as cool glass and will cause serious burns.

d. Learn and practice proper procedures when inserting glass tubing into rubber stoppers. (See Experiment 1)

12. **Learn the location and proper use of safety equipment:** fire extinguisher, eye wash, first aid kit, fire blanket, safety shower, spill kits, and other equipment available.

13. **Never work alone** in the laboratory area.

14. **Report all accidents** to the instructor, no matter how minor.

15. **Do not perform unauthorized experiments.**

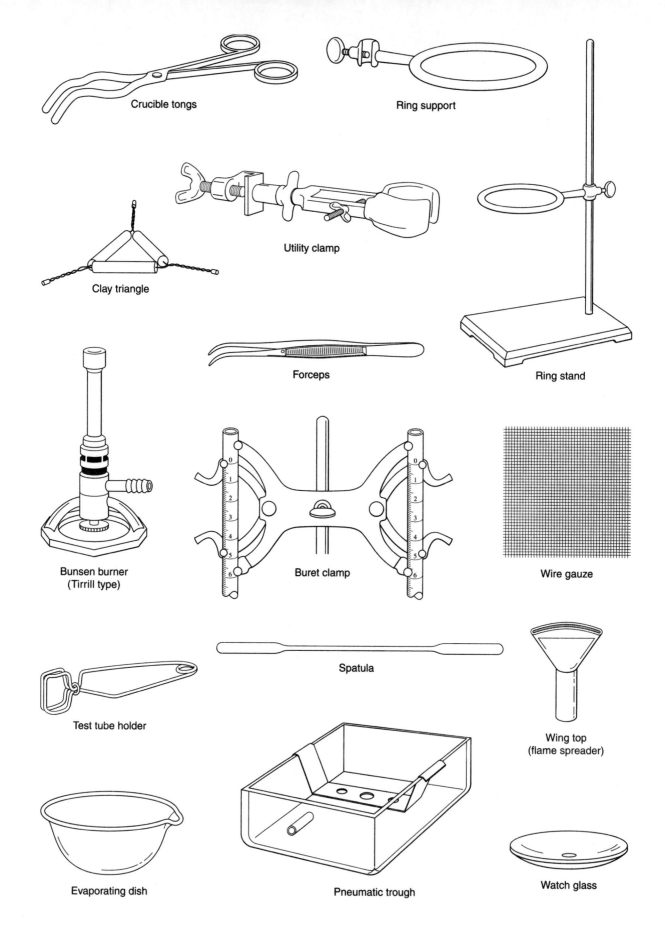

Crucible tongs

Ring support

Clay triangle

Utility clamp

Ring stand

Forceps

Bunsen burner
(Tirrill type)

Buret clamp

Wire gauze

Test tube holder

Spatula

Wing top
(flame spreader)

Evaporating dish

Pneumatic trough

Watch glass

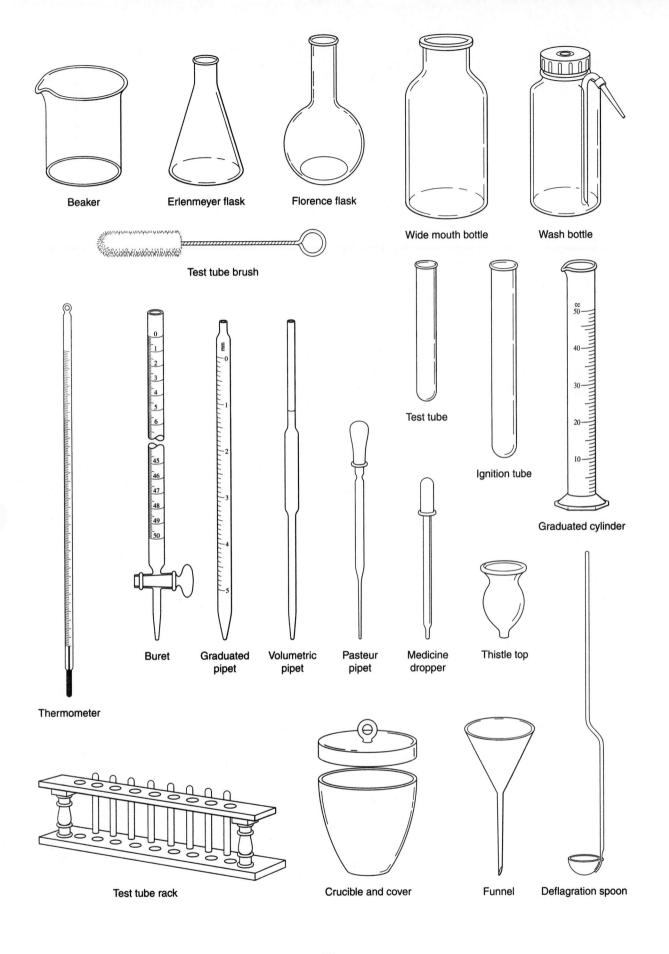

Beaker

Erlenmeyer flask

Florence flask

Wide mouth bottle

Wash bottle

Test tube brush

Test tube

Ignition tube

Graduated cylinder

Thermometer

Buret

Graduated pipet

Volumetric pipet

Pasteur pipet

Medicine dropper

Thistle top

Test tube rack

Crucible and cover

Funnel

Deflagration spoon

EXPERIMENT 1

Laboratory Techniques

MATERIALS AND EQUIPMENT

Solids: lead(II) chromate ($PbCrO_4$), potassium nitrate (KNO_3), and sodium chloride ($NaCl$). **Liquid:** glycerol. **Solutions:** 0.1 M lead(II) nitrate [$Pb(NO_3)_2$] and 0.1 M potassium chromate (K_2CrO_4). Ceramfab pad, 100 mL and 400 mL beakers, Bunsen burner, No. 1 evaporating dish, triangular file, funnel, wire gauze, filter paper, glass rod, clay triangle, 6 mm glass tubing, wing top (flame spreader).

DISCUSSION AND PROCEDURE

Wear protective glasses.

A. Laboratory Burners

Almost all laboratory burners used today are modifications of a design by the German chemist Robert Bunsen. In Bunsen's fundamental design, also widely used in domestic and industrial gas burners, gas and air are premixed by admitting the gas at relatively high velocity from a jet in the base of the burner. This rapidly moving stream of gas causes air to be drawn into the barrel from side ports and to mix with the gas before entering the combustion zone at the top of the burner.

The burner is connected to a gas cock by a short length of rubber or plastic tubing. With some burners the gas cock is turned to the **fully on** position when the burner is in use, and the amount of gas admitted to the burner is controlled by adjusting a needle valve in the base of the burner. In burners that do not have this needle valve, the gas flow is regulated by partly opening or closing the gas cock. With either type of burner **the gas should always be turned off at the gas cock when the burner is not in use** (to avoid possible dangerous leakage).

1. **Operation of the Burner.**   Examine the construction of your burner (Figure 1.1) and familiarize yourself with its operation. A burner is usually lighted with the air inlet ports nearly closed. The ports are closed by rotating the barrel of the burner in a clockwise direction. After the gas has been turned on and lighted, the size and quality of the flame is adjusted by admitting air and regulating the flow of gas. Air is admitted by rotating the barrel; gas is regulated with the needle valve, if present, or the gas cock. Insufficient air will cause a luminous yellow, smoky flame; too much air will cause the flame to be noisy and possibly blow out. A Bunsen burner flame that is satisfactory for most purposes is shown in Figure 1.2; such a flame is said to be "nonluminous." Note that the hottest region is immediately above the bright blue cone of a well-adjusted flame.

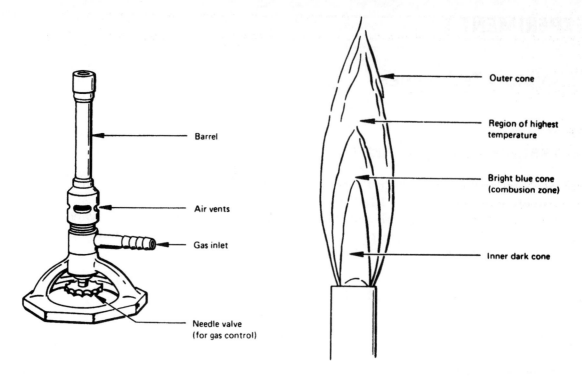

Figure 1. 1   Bunsen burner (Tirrill type)   **Figure 1.2   Bunsen burner flame**

B. Glassworking

Dispose of broken and nonusable glass in the container provided.

In laboratory work it is often necessary to fabricate simple items of equipment, making use of glass tubing and rubber stoppers. In working with glass tubing, improper techniques may result not only in an unsatisfactory apparatus but also in severe cuts and burns. Therefore the numbered instructions below should be studied carefully. Prepare the following list of items (illustrated in Figure 1.3), using 6 millimeter (mm) glass tubing and rod.

Two straight tubes, one 24 centimeters (cm) long, the other 12 cm (Figure 1.3A).

Two right-angle bends (Figure 1.3B).

One delivery tube (Figure 1.3C).

Two buret tips (Figure 1.3E). (Optional)

One stirring rod if there is none in your locker (Figure 1.3E).

This equipment will be used in future experiments. After it has been completed and approved by your instructor, store it in your locker.

1. **Cutting Glass Tubing.** (See Figure 1.4.)   Mark the tube with a pencil or ball-point pen at the point where it is to be cut. Grasp the tubing about 1 cm from the mark and hold it in position on the laboratory table. Hold the file by the tang (or handle) end and, pressing the edge of the file firmly against the glass at right angles to the tubing, make a scratch on the tubing by pushing the file away from you. If the file is in good condition a single stroke should suffice. Several strokes may be required if the file is dull, but if more than one stroke is needed, all must follow the same path so that only one scratch mark is present on the tubing. The scratch need not be very deep or very long, but it should be clearly defined.

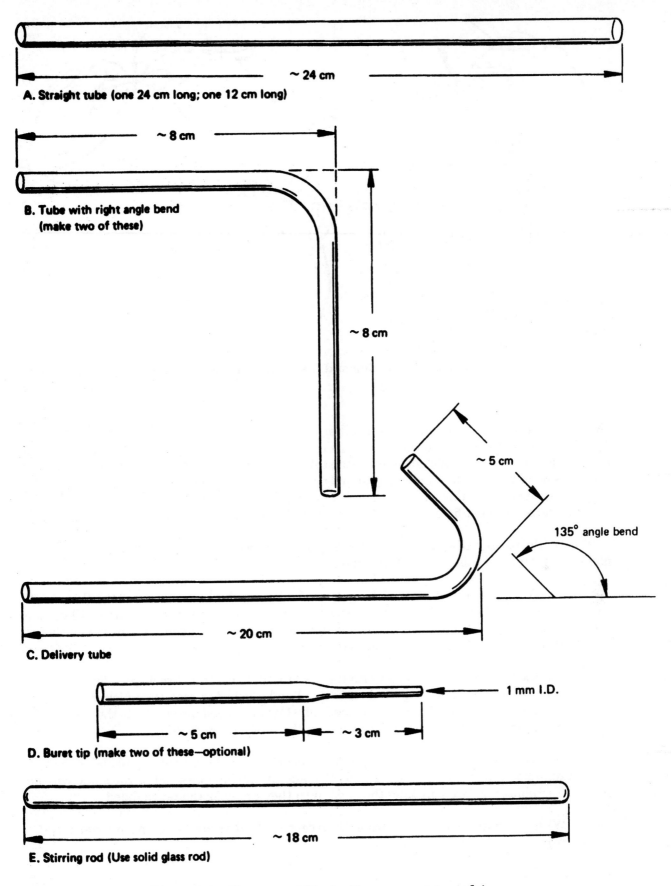

A. Straight tube (one 24 cm long; one 12 cm long)

~ 24 cm

~ 8 cm

B. Tube with right angle bend
   (make two of these)

~ 8 cm

~ 5 cm

135° angle bend

~ 20 cm

C. Delivery tube

1 mm I.D.

~ 5 cm

~ 3 cm

D. Buret tip (make two of these—optional)

~ 18 cm

E. Stirring rod (Use solid glass rod)

Figure 1.3   Glassware (Illustrations are not to scale)

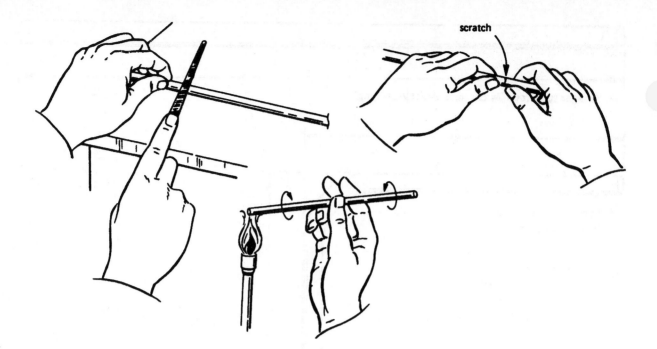

scratch

Figure 1.4   Cutting and fire-polishing glass tubing

Grasp the tubing with your thumbs together directly opposite the scratch mark (see Figure 1.4). Now apply pressure with the thumbs as though bending the ends toward your body while at the same time exerting a slight pull on the tubing. A straight, clean break should result. Use the flat side of your file to remove any sharp projections from the ends of the cut tubing. After cutting glass in this way, the ends of the cut glass, although clean and flat, are still very sharp and must be fire-polished in order to avoid personal injury.

2. **Fire-Polishing Glass.**   **Fire polishing** is the process of removing the sharp edges of glass by heating the tubing in a burner flame.

While continuously rotating the tubing, heat the end in the hottest part of the flame until the sharp edges are smooth. Be careful not to heat too much because the opening will become constricted. When the fire-polishing is completed, remember that the glass is **hot** even though it looks cool.

Put the hot glass tubing on a Ceramfab pad to cool. This is an excellent safety device. If hot objects are always placed on the pad and allowed to cool, then picked up with caution, one is less likely to get burned. The Ceramfab pad also protects the hot glass from sudden chilling (thermal shock) and the table top from injury.

Your instructor will have some examples, of properly fire-polished tubing available for your inspection. Laboratory stirring rods are easily made by cutting glass rod in the same way described for tubing and fire-polishing the ends until they are smooth and rounded.

Whenever glass is cut it must be fire-polished in order to avoid personal injury.

3. **Bending Glass Tubing.**   Put the wing top (flame spreader) on your burner and adjust the flame so that a sharply defined region of intense blue color is visible. Grasp the tubing to be bent at both ends and hold it in the flame lengthwise just above the zone of intense blue

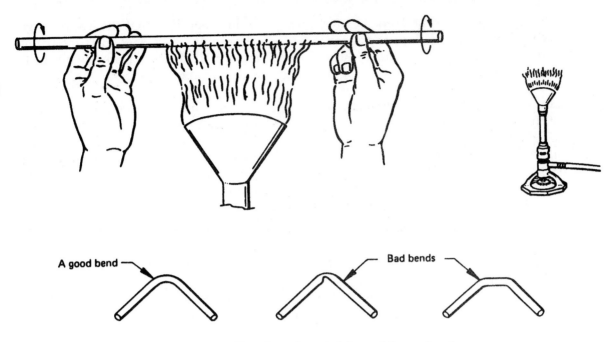

Figure 1.5   Heating glass tubing with a wing top

color. Continuously rotate the tubing in the flame until it has softened enough to bend easily (Figure 1.5). Remove the tubing from the flame, bend to the desired shape, and set aside to cool on the Ceramfab pad. If the bend is not satisfactory, discard the glass and repeat the work with a new piece.

If a bend is to be made where one arm of tubing is too short to hold in the hand while heating (Figure 1.3C), follow one of two procedures: (1) Proceed as in the above paragraph, using enough tubing to handle it from both ends; then cut to size and firepolish it after the bend is completed. (2) Heat a piece of tubing of proper size, holding it at one end and rotating it until it is soft in the region to be bent; then remove it from the flame and bend by grasping with tongs or by inserting the tang of a file into the hot end of the tubing.

4. **Preparing Buret Tips (Jets).**   A buret tip (or jet) (Figure 1.3D) is prepared as follows: Heat a small section in the center of a 14 cm length of tubing while rotating it in the hottest portion of the burner flame (without wing top) until the tubing is very soft. Remove from the flame and slowly pull the ends away from each other while holding the tubing in a vertical position. After cooling, cut the tips to the desired dimensions. Fire-polish all edges.

5. **Inserting Glass Tubing into a Stopper.**

⚠ If the glass tubing is held in the wrong place during its insertion into a stopper, serious injury can occur if the tubing shatters or breaks and pushes into the hand. Read the following instructions carefully.

Lubricate the hole in the stopper with glycerol (glycerine), using a stirring rod to make sure that the lubricant actually gets into the hole. Grasp the fire-polished tubing about 1 cm from the end to be inserted. Holding the lubricated stopper in the other hand, start the tubing into the hole by gently twisting it, and gradually work it all the way through the stopper. Be sure to grip the tubing at a point not more than 1 cm from the stopper at all times when making the insertion. Gripping at greater distances and twisting will break the tubing and probably cause personal injury. It is also good safety practice to protect your hands with a towel when inserting or removing glass tubing from rubber stoppers.

The end of the tube should protrude at least 5 mm from the stopper, so that the free passage of fluids is not prevented by flaps of rubber. After insertion make sure that the tube is not plugged, and remove the excess glycerol either by washing or with a towel. Note the lubricating properties of glycerol by rubbing a drop between your thumb and forefinger.

6. **Removing Glass Tubing from Stoppers.** Tubing should be removed from stoppers at the end of the laboratory period. Be careful to grip the tubing at a point very close to the stopper when twisting. If allowed to stand for several days the stopper may stick to the glass and be difficult to remove. In the event of sticking, do not use "strong arm" methods. A cork borer (No. 3 for 6 mm tubing) may be inserted between the rubber and glass to help remove the tubing from the stopper, but consult your instructor before attempting this procedure.

C. Evaporation

Evaporation is one of the processes used to separate a dissolved solid from a liquid.

1. Prepare the simple water bath illustrated in Figure 1.6. Before placing the beaker into position be sure the hottest part of the burner flame will reach the bottom of the beaker. Put the beaker of tap water on the wire gauze and begin heating.

2. **Preparing the Solution.** Cover the bottom of a 150 mm (standard) test tube with a small quantity of sodium chloride. Add distilled water until the test tube is about one-quarter full and stir with a glass rod until the salt is dissolved.

3. **Evaporating the Solution.** Pour the sodium chloride solution into an evaporating dish. Place the evaporating dish on the beaker of water being heated. Continue heating the water to maintain boiling—replenishing if necessary—until all of the water has evaporated from the solution in the evaporating dish, leaving the original solid as the residue.

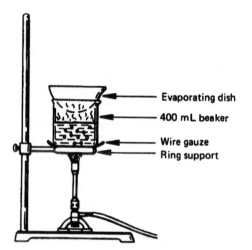

4. Dissolve the residue with tap water and flush the salt solution down the sink.

Figure 1.6   Evaporation on a simple water bath

D. Filtration

The process of separating suspended insoluble solids from liquids by means of filters is called **filtration.** Insoluble solids, called **precipitates,** are formed during some chemical reactions. In the laboratory these precipitates are generally separated from the solutions by filtering them out on a paper filter. The liquid that passes through the filter paper is the **filtrate;** the solid precipitate remaining on the filter paper is the **residue.**

1. **Forming a Precipitate.** Fill a test tube about one-half full of lead(II) nitrate solution. Fill a second tube about one-quarter full of potassium chromate solution. These solutions contain lead(II) nitrate and potassium chromate, each dissolved in water. Pour the lead(II) nitrate

solution into a 100 milliliter (mL) beaker. Slowly pour the potassium chromate solution into the beaker, stir, and observe the results. The chemical reaction that occurred formed potassium nitrate and lead(II) chromate. One of these products is a yellow precipitate.

2. **Filtering the Products.**   Prepare a filter as shown in Figure 1.7.

(a)   Fold a circle of filter paper in half. Fold in half again and open out into a cone. Tear off one corner of the outside folded edge. The top edge of the cone which is to touch the glass funnel should not be torn.

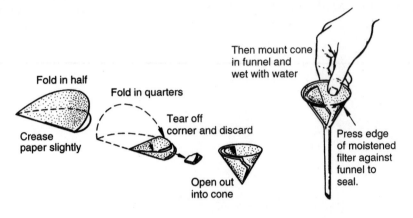

Figure 1.7   Folding and mounting filter paper

(b)   Fit the opened cone into a short-stemmed funnel, placing the torn edge next to the glass. Wet with distilled water and press the top edge of the paper against the funnel, forming a seal. Use one of the setups suggested in Figure 1.8 for supporting the funnel. Then, stir the mixture of products in the small beaker with a stirring rod and slowly pour it down the stirring rod into the filter paper in the funnel (see Figure 1.9). Do not to overfill the paper filter cone.

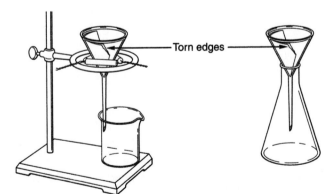

**Figure 1.8   Support the filter with a ring stand
or an Erlenmeyer flask**

**Figure 1.9   Pouring a solution
down a stirring rod**

3. **Identification of the Precipitate.** After the filtration is completed, compare the residue with the samples of solid potassium nitrate and lead(II) chromate provided by the instructor to determine which of these is the residue on the filter paper.

4. Use forceps to remove the filter paper with the precipitate and transfer it into the waste jar provided. Pour the filtrate into the waste bottle provided. Rinse the reaction beaker and the test tubes with water and pour the rinse solutions into the waste bottle.

WASTE
DISPOSE OF
PROPERLY

5. Review the safety procedures for chemical waste disposal in the preface to determine the specific reasons for putting the waste into special containers rather than the trash and the sink.

REPORT FOR EXPERIMENT 1

Laboratory Techniques

A, B. Laboratory Burners and Glassworking

Glassware shown in Figure 1.3

Articles	Instructor's Check and Comments
Straight tubes (2)	
Right-angle bends (2)	
Delivery tube (1)	
Buret tips (2)	(optional)
Stirring rod (1)	

Instructor's OK or grade on glass work _____

QUESTIONS AND PROBLEMS

1. Why is it necessary to turn off the gas with the gas cock rather than with the valve on the burner?

2. Why is air mixed with gas in the barrel of the burner before the gas is burned?

3. How would you adjust a burner which

    (a) has a yellow and smoky flame?

    (b) is noisy with a tendency to blow itself out?

4. Why are glass tubes and rods always fire-polished after cutting?

5. Explain briefly how to insert glass tubing into a rubber stopper.

6. Name the lubricant used for inserting glass tubing in rubber stoppers.

_____

C. Evaporation

Give the name and formula of the residue remaining after evaporation:

Name _____ Formula _____

D. Filtration

1. What is the name, formula, and color of the precipitate recovered by filtration?

Name _____ Formula _____ Color _____

2. Explain why the filter paper with the precipitate is collected in a jar instead of thrown into the trash can? (Refer to the section on waste disposal in Laboratory Rules and Safety Practices.)

3. Give the names and formulas of two compounds that must be present in the filtrate.

Name _____ Formula _____

Name _____ Formula _____

EXPERIMENT 2

Measurements

MATERIALS AND EQUIPMENT

Solids: sodium chloride (NaCl) and ice. Balance, ruler, thermometer, solid object for density determination, No. 1 or 2 solid rubber stopper.

DISCUSSION

Chemistry is an experimental science, and measurements are fundamental to most of the experiments. It is important to learn how to make and use these measurements properly.

The SI System of Units

The International System of Units (*Systeme Internationale*) or metric system is a decimal system of units for measurements used almost exclusively in science. It is built around a set of units including the meter, the gram, and the liter and uses factors of 10 to express larger or smaller multiples of these units. To express larger or smaller units, prefixes are added to the names of the units. Deci, centi, and milli are units that are 1/10, 1/100, and 1/1000, respectively, of these units. The most common of these prefixes with their corresponding values expressed as decimals and powers of 10 are shown in the table below.

Prefix	Decimal Equivalent	Power of 10
Deci	0.1	10^{-1}
Centi	0.01	10^{-2}
Milli	0.001	10^{-3}
Kilo	1000	10^{3}

Examples: A decigram is the equivalent of 0.1 gram, a milliliter is the equivalent of 10^{-3} liter, and a kilometer is 1000 meters. It is valuable to know that 1 milliliter (mL) equals 1 cubic centimeter (cm^3).

It will sometimes be necessary to convert from the American system to the metric system or vice versa. Many conversion factors are available and the following are common examples. A more complete list is given in Appendix 5.

1 pound (lb) = 453.6 grams (g)
1 inch (in.) = 2.54 centimeters (cm)
1 quart (qt) = 946 milliliters (ml)

Although the SI unit of temperature is the Kelvin (K), the Celsius (or centigrade) temperature scale is commonly used in scientific work and the Fahrenheit scale is commonly used in this country. On the Celsius scale the freezing point of water is designated 0°C, the boiling point 100°C.

Precision and Accuracy of Measurements

Scientific measurements must be as **precise** as possible. This means that every measurement will include one uncertain or estimated digit. When making measurements we normally estimate between the smallest scale divisions on the instrument being used. Then, only the uncertain digit should vary if the measurement is repeated using the same instrument, even if it is repeated by someone else. The **accuracy** of a measurement or calculated quantity refers to its agreement with some known value. For example, we need to make two measurements, volume and mass, to determine the density of a metal. This experimental density can then be compared with the density of the metal listed in a reference such as the *Handbook of Chemistry and Physics*. High accuracy means there is good agreement between the experimental value and the known value listed in the reference. Not all measurements can be compared with a known value.

Precision and Significant Figures

When a measured value is determined to the highest precision of the measuring instrument, the digits in the measurement are called **significant digits** or **significant figures**.

Suppose we are measuring two pieces of wire, using the metric scale on a ruler that is calibrated in tenths of centimeters as shown in Figures 2.1 a and b. One end of the first wire is placed at exactly 0.0 cm and the other end falls somewhere between 6.3 cm and 6.4 cm. Since the distance between 6.3 and 6.4 is very small, it is difficult to determine the next digit exactly. One person might estimate the length of the wire as 6.34 cm and another as 6.33 cm. The estimated digit is never ignored because it tells us that the ruler can be read to the 0.01 place. This measurement therefore has three significant digits (two certain and one uncertain figure).

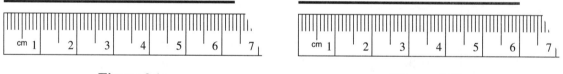

| Figure 2.1a | Figure 2.1b |

The second wire has a length which measures exactly 6 cm on the ruler as shown in Figure 2.1b. Reporting this length as 6 cm would be a mistake for it would imply that the 6 is an uncertain digit and others might record 5 or 7 as the measurement. Recording the measurement as 6.0 would also be incorrect because it implies that the 0 is uncertain and that someone else might estimate the length as 6.1 or 5.9. What we really mean is that, as closely as we can read it, the length is exactly 6 cm. So, we must write the number in such a way that it tells how precisely we can read it. In this example we can estimate to 0.01 cm so the length should be reported as 6.00 cm.

Significant Figures in Calculations

The result of multiplication, division, or other mathematical manipulation cannot be more precise than the least precise measurement used in the calculation. For instance, suppose we have an object that weighs 3.62 lb and we want to calculate the mass in grams.

$$(3.62 \text{ lb})\left(\frac{453.6 \text{ g}}{1 \text{ lb}}\right) = 1{,}642.032$$ when done by a calculator. To report 1,642.032 g as the mass is absurd, for it implies a precision far beyond that of the original measurement. Although the conversion factor has four significant figures, the mass in pounds has only three significant figures. Therefore the answer should have only three significant figures; that is, 1,640 g. In this case the zero cannot be considered significant. This value can be more properly expressed as 1.64×10^3 g. For a more comprehensive discussion of significant figures see Study Aid 1.

Precise Quantities versus Approximate Quantities

In conducting an experiment it is often unnecessary to measure an exact quantity of material. For instance, the directions might state, "Weigh about 2 g of sodium sulfite." This instruction indicates that the measured quantity of salt should be 2 g plus or minus a small quantity. In this example 1.8 to 2.2 g will satisfy these requirements. To weigh exactly 2.00 g or 2.000 g wastes time since the directions call for approximately 2 g.

Sometimes it is necessary to measure an amount of material precisely within a stated quantity range. Suppose the directions read, "weigh about 2 g of sodium sulfite to the nearest 0.001 g." This instruction does not imply that the amount is 2.000 g but that it should be between 1.8 and 2.2 g and measured and recorded to three decimal places. Therefore, four different students might weigh their samples and obtain 2.141 g, 2.034 g, 1.812 g, and 1.937 g, respectively, and each would have satisfactorily followed the directions.

Temperature

The simple act of measuring a temperature with a thermometer can easily involve errors. Not only does the calibration of the scale on the thermometer limit the precision of the measurement, but the improper placement of the thermometer bulb in the material being measured introduces a common source of human error. When measuring the temperature of a liquid, one can minimize this type of error by observing the following procedures:

1. Hold the thermometer away from the walls of the container.

2. Allow sufficient time for the thermometer to reach equilibrium with the liquid.

3. Be sure the liquid is adequately mixed.

When converting from degrees Celsius to Fahrenheit or vice versa, we make use of the following formulas:

$$°C = \frac{(°F - 32)}{1.8} \quad \text{or} \quad °F = (1.8 \times °C) + 32$$

Example Problem: Convert 70.0°F to degrees Celsius:

$$°C = \left(\frac{70.0°F - 32}{1.8}\right) = \frac{38.0}{1.8} = 21.11°C \text{ rounded to } 21.1°C$$

This example shows not only how the formula is used but also a typical setup of the way chemistry problems should be written. It shows how the numbers are used, but does not show the multiplication and division, which should be worked out by calculator. The answer was changed from 21.11°C to 21.1°C because the initial temperature, 70.0°F, has only three significant figures. The 1.8 and 32 in the formulas are exact numbers and have no effect on the number of significant figures.

Mass (Weight)

The directions in this manual are written for a 0.001 gram precision balance, but all the experiments can be performed satisfactorily using a 0.01 gram or 0.0001 gram precision balance. Your instructor will give specific directions on how to use the balance, but the following precautions should be observed:

1. The balance should always be "zeroed" before anything is placed on the balance pan. On an electronic digital balance, this is done with the "tare" or "T" button. Balances without this feature should be adjusted by the instructor.

2. Never place chemicals directly on the balance pan; first place them on a weighing paper, weighing "boat", or in a container. Clean up any materials you spill on or around the balance.

3. Before moving objects on and off the pan, be sure the balance is in the "arrest" position. When you leave the balance, return the balance to the "arrest" or standby position.

4. Never try to make adjustments on a balance. If it seems out of order, tell your instructor.

Volume

Beakers and flasks are marked to indicate only approximate volumes. Volume measurements are therefore made in a graduated cylinder by reading the point on the graduated scale that coincides with the bottom of the curved surface called the **meniscus** of the liquid (Figure 2.2). Volumes measured in your graduated cylinder should be estimated and recorded to the nearest 0.1 mL.

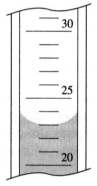

Figure 2.2 Read the bottom of the meniscus. The volume is 23.0 mL.

Density

Density is a physical property of a substance and is useful in identifying the substance. **Density** is the ratio of the mass of a substance to the volume occupied by that mass; it is the mass per unit volume and is given by the equations

$$\text{Density} = d = \frac{\text{Mass}}{\text{Volume}} = \frac{m}{V} \quad \frac{g}{mL} \quad \text{or} \quad \frac{g}{cm^3}$$

In calculating density it is important to make correct use of units and mathematical setups.

Example Problem: An object weighs 283.5 g and occupies a volume of 14.6 mL. What is its density?

$$d = \frac{m}{V} = \frac{283.5 \text{ g}}{14.6 \text{ mL}} = 19.4 \text{ g / mL}$$

Note that all the operations involved in the calculation are properly indicated and that all units are shown. If we divide grams by milliliters, we get an answer in grams per milliliter.

The volume of an irregularly shaped object is usually measured by the displacement of a liquid. An object completely submerged in a liquid displaces a volume of the liquid equal to the volume of the object.

Measurement data and calculations must always be accompanied by appropriate units.

PROCEDURE

Wear protective glasses.

Record your data on the report form as you complete each measurement.

A. Temperature

Record all temperatures to the **nearest 0.1°C.**

1. Fill a 400 mL beaker half full of tap water. Place your thermometer in the beaker. Give it a minute to reach thermal equilibrium. Keeping the thermometer in the water and holding the tip of the thermometer away from the glass, read and record the temperature.

2. Fill a 150 mL beaker half full of tap water. Set up a ring stand with the ring and wire gauze at a height so the hottest part of the burner flame will reach the bottom of the beaker. Heat the water to boiling. Read and record the temperature of the boiling water, being sure to hold the thermometer away from the bottom of the beaker.

3. Fill a 250 mL beaker one-fourth full of tap water and add a 100 mL beaker of crushed ice. Without stirring, place the thermometer in the beaker, resting it on the bottom. Wait at least 1 minute, then read and record the temperature. Now stir the mixture for about 1 minute. If almost all the ice melts, add more. Holding the thermometer off the bottom, read and record the temperature. Save the mixture for Part 4.

4. Weigh approximately 5 g of sodium chloride and add it to the ice-water mixture. Stir for 1 minute, adding more ice if needed. Read and record the temperature. Dispose of the salt water/ice mixture in the sink.

[WASTE DISPOSE OF PROPERLY]

B. Mass

Using the balance provided, do the following, recording all the masses to include one uncertain digit and all certain digits.

1. Weigh a 250 mL beaker.

2. Weigh a 125 mL Erlenmeyer flask.

3. Weigh a piece of weighing paper or a plastic weighing "boat."

4. Add approximately 2 g of sodium chloride to the weighing paper from step 3 and record the total mass. Calculate the mass of sodium chloride.

C. Length

Using a ruler, make the following measurements in both inches and centimeters; measure to the nearest 1/16th inch and to one uncertain digit in centimeters.

1. Measure the length of the arrow on the right. ⟵――――――――――――――⟶

2. Measure the external height of a 250 mL beaker.

3. Measure the length of a test tube.

D. Volume

Using the graduated cylinder most appropriate, measure the following volumes to the maximum precision possible, usually 0.1 mL. Remember to read the volume at the meniscus.

1. Fill a test tube to the brim with water and measure the volume of the water.

2. Fill a 125 mL Erlenmeyer flask to the brim with water and measure the volume of the water.

3. Measure 5.0 mL of water in a graduated cylinder and pour it into a test tube. With a ruler, measure the height (in cm) and mark the height with a marker.

4. Measure 10.0 mL of water in the graduated cylinder and pour it into a test tube like the one used in the previous step. Again, mark the height with a marker.

In the future, you will often find it convenient to estimate volumes of 5 and 10 mL simply by observing the height of the liquid in the test tube.

E. Density

Estimate and record all volumes to the highest precision, usually 0.1 mL. Make all weighing to the highest precision of the balance. Note that you must supply the units for the measurements and calculations in this section.

1. Density of Water. Weigh a clean, dry 50 mL graduated cylinder and record its mass. (Graduated cylinders should never by dried over a flame.) Fill the graduated cylinder with distilled water to 50.0 mL. Use a medicine dropper to adjust the meniscus to the 50.0 mL mark. Record the volume. Reweigh and calculate the density of water.

2. Density of a Rubber Stopper. Select a solid rubber stopper which is small enough to fit inside the 50 mL graduated cylinder. Weigh the dry stopper. Fill the 50 mL cylinder with tap water to approximately 25 mL. Read and record the exact volume. Carefully place the rubber stopper into the graduated cylinder so that it is submerged. Read and record the new volume. Calculate the volume of the rubber stopper and the density of rubber.

3. Density of a Solid Object.   Obtain a solid object from your instructor. Record the sample code on the report form. Determine the density of your solid by following the procedure given in Part 2 for the rubber stopper. To avoid the possibility of breakage, incline the graduated cylinder at an angle and slide, rather than drop, the solid into it.

Return the solid object to your instructor.

REPORT FOR EXPERIMENT 2

Measurements

A. Temperature

   1. Water at room temperature   _____ °C

   2. Boiling point   _____ °C

   3. Ice water

      Before stirring   _____ °C

      After stirring for 1 minute   _____ °C

   4. Ice water with salt added   _____ °C

B. Mass

   1. 250 mL beaker   _____ g

   2. 125 mL Erlenmeyer flask   _____ g

   3. Weighing paper or weighing boat   _____ g

   4. Mass of weighing paper/boat + sodium chloride   _____ g

      Mass of sodium chloride (show calculation setup)   _____ g

C. Length

   1. Length of ⟵——————————⟶   _____ in.   _____ cm

   2. Height of 250 mL beaker   _____ in.   _____ cm

   3. Length of test tube   _____ in.   _____ cm

D. Volume

   1. Test tube   _____ mL

   2. 125 mL Erlenmeyer flask   _____ mL

   3. Height of 5.0 mL of water in test tube   _____ cm

   4. Height of 10.0 mL of water in test tube   _____ cm

E. Density

1. Density of Water

Mass of empty graduated cylinder   _____

Volume of water   _____

Mass of graduated cylinder and water   _____

Mass of water (show calculation setup)   _____

Density of water (show calculation setup)   _____

2. Density of a Rubber Stopper

Mass of rubber stopper   _____

Initial volume in cylinder   _____

Final volume in cylinder   _____

Volume of rubber stopper (show calculation setup)   _____

Density of rubber stopper (show calculation setup)   _____

3. Density of a Solid Object

Number of solid object   _____

Mass of solid object   _____

Initial volume in graduated cylinder   _____

Final volume in graduated cylinder   _____

Volume of solid object (show calculation setup)   _____

Density of solid object (show calculation setup)   _____

QUESTIONS AND PROBLEMS

1. The directions state "weigh about 5 grams of sodium chloride". Give minimum and maximum amounts of sodium chloride that would satisfy these instructions.

2. Two students each measured the density of a quartz sample three times:

	Student A	Student B
1.	3.20 g/mL	2.82 g/mL
2.	2.58 g/mL	2.48 g/mL
3.	2.10 g/mL	2.59 g/mL
mean	2.63 g/mL	2.63 g/mL

The density found in the *Handbook of Chemistry and Physics* for quartz is 2.65 g/mL

a. Which student measured density with the greatest precision? Explain your answer.

b. Which student measured density with the greatest accuracy? Explain your answer.

Show calculation setups and answers for the following problems.

3. Convert 21°C to degrees Fahrenheit.   _____

4. Convert 101°F to degrees Celsius.   _____

5. An object is 9.6 cm long. What is the length in inches?   _____

6. An empty graduated cylinder weights 82.450 g. When filled to 50.0 mL with an unknown liquid it weighs 110.810 g. What is the density of the unknown liquid?

_____

EXPERIMENT 3

Preparation and Properties of Oxygen

MATERIALS AND EQUIPMENT

Solids: candles, magnesium (Mg) strips, manganese dioxide (MnO_2), fine steel wool (Fe), roll sulfur (S), wood splints. **Solution:** 9 percent hydrogen peroxide (H_2O_2). Deflagration spoon, pneumatic trough, 20 to 25 cm length rubber tubing, 25×200 mm ignition tube, five wide-mouth (gas-collecting) bottles, five glass cover-plates. **Demonstration supplies:** (1) cotton, sodium peroxide (Na_2O_2); (2) steel wool, 25×200 mm test tube; (3) Hoffman electrolysis apparatus.

DISCUSSION

Oxygen is the most abundant and widespread of all the elements in the earth's crust. It occurs both as free oxygen gas and combined in compounds with other elements. Free oxygen gas is diatomic and has the formula O_2. Oxygen is found combined with more elements than any other single element, and it will combine with all the elements except some of the noble gases. Water is 88.9 percent oxygen by mass, the atmosphere is about 21 percent oxygen by volume. Oxygen gas is colorless and odorless, and is only very slightly soluble in water, a property important to its collection in this experiment.

Oxygen may be obtained by decomposing a variety of oxygen-containing compounds. Some of these are mercury(II) oxide (HgO, mercuric oxide), lead(IV) oxide (PbO_2, lead dioxide), potassium chlorate ($KClO_3$), potassium nitrate (KNO_3), hydrogen peroxide (H_2O_2), and water (H_2O).

In this experiment oxygen is produced by decomposing hydrogen peroxide, and five bottles of oxygen will be collected by the downward displacement of water. After collection, some of the physical and chemical properties of oxygen will be observed.

A. Decomposition of Hydrogen Peroxide to Generate Oxygen

Hydrogen peroxide decomposes very slowly at room temperature. The rate of decomposition is greatly increased by adding a catalyst, manganese dioxide. Although manganese dioxide contains oxygen, it is not decomposed under conditions of this experiment. These equations represent the changes that occur.

Word Equation: Hydrogen peroxide $\longrightarrow$ Water + Oxygen

Formula Equation: $2\ H_2O_2(aq) \xrightarrow{\ MnO_2\ } 2\ H_2O(l) + O_2(g)$

B. Collection of Oxygen

The oxygen is collected by a method known as the downward displacement of water. The gas is conducted from a generator to a bottle of water inverted in a pneumatic trough (Figure 3.1). The oxygen, which is only very slightly soluble in the water, rises in the bottle and pushes the water down and out. Because oxygen is heavier than air, a glass plate is used to cover the opening of the bottle while it is inverted to a right-side-up position and placed on the benchtop until tested.

C. Properties of Oxygen

Like all kinds of matter, oxygen has both physical and chemical properties and you will observe both in this experiment. One outstanding and important chemical property of oxygen is its ability to support combustion. During combustion oxygen is consumed but does not burn and this ability to support combustion is one test for oxygen. Other substances (a wooden splint or a candle, for example) burn in oxygen producing a visible flame and heat. Compounds containing oxygen and one other element are known as oxides. Thus when elements such as sulfur, hydrogen, carbon, and magnesium burn in air or oxygen, they form sulfur dioxide, hydrogen oxide (water), carbon dioxide, and magnesium oxide, respectively. These chemical reactions may be represented by equations; for example:

Word Equation: Sulfur + Oxygen $\longrightarrow$ Sulfur dioxide

Formula Equation: $S(s) + O_2(g) \longrightarrow SO_2(g)$

See Study Aid 2 for discussion of writing formulas and chemical equations.

PROCEDURE

A. and B. Generation and Collection of Oxygen from Hydrogen Peroxide

⚠️ **Wear protective glasses.**
Wash hydrogen peroxide off skin with water immediately.

1. Assemble the apparatus shown in Figure 3.1. It consists of a 250 mL Erlenmeyer flask, two-hole stopper, thistle tube, glass right-angle bend (Figure 1.3B), glass delivery tube with 135 degree bend (Figure 1.3C), and a 20–25 cm length of rubber tubing. The thistle tube should be at least 24 cm (~10 in.) long and be inserted in the rubber stopper so that there is about 3 mm (1/8 in.) clearance between the end of the tube and the bottom of the flask with the stopper in place. Remember to use glycerol when inserting the glass tubing into the rubber stopper and to hold the glass tubing close to the point of insertion.

2. Fill a pneumatic trough with water until the water level is just above the removable shelf. Attach a piece of rubber tubing to the overflow spigot on the trough and put it in the sink so the water will not spill over the edges of the trough onto the counter. Completely fill five wide-mouth bottles with water. Transfer each bottle to the pneumatic trough by covering its mouth with a glass plate, inverting it, and lowering it into the water. Remove the glass plate below the water level. Place two bottles on the shelf in the trough (over the holes), leaving the other three standing for transfer to the shelf when needed.

3. Using a spatula, put a pea-sized quantity of manganese dioxide (MnO_2) in the generator flask. Replace the stopper, stabilize the flask on the ring stand with a clamp, and make sure that all glass-rubber connections are tight. Add 25 mL of water to the flask through the thistle tube. Make sure that the end of the thistle tube is covered with water (to prevent escape of oxygen gas through the thistle tube).

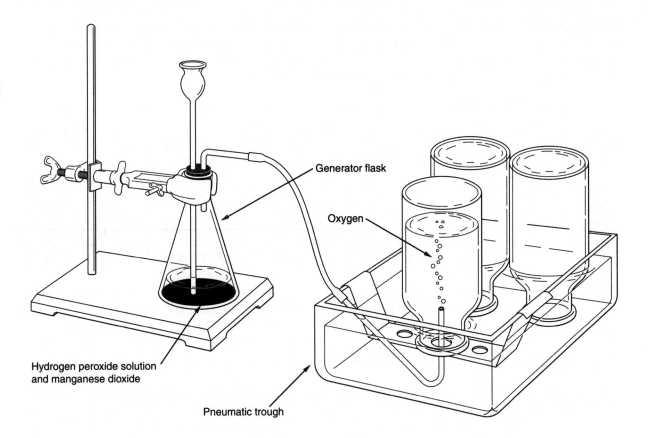

Figure 3.1 **Preparing oxygen by decomposing hydrogen peroxide**

4. Using a 50 mL graduated cylinder, measure out 50 mL of 9 percent hydrogen peroxide solution.

⚠ Reminder: If hydrogen peroxide gets on your skin, wash it off promptly with water.

To start the generation of oxygen, pour 5 to 10 mL of the peroxide solution into the thistle tube. If all the peroxide solution does not run into the generator, momentarily lift the delivery tube from the water in the trough. Immediately replace the end of the delivery tube under water and into the mouth of the first bottle to collect the gas. When one bottle is filled with gas, immediately start filling the next bottle. Continue generating oxygen by adding an additional 5 to 10 mL portion of hydrogen peroxide whenever the rate of gas production slows down markedly.

5. Cover the mouth of each gas-filled bottle with a glass cover-plate before removing it from the water. Store each bottle mouth upward without removing the glass plate; the oxygen will not readily escape since it is slightly more dense than air. Note which bottle of gas was collected first and continue until a total of five bottles of gas have been collected.

6. Allow the reaction to go to completion while you continue with the testing of the oxygen you collected. If you have any unreacted H_2O_2 remaining in the graduated cylinder, return it to the special bottle marked "9% unreacted H_2O_2." When you have completed the rest of this experiment, pour the material in the generator into the sink and flush generously with water.

C. Properties of Oxygen

Each of the following tests (except C.6) is conducted with a bottle of oxygen and, for comparison, with a bottle of air. Record your observations on the report form.

1. The **glowing splint test** is often used to verify the identity of oxygen. Ignite a wood splint, blow out the flame, and insert the still-glowing splint into the first bottle of oxygen collected. Repeat with a bottle of air. To ensure having a bottle of air, fill the bottle with water and then empty it, thus washing out other gases that may be present.

2. Take a small lump of sulfur, a deflagrating spoon, a bottle of oxygen, and a bottle of air to the fume hood. Light the burner in the fume hood and direct the flame directly into the spoon containing the sulfur. First the sulfur gets dark and melts, then it begins to burn with a blue flame that is barely visible. Lower the burning sulfur alternately into a bottle of oxygen and a bottle of air and compare combustions. Quench the excess burning sulfur in a beaker of water.

3. Stand a small candle (no longer than 5 cm) on a glass plate and light it. Lower a bottle of oxygen over the burning candle, placing the mouth of the bottle on the glass plate. Measure and record the time, in seconds, that the candle continues to burn. Repeat with a bottle of air. Note also the difference in the brilliance of the candle flame in oxygen and in air. Return the unused portion of the candle to the reagent shelf.

4. Invert a bottle of oxygen, covered with glass plate, and place it mouth to mouth over a bottle of air. Then remove the glass plate from between the bottles and allow them to stand mouth to mouth for 3 minutes. Cover each bottle with a glass plate and set the bottles down, mouths upward. Test the contents of each bottle by inserting a glowing splint.

5. Pour 25 mL of water into the fifth bottle of oxygen and replace the cover. Place the bottle close to (within 5 or 6 cm) the burner. Take a loose, 4 or 5 cm wad of steel wool (iron) in the crucible tongs and momentarily heat it in the burner flame until some of the steel wool first begins to glow. Immediately lower the glowing metal into the bottle of oxygen. (It is essential that some of the steel wool be glowing when it goes into the oxygen.) Repeat, using a bottle of air.

> NOTE: The 25 mL of water is to prevent breakage if the glowing steel wool is accidentally dropped into the bottle.

6. A small strip of magnesium ribbon will be burned next. Read the following precautions before proceeding. *Do not put burning magnesium into a bottle of oxygen.* There is enough oxygen in air for this reaction to proceed vigorously.

> Do not look directly at the burning magnesium ribbon. It is very bright and the light includes considerable ultraviolet light, which can cause damage to the retina of the eye.

Take a 2 to 5 cm strip of magnesium metal in a pair of crucible tongs and ignite it by heating it in the burner flame. After the burning is over, put the product on the Ceramfab plate and compare it to the metal from which it was produced.

C. Instructor Demonstrations (Optional)

1. **Sodium Peroxide as Source of Oxygen.** Spread some cotton on the bottom of an evaporating dish and sprinkle a small amount (less than 1 g) of fresh sodium peroxide on it. Sprinkle a few drops of water on the peroxide. Spontaneous combustion of the cotton will occur.

2. **Approximate Percentage of Oxygen in the Air.** Push a small wad of steel wool to the bottom of a 25 × 200 mm test tube. Wet the steel wool by covering with water; pour out the surplus water; and place the tube, mouth downward, in a 400 mL beaker half full of water. After the oxygen in the trapped air has reacted with the steel wool—at least three days are needed for complete reaction—adjust the water levels inside and outside the tube to the same height. Cover the mouth of the tube, remove from the beaker, and measure the volume of water in the tube. Alternatively, the height of the water column may be measured (in millimeters) without removing the tube from the beaker. The volume of water in the tube is approximately equal to the volume of oxygen originally present in the tube of air.

$$\% \text{ oxygen} = \frac{\text{Volume of water in tube}}{\text{Volume of tube}} \times 100$$

or

$$\% \text{ oxygen} = \frac{\text{Height of water column}}{\text{Length of tube}} \times 100$$

3. **Decomposition of Water.** Set up the Hoffman electrolysis apparatus, as shown in Figure 3.2. The solution used in the apparatus should contain about 2 mL of sulfuric acid per 100 mL of water. Direct current may be obtained from several 1.5 volt type A cells connected in series or from some other D.C. source.

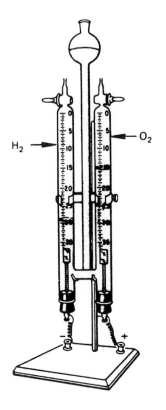

Figure 3.2   Hoffman electrolysis apparatus

NAME _____

SECTION _____ DATE _____

INSTRUCTOR _____

REPORT FOR EXPERIMENT 3

Preparation and Properties of Oxygen

A. Generation and Collection of Oxygen

1. What evidence did you observe that oxygen is not very soluble in water?

2. What is the source of oxygen in the procedure you used?

   Name _____ Formula _____

3. What purpose does the manganese dioxide serve in this preparation of oxygen?

4. What gas was in the apparatus before you started generating oxygen? Where did it go?

5. What is different about the composition of the first bottle of gas collected compared to the other four?

6. Why are the bottles of oxygen stored with the mouth up?

7. (a) What is the symbol of the element oxygen?   _____

   (b) What is the formula for oxygen gas?   _____

8. Which of the following formulas represent oxides? (Circle) MgO, $KClO_3$, SO_2, MnO_2, O_2, NaOH, PbO_2, Na_2O_2

9. Write the word and formula equations for the preparation of oxygen from hydrogen peroxide.

   Word Equation:

   Formula Equation:

10. What substances, other than oxygen, are in the generator when the decomposition of H_2O_2 is complete?

B. Properties of Oxygen

1. Write word equations for the chemical reactions that occurred. (See Study Aid 2.)

   B.1. Combustion of wood. Assume carbon is the combustible material.

   B.2. Combustion of sulfur.

   B.5. Combustion of steel wool (iron). (Call the product iron oxide.)

   B.6. Combustion of magnesium.

2. Write formula equations for these same combustions.

   B.1. (CO_2 is the formula for the oxide of carbon that is formed.)

   B.2 (SO_2 is the formula for the oxide of sulfur that is formed.)

B.5. (Fe_3O_4 is the formula for the oxide of iron that is formed.)

B.6. (MgO is the formula for the oxide of magnesium that is formed.)

3. Combustion of a candle.

   (a) Number of seconds that the candle burned in the bottle of oxygen. _____

   (b) Number of seconds that the candle burned in the bottle or air. _____

   (c) Explain this difference in combustion time.

   (d) Is it scientifically sound to conclude that all the oxygen in the bottle was reacted when the candle stopped burning? Explain.

4. What were the results of the experiment in which a bottle of oxygen was placed over a bottle of air? Explain the results.

5. (a) Describe the material that is formed when magnesium is burned in air.

   (b) What elements are in this product?

6. (a) What is your conclusion about the rate or speed of a chemical reaction with respect to the concentration of the reactants—for example, a combustion in a high concentration of oxygen (pure oxygen) compared to a combustion in a low concentration of oxygen (air)?

   (b) What evidence did you observe in the burning of sulfur to confirm your conclusion in 6(a)?

EXPERIMENT 4

Preparation and Properties of Hydrogen

MATERIALS AND EQUIPMENT

Solids: strips of copper, magnesium, and zinc; sodium metal; steel wool; mossy zinc; wood splints. **Solutions:** dilute (6 M) acetic acid ($HC_2H_3O_2$), 0.1 M copper(II) sulfate ($CuSO_4$), dilute (6 M) hydrochloric acid (HCI), dilute (3 M) phosphoric acid (H_3PO_4), 9 M sulfuric acid (H_2SO_4) and dilute (3 M) sulfuric acid (H_2SO_4), phenolphthalein solution. Pneumatic trough, five wide-mouth (gas-collecting) bottles.

DISCUSSION

Hydrogen, having atomic number 1 and atomic mass 1.008, is the simplest element. It is the ninth most abundant element in the earth's crust (about 0.9 percent by mass). At ordinary temperatures and pressures it is a gas, composed of diatomic molecules, H_2, and is only very slightly soluble in water. Hydrogen is usually found combined with other elements. Water is the most common and probably the most important compound of hydrogen. Hydrogen will not support combustion, but in the presence of oxygen it burns readily to form water:

$$2 H_2(g) + O_2(g) \longrightarrow 2 H_2O(g)$$

This reaction is used as a simple test for hydrogen, for mixtures of hydrogen and air (or oxygen) burn explosively with a distinctive "popping" or "barking" sound.

A. Methods of Preparing Hydrogen

There are several ways of producing hydrogen gas. Two methods are demonstrated in this experiment.

1. **Active Metal with Water.** Several of the most active metals—such as lithium, sodium, potassium, rubidium, cesium, and calcium—will react with cold water; magnesium will react with hot water. An example of such a reaction is

$$\text{Sodium} + \text{Water} \longrightarrow \text{Sodium hydroxide} + \text{Hydrogen}$$

$$2 Na(s) + 2 H_2O(l) \longrightarrow 2 NaOH(aq) + H_2(g)$$

Besides hydrogen, this reaction also produces sodium hydroxide, which causes the solution to become basic (alkaline). Basic solutions can be differentiated from neutral or acidic solutions by using acid-base indicators such as litmus or phenolphthalein.

2. **Active Metal with Dilute Acid.** In general, metals that are more active than hydrogen will react with dilute acids to produced hydrogen gas and a salt. For example, the salts produced in the following reactions are magnesium chloride, zinc sulfate, and zinc phosphate, respectively.

$$Mg(s) + 2\ HCl(aq) \longrightarrow MgCl_2(aq) + H_2(g)$$
$$Zn(s) + H_2SO_4(aq) \longrightarrow ZnSO_4(aq) + H_2(g)$$
$$3\ Zn(s) + 2\ H_3PO_4(aq) \longrightarrow Zn_3(PO_4)_2(aq) + 3\ H_2(g)$$

The strong oxidizing acids, such as nitric acid (HNO_3) and concentrated sulfuric acid, also react with metals but do not produce hydrogen gas.

B. Collection of Hydrogen

Hydrogen, like other gases such as oxygen, which are insoluble in water, is collected by the method known as the downward displacement of water. The gas is conducted from a generator to a bottle of water inverted in a pneumatic trough (Figure 4.1). The insoluble hydrogen rises in the bottle and pushes the water down and out. Because hydrogen gas is lighter than air, bottles containing hydrogen are stored upside down until the gas is ready to be tested.

C. Properties of Hydrogen

During this experiment you will collect several bottles of hydrogen and observe some of its physical and chemical properties. The physical properties tested are the density of hydrogen relative to air and its ability to spontaneously mix with other gases (diffusion). When observing the chemical properties of hydrogen be sure to keep in mind that there is a distinction between combustion and the ability to support combustion.

We can observe three different situations: (1) A flame inserted into a bottle of pure hydrogen will go out, because there is no oxygen to support combustion. (2) A flame inserted into a mixture of air and hydrogen will set off a very rapid combustion (burning) of the hydrogen, causing a small explosion. (3) A flame brought to the mouth of a bottle of pure hydrogen will cause the hydrogen to burn, but only at the mouth, where the hydrogen is in contact with the air. Situation (3) is the hardest to detect because the reaction is not explosive, there is little noise, and hydrogen burns with a colorless flame.

PROCEDURE

⚠ PRECAUTIONS:

1. **Wear protective glasses.**

2. Mixtures of hydrogen and air may explode when ignited. Keep the burner away from the generator tube and the delivery tube. Wrap the generator with a towel to prevent flying glass.

A. Preparing Hydrogen from Water

Record your observations on the report form.

 Do not put your head over the tube while the sodium is reacting.

1. Fill a test tube half full of water and place it in the test tube rack. Obtain a piece of sodium (no larger than a 4 mm cube) from your instructor and place it on a piece of filter paper; do not touch the sodium with your fingers. Fold the filter paper over the sodium and press out the kerosene, noting how soft the sodium metal is. (Sodium is kept in kerosene because it

reacts rapidly with water or with the oxygen in air.) Pick up the sodium with tweezers or tongs and, holding it at arm's length, drop it into the test tube.

Immediately bring a burning splint to the mouth of the test tube and observe the results. When the reaction has ceased, use a clean stirring rod to place drops of the solution on pieces of red and blue litmus paper to determine whether the solution is acidic or basic. Acids turn litmus red, bases turn litmus blue. Add a drop of phenolphthalein to the solution in the test tube. A bright pink color indicates the presence of a base.

 2. Empty the solution in the test tube into the sink and flush generously with water.

B. Preparing Hydrogen from Acids

1. **Using Various Metals.** Set up four test tubes in a rack. Place one small strip of zinc into the first tube. In like manner place samples of copper, steel wool (iron), and magnesium, in this order, into the other three tubes. In rapid succession add a few milliliters of dilute (6 M) hydrochloric acid to each test tube. Note whether gas is evolved in each case and also note its relative rate of evolution. Test any gas evolved for evidence of hydrogen by bringing a flame to the mouth of the test tube.

 Carefully pour the liquids in the test tubes into the sink keeping the metal strips in the test tubes. Dispose of the unreacted metal strips by rinsing with water and putting them into the container provided by the instructor. Waste solids are never put into the sink.

2. **Using Various Acids.** In the preceding section you found that zinc is one of the metals capable of releasing hydrogen from hydrochloric acid. In this section you will compare the relative ease with which zinc displaces hydrogen from a variety of acids. The stronger acids react at a faster rate than the weaker acids.

Set up four test tubes in a rack. Place several milliliters of dilute (6 M) hydrochloric acid into the first tube, dilute (6 M) acetic acid into the second, dilute (3 M) sulfuric acid into the third, and dilute (3 M) phosphoric acid into the fourth. Now drop a small strip of zinc into each of the four tubes. The variation in the rates of evolution of hydrogen gas is a measure of the relative strengths of the acids. Let the reactions proceed for three minutes before making your evaluation.

Carefully pour the liquids into the sink, keeping the zinc in the test tube. Rinse the zinc with water and dispose of the unreacted zinc in the container provided, not in the sink or the trash.

C. Generation and Collection of Hydrogen

1. Assemble the generator shown in Figure 4.1, using a wide-mouth bottle or a 250 mL Erlenmeyer flask, a 2-hole rubber stopper equipped with a thistle tube reaching to within 1 cm of the bottom of the bottle, and a delivery tube. Clamp the generator to the ring stand. All connections must be airtight.

2. Place approximately 10 g of mossy zinc in the bottle; add 2 mL of 0.1 M copper(II) sulfate solution (as a catalyst) and 50 mL of water and make sure that the bottom end of the thistle tube is under water. Fill four wide-mouth bottles with water and invert them in the pneumatic trough with two over the holes in the shelf. Put the delivery tube into the opening of one of the bottles.

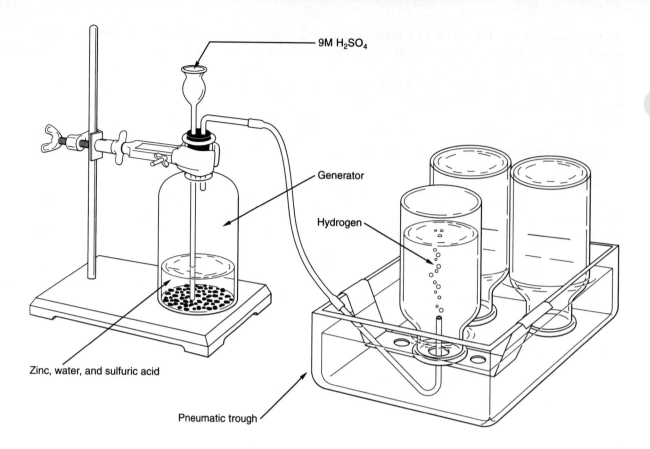

9M H₂SO₄

Generator

Hydrogen

Zinc, water, and sulfuric acid

Pneumatic trough

Figure 4.1   Preparing hydrogen from zinc and sulfuric acid

3. When you are ready to collect the gas, pour about 4–5 mL of 9 M sulfuric acid, H_2SO_4, through the thistle tube. Add more acid in 4–5 mL increments, as needed, to keep the reaction going. As each bottle becomes filled with gas, place a glass plate over its mouth while the bottle is still under water; then remove the bottle from the water and store it mouth downward without removing the glass plate.

4. Set aside—but do not discard—the first bottle of gas collected, and fill the other three with hydrogen.

5. Since no more hydrogen gas is needed, fill the generator with water to dilute the acid and quench the reaction. Open the generator, pour out the diluted acid into the sink and flush generously with water. Be careful. The leftover zinc should NOT go into the sink. Rinse the remaining zinc thoroughly with water, and return this zinc to the container provided by the instructor.

D. Properties of Hydrogen

NOTE: Unless otherwise directed, keep the bottles mouth downward while performing the following tests.

1. Raise the first bottle of gas collected a few inches straight up from the table, and immediately apply a burning splint to the mouth of the bottle.

2. Raise the second bottle straight up a few inches from the table. Then without delay slowly insert a burning splint halfway into the bottle. Continue with the insertion of the splint even though you hear a muffled report as the flame approaches the mouth of the bottle. Slowly withdraw the splint until the charred end is in the neck of the bottle; hold the splint there a few seconds, but do not withdraw it completely. Repeat inserting and withdrawing the splint several times.

3. Place the third bottle mouth upward and remove the glass plate. After one minute, bring a burning splint to its mouth.

4. Keeping the cover plate over its mouth, place the fourth bottle of hydrogen (still upside down) mouth to mouth over a bottle of air. Then remove the cover plate from between them. Let the two bottles remain in this position for three minutes, then replace the cover plate between them. Leaving the cover plate on the mouth of the lower bottle, raise the top bottle straight up, at least 6 inches, and immediately bring a burning splint to its mouth. Turn the lower bottle mouth downward, with the cover plate in place. Bring a burning splint to the mouth of this bottle as you lift it straight up. Compare the results.

NAME _____

SECTION _____ DATE _____

INSTRUCTOR _____

REPORT FOR EXPERIMENT 4

Preparation and Properties of Hydrogen

A. Preparing Hydrogen from Water

1. Describe what you observed when sodium was dropped into water.

2. Describe what you observed when a flame was brought to the mouth of the test tube.

3. (a) What color did the litmus papers turn?   _____

   (b) What color did the phenolphthalein turn?   _____

4. Did the reaction make the solution acidic or basic?   _____

5. Complete and balance the following word and formula equations:

   Sodium + Water $\longrightarrow$

   Na + H_2O $\longrightarrow$

B. Preparing Hydrogen from Acids

1. (a) Write the symbols of the metals that reacted with dilute hydrochloric acid.

   _____

   (b) Write the symbols of the four metals in order of decreasing ability to liberate hydrogen from hydrochloric acid.

   _____ _____ _____ _____

2. Write the formulas of the four acids in order of decreasing strength based on the rates at which they liberate hydrogen when reacting with zinc.

   _____ _____ _____ _____

C. Generation and Collection of Hydrogen

1. Why was water added to cover the bottom of the thistle tube?

2. What do we call this method of collecting gas?

3. What physical property of hydrogen, other than that it is less dense than water, allows it to be collected in this manner?

D. Properties of Hydrogen

1. What happened when the splint was brought to the mouth of the first bottle of gas collected?

2. Describe fully the results of testing the second bottle of gas.

3. (a) Is hydrogen combustible?   _____

   (b) What evidence do you have from testing the second bottle of gas?

4. (a) Does hydrogen support combustion?   _____

   (b) What evidence do you have of this from testing the second bottle of gas?

5. What compound was formed during the testing of the first and second bottles of gas?

                                                                  _____

6. Why did the first bottle of gas behave differently from the second bottle?

7. (a) What happened when the splint was brought to the mouth of the third bottle of gas?

(b) How do you account for this?

8. When testing the fourth bottle:

(a) What was the result with the top bottle?

(b) What was the result with the lower bottle?

(c) How do you account for these results?

9. Complete the following word equations:

Zinc + Sulfuric acid $\longrightarrow$

Magnesium + Hydrochloric acid $\longrightarrow$

Hydrogen + Oxygen $\longrightarrow$

10. Complete and balance the following corresponding formula equations:

$Zn(s)$ + $H_2SO_4(aq)$ $\longrightarrow$

$Mg(s)$ + $HCl(aq)$ $\longrightarrow$

$H_2(g)$ + $O_2(g)$ $\longrightarrow$

EXPERIMENT 5

Calorimetry and Specific Heat

MATERIALS AND EQUIPMENT

Styrofoam cups, 6 oz; thermometers, metal samples, test tube with a diameter of at least 22 mm, bunsen burner, wire gauze, 400 ml beaker, cardboard cut into 4" squares with small thermometer hole in middle.

DISCUSSION

Calorimetry is the science of measuring a quantity of heat. Heat is a form of energy associated with the motion of atoms or molecules of a substance. Heat (often represented as "q") is measured in energy units such as joules or calories. Temperature (often represented as "t") is measured in degrees (usually Celsius). The measurement of temperature is already familiar to you. The same temperature is obtained for the water in a lake and for a thermos of water taken from the lake. But the heat content of the whole lake is much more than the heat content in that thermos of water even though both are exactly the same temperature.

Temperature and heat are related to each other by the specific heat (sp ht) of a substance, defined as the quantity of heat needed to raise one gram of a substance by one degree Celsius (J/g°C). The relationship between quantity of heat (q), specific heat (sp ht), mass (m) and temperature change (Δt) is mathematically expressed by the equation:

$$q = (m)\,(sp\ ht)\,(\Delta t) \quad \text{or} \quad \text{Joules} = (g)\left(\frac{J}{g°C}\right)(°C)$$

Since the mass and temperature can be measured by a balance and a thermometer, respectively, q can be calculated if the sp ht for a substance is known. Also, sp ht can be calculated if the heat content (q) of the substance is known. The amount of heat needed to raise the temperature of 1 g of water by 1 degree Celsius is the basis of the calorie. Thus, the specific heat of water is exactly 1.00 cal/g°C. The SI unit of energy is the joule and it is related to the calorie by 1 calorie = 4.184 J. Thus, the specific heat of water is also 4.184 J/g°C. The specific heat of a substance relates to its capacity to absorb heat energy. The higher the specific heat, the more energy required by a substance to change its temperature.

The specific heat of metals generally varies with their atomic masses. You will see this relationship later when the data in Table 5.1 is graphed. For this data, atomic mass is the independent variable and specific heat is the dependent variable because we are examining how specific heat changes as a function of atomic mass. For a review of variables and graphing techniques, see Study Aid 3.

In this experiment, we will use calorimetry to determine the specific heat of a metal. Heat energy is transferred from a hot metal to water until the metal and the water have reached the same temperature. This transfer is done in an insulated container to minimize heat losses to

Table 5.1   Specific Heat of Metals

Name of Metal	Atomic Mass, amu	Specific Heat, J/g°C
Aluminum	26.98	.900
Copper	63.55	.385
Gold	197.0	.131
Iron	55.85	.451
Lead	207.2	.128
Silver	107.9	.237
Tin	118.7	.222

the surroundings. We then make the assumption that all the heat lost by the metal (q_x) was absorbed by the water and is equal to the heat gained by the water, (q_w). Since we know the specific heat of water, we have all the variables needed to calculate q_w using the equation:

$$q_w = (m_w)(sp\ ht_w)(\Delta t_w)$$

Since q_w is equal to q_x we can say that

$$q_w = q_x = (m_x)(sp\ ht_x)(\Delta t_x)$$

This relationship can be used to calculate $sp\ ht_x$ of the metal because both m_x and Δt_x can be measured.

Sample Calculation:

A metal sample weighing 68.3820 g was heated to 99.0°C, then quickly transferred into a styrofoam calorimeter containing 62.5515 g of distilled water at a temperature of 18.0°C. The temperature of the water in the styrofoam cup increased and stabilized at 20.6°C. Calculate the $sp\ ht_x$ and identify the metal using 4.184 J/g°C for the $sp\ ht_w$.

$$\Delta t_w = 20.6°C - 18.0°C = 2.6°C$$

$$q_w = (m_w)(sp\ ht_w)(\Delta t_w)$$

$$= (62.5515g)\ (4.184\ J/g°C)\ (2.6°C)$$

$$= 680\ J \qquad \text{(heat absorbed by the water)}$$

Let x be the metal. Then, since all of the heat absorbed by the water came from the hot metal, we can say that

$$q_w = q_x = (\ m_x)\ (sp\ ht_x)\ \Delta t_x$$

$$\Delta t_x = 99.0°C - 20.6°C = 78.4°C$$

$$680\ J = (68.3820g)\ (sp\ ht_x)\ (78.4°C)$$

$$sp\ ht_x = 0.13\ J/g°C$$

Refer to Table 5.1 and determine that the unknown metal is lead or gold.

PROCEDURE: TRIAL 1

Wear protective glasses.

 No waste generated by this experiment.

1. Weigh a dry metal sample and record the mass and the name of the metal on the report form.

2. Carefully slide the sample into a large dry test tube and put a thermometer beside it in the test tube.

3. Attach the test tube to a ring stand and place it into an empty 400 mL beaker as shown in Figure 5.1. Be sure the height of the beaker is adjusted so the hottest part of the burner flame will be on the bottom of the beaker. Do not heat the dry beaker while making this adjustment. The bottom of the test tube should be at least one-half inch above the bottom of the beaker.

4. Fill the beaker with tap water so the height of the water in the beaker is about two inches higher than the top of the metal sample. There should be no water inside the test tube.

5. Begin heating the water in the beaker and continue with the next step(s). As you are working, check the water and note when it starts to boil. Turn down the burner but keep the water gently boiling. Do not do step 11 until the water has been boiling for about ten minutes and the temperature in the test tube has stabilized.

6. Nest two dry styrofoam cups together, weigh them, and record the mass on your report form.

7. Take another metal sample similar to the one you are heating. It does not matter if it is not the same metal. It is a "stand in" for the metal sample you are heating and will not be used during the experiment. Put this "stand in" into the styrofoam cup and add enough distilled water to cover the metal by no more than one-half inch.

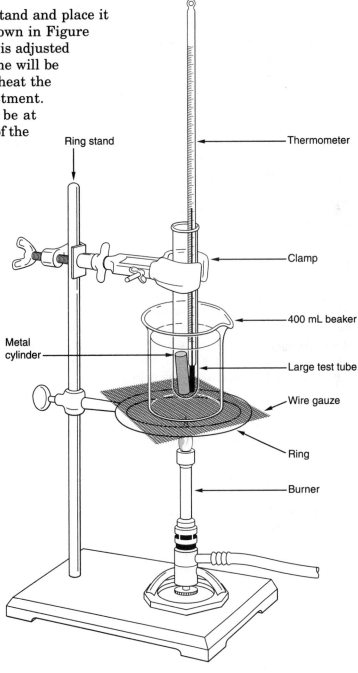

Figure 5.1   The boiler

8. Remove the "stand in" and weigh and record the mass of the styrofoam cups plus the water that you added.

9. Take a cardboard cover for the styrofoam cup and insert a thermometer through the hole. The nested cups with the cardboard cover and thermometer are referred to as a calorimeter, Figure 5.2. If you just leave the calorimeter on the benchtop it might fall over and break the thermometer so put the whole setup into a small beaker to stabilize it. The cardboard cover must rest directly on top of the styrofoam cup and not on the beaker.

10. Measure and record the temperature of the water in the styrofoam cup. Leave the thermometer in the cover until you are ready to transfer the hot metal into the calorimeter.

11. After the water in the beaker has been boiling for 10 minutes and the temperature inside the test tube with the metal has been stable for 5 minutes record the temperature on your report form. Remove the thermometer from the test tube and set it aside so it does not get mixed up with the thermometer used in the calorimeter.

12. Now, you are going to transfer the metal from the test tube to the water in the calorimeter. It is important that the transfer take place quickly and carefully to minimize heat loss to the surroundings and to avoid

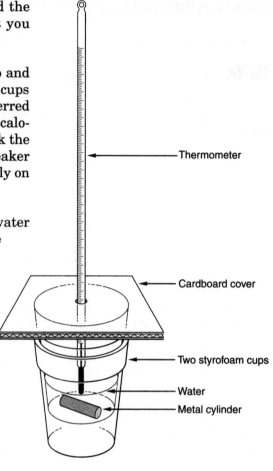

Figure 5.2   The calorimeter

splashing. Remove the cardboard cover and thermometer from the calorimeter. Loosen the clamp on the boiler ring stand, lift the clamp and test tube out of the boiler, and quickly slide the metal into the water in the calorimeter.

13. Immediately, put the cardboard cover with the thermometer back on the styrofoam cup. Stir gently for 2–3 minutes while monitoring the temperature. Record the temperature after it has remained constant for about one minute.

14. Unless instructed otherwise, repeat the experiment with Trial 2.

PROCEDURE: TRIAL 2

15. Repeat this experiment using the same metal sample, but this time use colder distilled water (5–10°C). Do not record the initial temperature of the cold water in the calorimeter until immediately before you add the hot metal. Fill out the report form in the column labeled Trial 2.

16. Look up and record the theoretical value for the specific heat for your metal sample. This can be found in Table 5.1 or in the *Handbook of Chemistry and Physics* if it is a metal not listed in the table.

REPORT FOR EXPERIMENT 5

Calorimetry and Specific Heat

Measurements and Calculations

		TRIAL 1	TRIAL 2
1.	Mass of metal sample	_____	_____
2.	Mass of calorimeter (styrofoam cups)	_____	_____
3.	Mass of styrofoam cups + water	_____	_____
4.	Mass of water (show setup for calculation)	_____	_____
5.	Initial water temperature	_____	_____
6.	Temperature of heated metal sample	_____	_____
7.	Final temperature of water and metal	_____	_____
8.	Change in temperature, Δt_w of the water in the calorimeter (show setup for calculation)	_____	_____
9.	Change in temperature, Δt_x of the metal sample (show setup for calculation)	_____	_____
10.	Specific heat ($sp\ ht_w$) of water	*4.184 J/g°C*	*4.184 J/g°C*
11.	Heat (q_w) gained by water (show setup for calculation)	_____	_____
12.	Heat (q_x) lost by metal sample	_____	_____
13.	Specific heat ($sp\ ht_x$) of the metal (experimental value) (show setup for calculation)	_____	_____

14. Name of metal _____   Theoretical Specific Heat _____

QUESTIONS AND PROBLEMS

1. Why is it important for there to be enough water in the calorimeter to completely cover the metal sample?

2. Why did we heat the metal in a dry test tube rather than in the boiling water?

3. The water in the beaker gets its heat energy from the _____ and the water in the calorimeter gets its heat energy from the _____.

4. What is the specific heat in J/g°C for a metal sample with a mass of 95.6 g which absorbs 841 J of energy when its temperature increases from 30.0°C to 98.0°C?

5. What effect does the initial temperature of the water have on the change in temperature of the water after the hot metal is added? Explain your answer.

> Results of scientific experiments must be reproducible when repeated or they do not mean anything. When results are repeated, the experiment is said to have good **precision**. When the results agree with a theoretical value they are described as **accurate**.

6. Which is better, the precision or the accuracy of the experimental specific heats determined for your metal sample? Support your answer with your data.

> Use the data presented in Table 5.1 to answer questions 7–9 and complete the graph on the next page to show the relationship between the atomic mass and the specific heat of the seven metals listed. Make the graph following the guidelines provided in Study Aid 3.

7. a. What is the independent variable? _____

   b. What is the dependent variable? _____

8. In Table 5.1, what is the range of values for atomic mass? _____

9. In Table 5.1, what is the range of values for specific heat? _____

10. Plot the data in Table 5.1 on the graph below. Be sure to include the following:

   a. A title

   b. Placement of the independent and dependent variables on the appropriate axes

   c. Increments for each axis

   d. Labels for each axis

   e. Plotting the data points

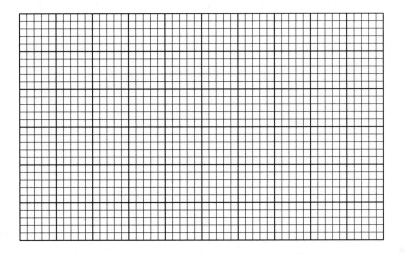

11. Use your graph to summarize the relationship between the atomic mass of metal atoms and the specific heat of a metal.

12. The specific heat was measured for two unknown metal samples. The first sample tested had a specific heat of 0.54 J/g°C. Use your graph to estimate the atomic mass of the metal.

   The second metal had a specific heat of 0.24 J/g°C. Use your graph again and estimate the atomic mass of this metal.

EXPERIMENT 6

Freezing Points—Graphing of Data

MATERIALS AND EQUIPMENT

Solids: benzoic acid (C_6H_5COOH) and crushed ice. **Liquid:** glacial acetic acid ($HC_2H_3O_2$). Thermometer, watch or clock with second hand, slotted corks or stoppers.

DISCUSSION

All pure substances, elements and compounds, possess unique physical and chemical properties. Just as one human being can be distinguished from all others by certain characteristics—fingerprints, for example—it is also possible, through knowledge of its properties, to distinguish any given compound from among the many hundreds of thousands that are known.

A. Melting and Freezing Points of Pure Substances.

The melting point and the boiling point are easily determined physical properties that are very useful in identifying a substance. Consequently, these properties are almost always recorded when a compound is described in the chemical literature (textbooks, handbooks, journal articles, etc.). The freezing and melting of a pure substance occurs at the same temperature, measured when the liquid and solid phases of the substance are in equilibrium. When energy is being removed from a liquid in equilibrium with its solid, the process is called freezing; when energy is being added to a solid in equilibrium with its liquid, the process is called melting.

$$\text{liquid} \underset{\substack{+ \text{ energy} \\ \text{melting}}}{\overset{\substack{\text{freezing} \\ - \text{ energy}}}{\rightleftharpoons}} \text{solid}$$

In this experiment, we will determine the freezing point of a pure organic compound, glacial acetic acid ($HC_2H_3O_2$). When the experimental freezing point has been determined, it will be compared with the melting point temperature listed in the *Handbook of Chemistry and Physics*.

When heat is removed from a liquid, the liquid particles lose kinetic energy and move more slowly causing the temperature of the liquid to decrease. Finally enough heat is removed and the particles move so slowly that the liquid becomes a solid, often a crystalline solid. The temperature when this happens (the freezing point) is different for different substances.

The amount of energy removed from a quantity of liquid to freeze it is equal to the amount of energy added to the same quantity of its solid to melt it. Thus, depending on the direction of energy flow, this equilibrium temperature is called the melting point or the freezing point.

B. Freezing Point of Impure Substances

When a substance (solvent) is uniformly mixed with a small amount of another substance (solute), the freezing point of the resulting solution (an "impure substance") will be lower than that of the pure solvent. For example, the accepted freezing point for pure water is 0.0°C. Solutions of salt in water may freeze at temperatures as low as –21°C depending on the amount of salt added to the water. Antifreeze is added to the water in a car radiator to lower the freezing point of the water.

Melting point/freezing point data are of great value in determining the identity and/or purity of substances, especially in the field of organic chemistry. If a sample of a compound melts or freezes appreciably below the known melting point of the pure substance, we know that the sample contains impurities which have lowered the melting point. If the melting point of an unknown compound agrees with that of a known compound, the identity can often be confirmed by mixing the unknown compound with the known and determining the melting point of the mixture. If the melting point of the mixture is the same as that of the known compound, the compounds are identical. On the other hand, a lower melting point for the mixture indicates that the two compounds are not identical.

C. Supercooling During Freezing

Frequently when a substance is being cooled, the temperature will fall below the true freezing point before crystals begin to form. This phenomenon is known as supercooling because the substance is cooled below its freezing point without forming a solid. Supercooling is more likely to occur if the liquid remains very still and undisturbed as its temperature is lowered. When the system is disturbed in any way, by stirring or jarring, crystallization occurs rapidly throughout the system. As the crystals form, heat is released (called the heat of crystallization) and the temperature rises quickly to the freezing point. Thus, supercooling does not change the freezing point of the substance.

D. Freezing Point Determinations

You will do three freezing point determinations during this experiment.

Trial 1.   Freezing point determination of pure glacial acetic acid WITH STIRRING. This will usually eliminate supercooling.

Trial 2.   Freezing point determination of pure glacial acetic acid WITHOUT STIRRING. This should enhance the possibility of supercooling but does not guarantee it.

Trial 3.   Freezing point determination of acetic acid (the solvent) after benzoic acid (a solute) has been dissolved in it. This will be done WITHOUT STIRRING to enhance supercooling again.

The time/temperature data will be graphed and the freezing point for each trial read from the graph.

PROCEDURE

Wear protective glasses.

> Notes: Since water and other contaminants will influence the freezing points in this experiment, use only clean, dry equipment.
>
> Read and record all temperatures to the nearest 0.1°C.

A. Freezing Point Determination of Pure Glacial Acetic Acid

Trial 1: With stirring

1. Fasten a utility clamp to the top of a clean, dry test tube. Position this clamp-tube assembly on a ring stand so that the bottom of the tube is about 20 cm above the ring stand base.

2. Obtain a slotted one-hole cork (or stopper) to fit the test tube (see Figure 6.1). Insert a thermometer in the cork and position it in the test tube so that the end of the bulb is about 1.5 cm from the bottom of the test tube. Turn the thermometer so that the temperature scale can be read in the slot.

3. Take your test tube, the cork/ thermometer and a graduated cylinder to the fume hood. Measure out 10. mL of glacial acetic acid. Pour it into the test tube and close the test tube with the cork/ thermometer. Glacial acetic acid is irritating and harmful if inhaled so keep the test tube stoppered while you work outside the hood at your bench. Rinse the graduated cylinder immediately.

4. Reclamp the test tube to your ring stand to minimize the risk of spilling. Make sure the thermometer bulb is covered by the acid and adjust the temperature of the acetic acid to approximately 25°C by warming or cooling the tube in a beaker of water.

5. Fill a 400 mL beaker about three-quarters full of crushed ice; add cold water until the ice is almost covered. Position the beaker of ice and water on the ring stand base under the clamped tube-thermometer assembly.

6. Read the temperature of the acetic acid and record as the 0.0 minute time reading in the Data Table. Now loosen the clamp on the ring stand and observe the second hand of your watch or clock. As the second hand crosses 12, lower the clamped tube-thermometer assembly so that all of the acetic acid in the tube is below the surface of the ice water. Fasten the clamp to hold the tube in this position.

7. Loosen the cork on the tube and stir (during Trial 1 only) the acid with the thermometer, keeping the bulb of the thermometer completely immersed in the acid. Take accurate temperature readings at 30-second intervals as the acid cools. (Zero time was when the second hand crossed 12.) Stop stirring and center the thermometer bulb in the tube as soon as you are sure that crystals are forming in the acid (three to four minutes). Circle the temperature reading when the crystals were first observed.

8. Continue to take temperature readings at 30-second intervals until a total time of 12 minutes has elapsed or until the entire volume of liquid becomes solidified. After that occurs, read the temperature for an additional 2 minutes (4 time intervals) and continue with the next step.

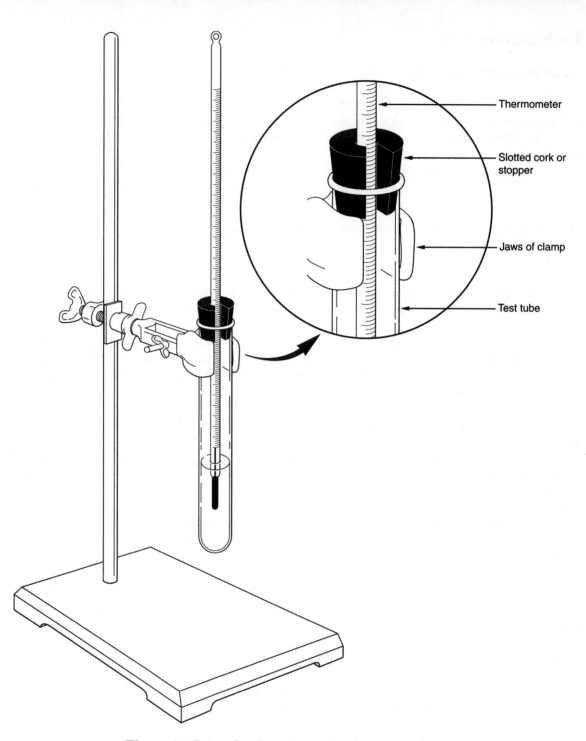

Figure 6.1 Setup for freezing-point determination

9. After completing the temperature readings, remove the test tube-thermometer assemble from the ice bath, keeping the thermometer in place. Immerse the lower portion of the test tube in a beaker of warm water to melt the frozen acetic acid. Do not discard this acid; it will be used in Trials 2 and 3.

Trial 2: Without stirring

10. Repeat steps 4–9 with the following changes:

    a. Replenish the ice bath as in step 5.

    b. After submerging the tube in the ice bath, do NOT stir. Do NOT touch or move the apparatus in any way.

    c. If the temperature goes down to about 4°C or lower without the formation of acetic acid crystals and remains there, touch the thermometer and move it until crystals form which usually happens quickly. When you do this be very observant of the temperature changes. Continue to record temperature readings for the full 12 minutes or until the temperature stabilizes after crystallization for 5 minutes.

B. Freezing Point Determination of An Acetic Acid/Benzoic Acid Solution

Trial 3: Without stirring

11. Weigh approximately 0.50 g (between 0.48 and 0.52 g) of benzoic acid crystals. Now remove the thermometer from the test tube of acetic acid and lay it on the table) being careful not to contaminate the thermometer or lose any acid. Carefully add all of the benzoic acid to the acetic acid. Stir gently with the thermometer until all of the crystals have dissolved. Stir for an additional minute or two to ensure a uniform solution. Adjust the temperature of the solution to approximately 25°C.

12. Repeat step 10.

13. Dispose of the acetic acid/benzoic acid solution in the waste container provided.

C. Graphing Temperature Data

Graph the three sets of data using the graph paper in the report form or prepare a computer graph. If necessary, review the instructions for preparing a graph in Study Aid 3.

REPORT FOR EXPERIMENT 6

Freezing Points—Graphing of Data

Data Table

time, minutes	Pure Acetic Acid temp, °C WITH STIRRING	Pure Acetic Acid temp, °C WITHOUT STIRRING	Impure Acetic Acid temp, °C WITHOUT STIRRING
0.0			
0.5			
1.0			
1.5			
2.0			
2.5			
3.0			
3.5			
4.0			
4.5			
5.0			
5.5			
6.0			
6.5			
7.0			
7.5			
8.0			
8.5			
9.0			
9.5			
10.0			
10.5			
11.0			
11.5			
12.0			

Graphing of Freezing Point Data

Plot your data on the graph paper or the computer using a legend as follows:

⊙ = Pure acetic acid with stirring

⚠ = Pure acetic acid without stirring

⊡ = Acetic acid/benzoic acid solution without stirring

Draw rectangles around the portions of your curves that show supercooling.

Questions

Use your graph to answer the questions 1–3.

1. a. At what temperature did crystals first form in Trial 1?   _____

     b. Where did the temperature stabilize after supercooling in Trial 2?   _____

     c. What is your experimental freezing point of glacial acetic acid?   _____

     d. What is the theoretical freezing point of glacial acetic acid?   _____
        (Consult the *Handbook of Chemistry and Physics*)

2. How many degrees was the freezing point depressed by the benzoic acid?   _____

Do this by estimating to the nearest 0.1 degree the number of degrees between the flattest (most nearly horizontal) portions of the curves. Mark the area on the graph with an arrow (↓) to show where this temperature difference estimate was made.

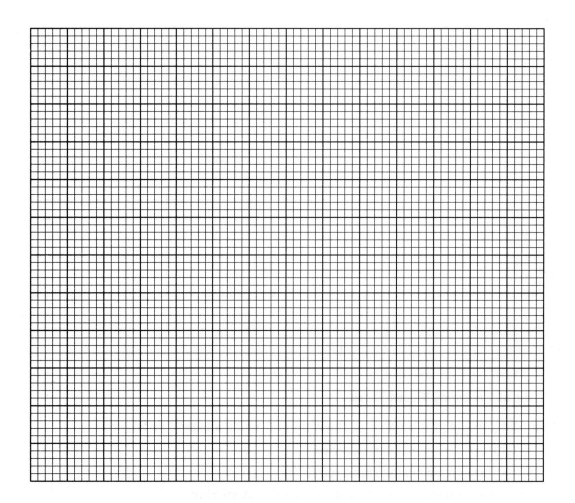

3. a. What is the effect of stirring on the freezing point of pure acetic acid?

    b. What is the effect of stirring on supercooling?

4. a. What do the melting point and freezing point of a substance have in common?

    b. What is the difference between the melting and freezing of a substance?

5. When the solid and liquid phases are in equilibrium, which phase, solid or liquid contains the greater amount of energy? Explain the rationale for your answer.

EXPERIMENT 7

Water in Hydrates

MATERIALS AND EQUIPMENT

Solids: finely ground copper(II) sulfate pentahydrate ($CuSO_4 \cdot 5\ H_2O$), and unknown hydrate. Cobalt chloride test paper, clay triangle, crucible and cover, 25×200 mm hard-glass test tube.

DISCUSSION

Many salts form compounds in which a definite number of moles of water are combined with each mole of the anhydrous salt. Such compounds are called **hydrates**. The water which is chemically combined in a hydrate is referred to as **water of crystallization** or **water of hydration**. The following are representative examples:

$$CaSO_4 \cdot 2H_2O, \quad CoCl_2 \cdot 6H_2O, \quad MgSO_4 \cdot 7H_2O, \quad Na_2CO_3 \cdot 10H_2O$$

In a hydrate the water molecules are distinct parts of the compound but are joined to it by bonds that are weaker than either those forming the anhydrous salt or those forming the water molecules. In the formula of a hydrate a dot is commonly used to separate the formula of the anhydrous salt from the number of molecules of water of crystallization. For example, the formula of calcium sulfate dihydrate is written $CaSO_4 \cdot 2H_2O$ rather than H_4CaSO_6.

Hydrated salts can usually be converted to the anhydrous form by heating:

$$\text{Hydrated salt} \xrightarrow{\Delta} \text{Anhydrous salt} + \text{water}$$

Hence it is possible to determine the percentage of water in a hydrated salt by determining the amount of mass lost (water driven off) when a known mass of the hydrate is heated:

$$\text{Percentage water} \ = \ \frac{\text{Mass lost}}{\text{Mass of sample}} \times 100$$

It is possible to condense the vapor driven off the hydrate and demonstrate that it is water by testing it with anhydrous cobalt(II) chloride ($CoCl_2$). Anhydrous cobalt(II) chloride is blue but reacts with water to form the red hexahydrate, $CoCl_2 \cdot 6H_2O$.

PROCEDURE

Wear protective glasses.

A. Qualitative Determination of Water

1. Fold a 2.5×20 cm strip of paper lengthwise to form a V-shaped trough or chute. Load about 4 g of finely ground copper(II) sulfate pentahydrate in this trough, spreading it evenly along the length of the trough.

2. Clamp a **dry** 25 x 200 mm hard-glass test tube so that its mouth is 15–20 degrees **above the horizontal** (Figure 7.1a). Insert the loaded trough into the tube. Rotate the tube to a nearly vertical position (Figure 7.1b) to deposit the copper(II) sulfate in the bottom of the tube. Tap the paper chute gently if necessary, but make sure that no copper sulfate is spilled and adhering to the sides of the upper part of the tube.

3. Remove the chute and turn the tube until it slants mouth downward at an angle of 15–20 degrees **below the horizontal** (Figure 7.1c). Make sure that all of the copper(II) sulfate remains at the bottom of the tube. To obtain a sample of the liquid that will condense in the cooler part of the tube, place a small clean, dry test tube, held in an upright position in either a rack of an Erlenmeyer flask, just below the mouth of the tube containing the hydrate.

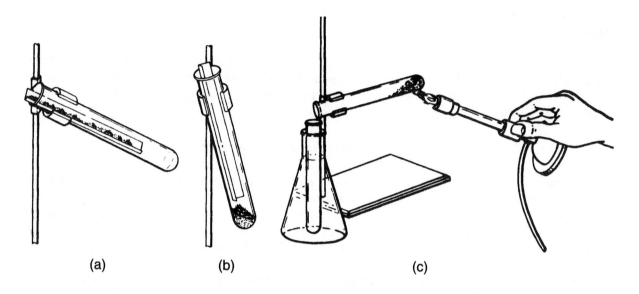

(a)　　　　　　(b)　　　　　　(c)

Figure 7.1　Setup for dehydration of a hydrate

4. Heat the hydrate gently at first to avoid excessive spattering. Gradually increase the rate of heating, noting any changes that occur and collecting some of the liquid that condenses in the cooler part of the tube. Continue heating until the blue color of the hydrate has disappeared, but do not heat until the residue in the tube has turned black. Finally warm the tube over its entire length—without directly applying the flame to the clamp—for a minute or two to drive off most of the liquid that has condensed on the inner wall of the tube. Allow the tube and contents to cool.

> **Note:** At excessively high temperatures (above 600°C) copper(II) sulfate decomposes; sulfur trioxide is driven off and the black copper(II) oxide remains as a residue.

Observe and record the appearance and odor of the liquid that has been collected.

5. While the tube is cooling, dry a piece of cobalt chloride test paper by holding it about 20 to 25 cm above a burner flame; that is, close enough to heat but not close enough to char or ignite the paper. When properly dried, the test paper should be blue. Using a clean stirring rod, place a drop of the liquid collected from the hydrate on the dried cobalt chloride test paper. For comparison place a drop of distilled water on the cobalt chloride paper. Record your observations.

6. Empty the anhydrous salt residue in the tube onto a watch glass and divide it into two portions. Add 3 or 4 drops of the liquid collected from the hydrate to one portion and 3 or 4 drops of distilled water to the other. Compare and record the results of these tests.

 Dispose of solid residues in the container provided.

B. Quantitative Determination of Water in a Hydrate

> NOTES:
>
> 1. **Weigh crucible and contents to the highest precision with the balance available to you.**
>
> 2. Since there is some inaccuracy in any balance, use the same balance for successive weighings of the same sample. When subtractions are made to give mass of sample and mass lost, the inaccuracy due to the balance should cancel out.
>
> 3. Handle crucibles and covers with tongs only, after initial heating.
>
> 4. Be sure crucibles are at or near room temperature when weighed.
>
> 5. **Record all data directly on the report form as soon as you obtain them.**

1. Obtain a sample of an unknown hydrate, as directed by your instructor. Be sure to record the identifying number.

2. Weigh a clean, dry crucible and cover to the highest precision of the balance.

3. Place between 2 and 3 g of the unknown into the weighed crucible, cover, and weigh the crucible and contents.

4. Place the covered crucible on a clay triangle; adjust the cover so that is slightly ajar, to allow the water vapor to escape (see Figure 7.2); and **very gently** heat the crucible for about 5 minutes. Readjust the flame so that a sharp, inner-blue cone is formed. Heat for another 12 minutes with the tip of the inner-blue cone just touching the bottom of the crucible. The crucible bottom should become dull red during this period.

5. After this first heating is completed, close the cover, cool (about 10 minutes), and weigh.

6. Heat the covered crucible and contents for an additional 6 minutes at maximum temperature; cool and reweigh. The results of the last two weighings should agree within 0.05 g. If the

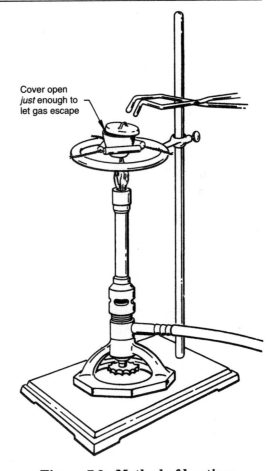

Cover open *just* enough to let gas escape

Figure 7.2 Method of heating a crucible

decrease in mass between the two weighings is greater than 0.05 g, repeat the heating and weighing until the results of two successive weighings agree to within 0.05 g.

7. Calculate the percentage of water in your sample on the basis of the final weighing.

 Dispose of the solid residue in the container provided. Return the unused portion of your unknown to the instructor.

REPORT FOR EXPERIMENT 7

Water in Hydrates

A. Qualitative Determination of Water

1. Describe the appearance and odor of the liquid obtained by heating copper(II) sulfate pentahydrate.

2. Compare the results obtained when the liquid from the hydrate and the distilled water were tested with cobalt chloride paper.

3. Describe what happened when the liquid from the hydrate and distilled water were added to the anhydrous salt obtained by heating the hydrate.

B. Quantitative Determination of Water in a Hydrate

Mass of crucible and cover	
Mass of crucible, cover, and sample	
Mass of crucible, cover, and sample after first heating	
Mass of crucible, cover, and sample after second heating	
Mass of crucible, cover, and sample after third heating (if needed)	

CALCULATIONS

Make all calculations needed to complete the following table. Be sure to carry the proper number of significant figures in your work. Show numerical setups in the spaces provided; do not show longhand division and multiplication.

Calculate mass of original sample Numerical setup:	
Calculate mass lost from sample by heating Numerical setup:	
Calculate percentage of water in sample Numerical setup:	Sample No. _____ _____

QUESTIONS AND PROBLEMS

1. What evidence did you see that indicated the liquid obtained from the copper(II) sulfate pentahydrate was water?

    (a)

    (b)

2. What was the evidence of a chemical reaction when the anhydrous salt samples were treated with the liquid obtained from the hydrate and with water?

    (a)

    (b)

3. Write a balanced chemical equation for the decomposition of copper(II) sulfate pentahydrate.

4. When the unknown was heated, could the decrease in mass have been partly due to the loss of some substance other than water? Explain.

EXPERIMENT 8

Properties of Solutions

MATERIALS AND EQUIPMENT

Solids: ammonium chloride (NH_4Cl), barium chloride ($BaCl_2$), barium sulfate ($BaSO_4$), fine and coarse crystals of sodium chloride ($NaCl$), and sodium sulfate (Na_2SO_4). **Liquids:** decane ($C_{10}H_{22}$), isopropyl alcohol (C_3H_7OH), and kerosene. **Solutions:** saturated iodine-water (I_2), and saturated potassium chloride (KCl).

DISCUSSION

The term **solution** is used in chemistry to describe a system in which one substance (or more) is dissolved in another substance.

A solution has two components, a solute and a solvent. The **solute** is the substance that is dissolved. The **solvent** is the dissolving medium; it is the substance present in greater quantity. The name of the solution is taken from the name of the solute. Thus when sodium chloride is dissolved in water, sodium chloride is the solute, water is the solvent, and the solution is called a sodium chloride solution.

Like other mixtures, a solution has variable composition, since more or less solute may be dissolved in a given quantity of a solvent. But unlike other mixtures, a solution is homogeneous, since the solute remains uniformly dispersed throughout the solution after mixing. The size of the dissolved particles (molecules or ions) is of the order of 10^{-8} to 10^{-7} cm (1-10Å).

For many substances whose reactions are those of their ions, it is necessary to use solutions in order to obtain satisfactory reactions. For example, when the two solids sodium chloride ($NaCl$) and silver nitrate ($AgNO_3$) are mixed, no detectable reaction is observed. However, when water solutions of these salts are mixed, a white precipitate of silver chloride ($AgCl$) is formed immediately.

The formation of a solution depends on the nature of the solute and the solvent. In general, water, which is polar, is a better solvent for inorganic than for organic substances. On the other hand, nonpolar solvents such as benzene, decane, and ether are good solvents for many organic substances that are practically insoluble in water.

The rate of dissolving of a solute depends on:

1. The particle size of the solute.

2. Agitation or stirring of the solution.

3. The temperature of the solution.

4. The concentration of the solute in solution.

In addition, the **amount of solute** that will dissolve per unit mass or unit volume of solvent depends on the temperature of the solution.

Solubility expresses the amount of solute that will dissolve in a given amount of solvent. Terms used to describe the relative solubility of a solute are: **insoluble, slightly soluble, moderately soluble,** and **very soluble.** When two liquids are **immiscible** they do not form a solution or are generally insoluble in each other.

The **concentration** of a solution expresses the relative content of dissolved solute. Common ways to express concentration are as follows:

1. **Dilute Solution:** A solution that contains a relatively small amount of solute per unit volume of solution.

2. **Concentrated Solution:** A solution that contains a relatively large amount of solute per unit volume of solution.

3. **Saturated Solution:** A solution that contains as much solute as can dissolve at a given temperature and pressure; the dissolved solute is in equilibrium with undissolved solute:

$$\text{Solute (solid)} \rightleftharpoons \text{Solute (dissolved)}$$

A saturated solution can be dilute or concentrated.

4. **Unsaturated Solution:** A solution containing less solute per unit volume than the corresponding saturated solution.

5. **Supersaturated Solution:** A solution containing more dissolved solute than is normally present in the corresponding saturated solution. However, a supersaturated solution is in a metastable condition and will form a saturated solution if disturbed. For example, when a small crystal of a dissolved salt is dropped into a supersaturated solution, crystallization begins at once, and salt precipitates until a saturated solution in formed.

6. **Mass-percent Solution:** The percent by mass of the solute in a solution. Thus a 10 percent sodium hydroxide solution contains 10 g of NaOH in 100 g of solution (10 g NaOH + 90 g H_2O; 2 g NaOH + 18 g H_2O; etc.). The formula for calculating mass percent is

$$\text{Mass percent} = \frac{\text{g solute}}{\text{g solute} + \text{g solvent}} \times 100$$

7. **Molarity:** The number of moles (molar mass) of solute per liter of solution. Thus, a solution containing 1 mole of NaOH (40.00 g) per liter is 1 molar (abbreviated 1 M). The concentration of a solution containing 0.5 mole in 500 mL of solution is also 1 molar, etc. The equation is

$$\text{Molarity} = \frac{\text{moles of solute}}{\text{liter of solution}}$$

PROCEDURE

Wear protective glasses.

A. Concentration of a Saturated Solution

> Use the same balance for all weighings.

1. Prepare a water bath with a 400 mL beaker half full of water and heat to boiling. (See Figure 1.6.)

2. Weigh an evaporating dish to the highest precision of the balance. Obtain 6 mL of saturated potassium chloride solution and pour it into the dish. Weigh the dish with the solution in it and record these masses on the report form.

3. Place the evaporating dish on the beaker of boiling water and continue to boil until the potassium chloride solution has evaporated almost to dryness (about 25 to 30 minutes), **adding more water to the beaker as needed.**

> While the evaporation is proceeding, continue with other parts of the experiment.

4. Remove the evaporating dish and beaker from the wire gauze and dry the bottom of the dish with a towel. Put the dish on the wire gauze and heat gently to evaporate the last traces of water. Do not heat too strongly because at high temperatures there is danger of sample loss by spattering.

5. Allow the dish to cool for 5 to 10 minutes on the Ceramfab pad and reweigh. Add water to the residue in the dish to redissolve the potassium chloride. Pour the solution into the sink and flush generously with water.

B. Relative Solubility of a Solute in Two Solvents

1. Add about 2 mL of decane and 5 mL of water to a test tube, stopper it, and shake gently for about 5 seconds. Allow the liquid layers to separate and note which liquid has the greatest density.

2. Now, add 5 mL of saturated iodine-water to the test tube, note the color of each layer, insert the stopper, and shake gently for about 20 seconds. Allow the liquids to separate and again note the color of each layer.

3. Dispose of the mixture in this test tube in the bottle labeled **Decane Waste.**

C. Miscibility of Liquids

1. Take three dry test tubes and add liquids to each as follows:

    a. 1 mL kerosene and 1 mL isopropyl alcohol

    b. 1 mL kerosene and 1 mL water

    c. 1 mL water and 1 mL isopropyl alcohol

2. Stopper each tube and mix by shaking for about 5 seconds. Note which pairs are miscible.

 Dispose of the kerosene mixtures (a and b) in the bottle labeled **Kerosene Waste.**

D. Effect of Particle Size on Rate of Dissolving

1. Fill a dry test tube to a depth of about 0.5 cm with fine crystals of sodium chloride. Fill another dry tube to the same depth with coarse sodium chloride crystals. Add 10 mL of tap water to each tube and stopper. Shake both tubes at the same time, noting the number of seconds required to dissolve the salt in each tube. (Don't shake the tubes for more than two minutes.)

 2. Dispose of these solutions in the sink.

E. Effect of Temperature on Rate of Dissolving

1. Weigh two 0.5 g samples of fine sodium chloride crystals.

2. Take a 100 mL and a 150 mL beaker and add 50 mL tap water to each. Heat the water in the 150 mL beaker to boiling and allow it to cool for about 1 minute.

3. Add the 0.5 g samples of salt to each beaker and observe the time necessary for the crystals to dissolve in the hot water.

4. As soon as the crystals are dissolved in the hot water, take the beaker containing the hot solution in your hand, slowly tilt it back and forth, and observe the layer of denser salt solution on the bottom. Repeat with the cold-water solution.

 5. Dispose of these solutions in the sink.

F. Solubility versus Temperature; Saturated and Unsaturated Solutions

1. Label four weighing boats or papers as follows and weigh the stated amounts onto each one.

     a. 1.0 g NaCl   b. 1.4 g NaCl   c. 1.0 g NH_4Cl d. 1.4 g NH_4Cl

2. Place each of the 1.0 g samples into separate, labeled test tubes, add 5 mL of water, stopper, and shake until each salt is dissolved.

3. Add 1.4 g NaCl to the NaCl solution. Add 1.4 g NH_4Cl to the NH_4Cl solution. Stopper each tube and shake for about 3 minutes. Note whether all of the crystals have dissolved.

4. Place both tubes (unstoppered) into a beaker of boiling water, shake occasionally, and note the results after about 5 minutes.

5. Remove the tubes and cool in running tap water for about 1 minute. Let stand for a few minutes and record what you observe. Dispose of these solutions in the sink.

G. Ionic Reactions in Solution

1. Into four labeled test tubes, place pea-sized quantities of the following salts, one salt in each tube: (a) barium chloride, (b) sodium sulfate, (c) sodium chloride, (d) barium sulfate.

2. Add 5 mL of water to each tube, stopper, and shake to dissolve. One of the four salts does not dissolve.

3. Mix the barium chloride and sodium sulfate solutions together. Note the results. (Sodium chloride and barium sulfate are the products of this reaction.)

 Dispose of all tubes containing barium in the waste bottle provided. The remaining tubes can be rinsed in the sink.

NAME ____________________

SECTION ______ DATE ____________

REPORT FOR EXPERIMENT 8

INSTRUCTOR ______________

Properties of Solutions

A. Concentration of a Saturated Solution

1. (a) Mass of evaporating dish. ____________

   (b) Mass of dish and potassium chloride solution. ____________

   (c) mass of dish and potassium chloride (after evaporation). ____________

2. Calculate: (Show numerical setups.)

   (a) Mass of the saturated potassium chloride solution.

   ____

   (b) Mass of the potassium chloride in the saturated solution.

   ____

   (c) Mass of water in the saturated solution.

   ____

   (d) Mass percent of potassium chloride in the saturated solution.

   ____

   (e) The grams of potassium chloride per 100 g of water in the saturated solution.

   ____

B.   Relative Solubility of a Solute in Two Solvents

1.   (a) Which liquid is denser, decane or water?   _____

     (b) What experimental evidence supports your answer?

2.   Color of iodine in water: _____

     Color of iodine in decane: _____

3.   (a) In which of the two solvents used is iodine more soluble?   _____

     (b) Cite experimental evidence for your answer.

C.   Miscibility of Liquids

1.   Which liquid pairs tested are miscible?

2.   How do you classify the liquid pair decane – H_2O, miscible or immiscible?

D.   Rate of Dissolving versus Particle Size

1.   Time required for fine salt crystals to dissolve: _____

     Time required for coarse salt crystals to dissolve: _____

2.   What do you conclude from this experiment?

E.   Rate of Dissolving versus Temperature

1.   Under which condition, hot or cold, did the salt dissolve faster?   _____

2.   Length of time for the salt to dissolve in hot water.   _____

F. Solubility versus Temperature; Saturated and Unsaturated Solutions

1. (a) Which solution(s) containing 1.0 g salt per 5 mL of water are unsaturated?

   _____

   (b) Evidence:

2. (a) Which solution(s) containing 2.4 g salt per 5 mL of water are saturated at room temperature?

   _____

   (b) Evidence:

3. (a) Which salt is the least soluble at elevated temperatures? _____

   (b) Classify the salt solutions at elevated temperatures as either saturated or unsaturated.

   Ammonium chloride: _____

   Sodium chloride: _____

4. (a) Did the salt solution(s) that were unsaturated at the elevated temperature become saturated when cooled?

   _____

   (b) Evidence:

G. Ionic Reactions in Solution

1. Write the word and formula equations representing the chemical reaction that occurred between barium chloride and sodium sulfate.

   Word Equation:

   Formula Equation:

2. (a) Which of the products is the white precipitate? _____

   (b) What experimental evidence leads you to this conclusion?

SUPPLEMENTARY QUESTIONS AND PROBLEMS

1. Use the solubility table below in answering the following:

(a) What is the percentage by mass of NaCl in a saturated solution of sodium chloride at 50°C?

_____

(b) Calculate the solubility of potassium bromide at 23°C. Hint: Assume that the solubility increases by an equal amount for each degree between 20°C and 30°C.

_____

(c) A saturated solution of barium chloride at 30°C contains 150 g water. How much additional barium chloride can be dissolved by heating this solution to 60°C?

_____

Solubility of Salts in Water*

	0°	10°	20°	30°	40°	50°	60°	70°	80°	90°	100°
NaCl	35.7	35.8	36.0	36.3	36.6	37.0	37.3	37.8	38.4	39.0	39.8
KBr	53.5	59.5	65.2	70.6	75.5	80.2	85.5	90.0	95.0	99.2	104.0
$BaCl_2$	31.6	33.3	35.7	38.2	40.7	43.6	46.6	49.4	52.6	55.7	58.8

*This table shows the mass in grams of each salt that can be dissolved in 100 g of water at the temperatures (°C) indicated.

EXPERIMENT 9

Composition of Potassium Chlorate

MATERIALS AND EQUIPMENT

Solids: Reagent Grade potassium chlorate ($KClO_3$) and potassium chloride (KCl). **Solutions:** dilute (6 M) nitric acid (HNO_3) and 0.1 M silver nitrate ($AgNO_3$). Two No. 0 crucibles with covers; Ceramfab pad.

DISCUSSION

The **percentage composition** of a compound is the percentage by mass of each element in the compound. If the formula of a compound is known, the percentage composition can be calculated from the molar mass and the total mass of each element in the compound. The **molar mass** of a compound is determined by adding up the atomic masses of all the atoms making up the formula. The **total mass** of an element in a compound is determined by multiplying the atomic mass of that element by the number of atoms of that element in the formula. The percentage of each element is then calculated by dividing its total mass in the compound by the molar mass of the compound and multiplying by 100.

The percentage composition of many compounds may be directly determined or verified by experimental methods. In this experiment the percentage composition of potassium chlorate will be determined both experimentally and from the formula.

When potassium chlorate is heated to high temperatures (above 400°C) it decomposes to potassium chloride and elemental oxygen, according to the following equation:

$$2\ KClO_3(s) \longrightarrow 2\ KCl(s) + 3\ O_2(g)$$

The relative amounts of oxygen and potassium chloride are measured by heating a weighed sample of potassium chlorate and determining the amount of residue (potassium chloride) remaining. The decrease in mass brought about by heating represents the amount of oxygen originally present in the sample.

From the experiment we obtain the following three values:

1. Mass of original sample ($KClO_3$).

2. Mass lost when sample was heated (Oxygen).

3. Mass of residue (KCl).

From these experimental values (and a table of atomic masses) we can calculate the following:

4. Percentage oxygen in sample (Experimental value)

$$= \frac{\text{Mass lost by sample}}{\text{Original sample mass}} \times 100$$

5. Percentage KCl in sample (Experimental value)

$$= \frac{\text{Mass of residue}}{\text{Original sample mass}} \times 100$$

6. Percentage oxygen in $KClO_3$ from formula (Theoretical value)

$$= \frac{3 \text{ at. masses of oxygen}}{\text{Molar mass of } KClO_3} \times 100 = \frac{3 \times 16.00 \text{ g}}{122.6 \text{ g}} \times 100$$

7. Percentage KCl in $KClO_3$ from formula (Theoretical value)

$$= \frac{\text{Molar mass of KCl}}{\text{Molar mass of } KClO_3} \times 100 = \frac{74.55 \text{ g}}{122.6 \text{ g}} \times 100$$

8. Percentage error in experimental oxygen determination

$$= \frac{(\text{Experimental value}) - (\text{Theoretical value})}{\text{Theoretical value}} \times 100$$

PROCEDURE

PRECAUTIONS: Since potassium chlorate is a strong oxidizing agent it may cause fires or explosions if mixed or heated with combustible (oxidizable) materials such as paper. Observe the following safety precautions when working with potassium chlorate:

1. **Wear protective glasses.**

2. Use clean crucibles that have been heated and cooled prior to adding potassium chlorate.

3. Use Reagent Grade potassium chlorate.

4. **Dispose of any excess or spilled potassium chlorate as directed by your instructor. (Potassium chlorate may start fires if mixed with paper or other solid wastes.)**

5. Heat samples slowly and carefully to avoid spattering molten material—and to avoid poor experimental results.

NOTES:

1. Make all weighings to the highest precision possible with the balance available to you. Use the same balance to make all weighings for a given sample. Record all data directly on the report sheet as they are obtained.

2. Duplicate samples of potassium chlorate are to be analyzed, if two crucibles are available.

3. For utmost precision, handle crucibles with tongs after the initial heating.

A. Determining Percentage Composition

Place a clean, dry crucible (uncovered) on a clay triangle and heat for 2 or 3 minutes at the maximum flame temperature. The tip of the sharply defined inner-blue cone of the flame should almost touch and heat the crucible bottom to redness. Allow the crucible to cool. If two crucibles are being used, carefully transfer the first to a Ceramfab pad and heat the second while the first crucible is cooling.

Weigh the cooled crucible and its cover; add between 1 and 1.5 g of potassium chlorate; weigh again.

> **NOTE:** The crucible must be covered when potassium chlorate is being heated in it.

Place the covered crucible on the clay triangle and heat gently for 8 minutes with the tip of the inner-blue cone of the flame 6 to 8 cm (about 2.5 to 3 in.) below the crucible bottom. Then carefully lower the crucible or raise the burner until the tip of the sharply defined inner-blue cone just touches the bottom of the crucible, and heat for an additional 10 minutes. The bottom of the crucible should be heated to a dull red color during this period.

Grasp the crucible just below the cover with the concave part of the tongs and very carefully transfer it to a Ceramfab pad. Allow to cool (about 10 minutes) and weigh. Begin analysis of a second sample while the first is cooling.

After weighing, heat the first sample for an additional 6 minutes at the maximum flame temperature (bottom of the crucible heated to a dull red color); cool and reweigh. The last two weighings should be in agreement. If the mass decreased more than 0.05 g between these two weighings, repeat the heating and weighing until two successive weighings agree within 0.05 g.

Complete the analysis of the second sample following the same procedure used for the first.

B. Qualitative Examination of Residue

This part of the experiment should be started as soon as the final heating and weighing of the first sample is completed and while the second sample is in progress.

Number and place three clean test tubes in a rack. Put a pea-sized quantity of potassium chloride into tube No. 1 and a like amount of potassium chlorate into tube No. 2. Add 10 mL of distilled water to each of these two tubes and shake to dissolve the salts. Now add distilled water to the crucible containing the residue from the first sample so it is one-half full. Heat the uncovered crucible very gently for about 1 minute; transfer 1 to 2 mL of the resulting solution from the crucible to tube No. 3; add about 10 mL of distilled water and mix.

Test the solution in each tube as follows: Add 5 drops of dilute (6 M) nitric acid and 5 drops of 0.1 M silver nitrate solution. Mix thoroughly. Record your observations. This procedure using nitric acid and silver nitrate is a general test for chloride ions. The formation of a white precipitation is a positive test and indicates the presence of chloride ions. A positive test is obtained with any substance that produces chloride ions in solution.

 Dispose of solutions and precipitates containing silver in the container provided.

NAME ______________________

SECTION ______ DATE ______________

REPORT FOR EXPERIMENT 9

INSTRUCTOR ______________________

Composition of Potassium Chlorate

A. Determining Percentage Composition

Data Table	Sample 1	Sample 2
(a) Mass of crucible		
(b) Mass of crucible and sample		
(c) Mass of crucible and sample after first heating		
(d) Mass of crucible and sample after second heating		
(e) Mass of crucible and sample after third heating (if needed)		

Make all calculations necessary to complete the following. Be sure to carry the proper numbers of significant figures in your computations. Show calculation setups in the spaces provided in factor form (do not show longhand division or multiplication). Place the calculated answers in the spaces provided at the right.

1. **Mass of Original Sample**

   Calculation setups

   Sample 1:

   Sample 1 ______________

   Sample 2:

   Sample 2 ______________

2. **Mass Loss upon Heating**

   Calculation setups

   Sample 1:

                                                    Sample 1 _____

   Sample 2:

                                                    Sample 2 _____

---

3. **Mass of Residue**

   Calculation setups

   Sample 1:

                                                    Sample 1 _____

   Sample 2:

                                                    Sample 2 _____

---

4. **Percentage of Oxygen in $KClO_3$ (Experimental Value)**

   Calculation setups

   Sample 1:

                                                    Sample 1 _____

   Sample 2:

                                                    Sample 2 _____

---

5. **Percentage of KCl in KClO$_3$ (Experimental Value)**

    Calculation setups

    Sample 1:

    Sample 1 _____

    Sample 2:

    Sample 2 _____

---

6. **Percentage of Oxygen in KClO$_3$ from Formula (Theoretical Value)**

    Calculation setup

    _____

---

7. **Percentage of KCl in KClO$_3$ from Formula (Theoretical Value)**

    Calculation setup

    _____

---

8. **Percentage Error in Experimental Oxygen Determinations**

    Calculation setups

    Sample 1:

    Sample 1 _____

    Sample 2:

    Sample 2 _____

B. Qualitative Examination of Residue

1. Record what you observed when silver nitrate was added to the following:

    (a) Potassium chloride solution

    (b) Potassium chlorate solution

    (c) Residue solution

2. Questions

    (a) What evidence did you observe that would lead you to believe that the residue was potassium chloride?

    (b) Did the evidence obtained in the silver nitrate tests of the three solutions prove conclusively that the residue actually was potassium chloride? Explain.

SUPPLEMENTARY QUESTIONS AND PROBLEMS

1. Calculate the percentage of Cl in $Al(ClO_3)_3$.

    _____

2. Other metal chlorates when heated show behavior similar to that of potassium chlorate yielding metal chlorides and oxygen. Write the balanced formula equation for the reaction to be expected when calcium chlorate, $Ca(ClO_3)_2$, is heated.

EXPERIMENT 10

Double Displacement Reactions

MATERIALS AND EQUIPMENT

Solid: sodium sulfite (Na_2SO_3). **Solutions:** dilute (6 M) ammonium hydroxide (NH_4OH), 0.1 M ammonium chloride (NH_4Cl), 0.1 M barium chloride ($BaCl_2$), 0.1 M calcium chloride ($CaCl_2$), 0.1 M copper(II) sulfate ($CuSO_4$), dilute (6 M) hydrochloric acid (HCl), concentrated (12 M) hydrochloric acid (HCl), 0.1 M iron(III) chloride ($FeCl_3$), dilute (6 M) nitric acid (HNO_3), 0.1 M potassium nitrate (KNO_3), 0.1 M silver nitrate ($AgNO_3$), 0.1 M sodium carbonate (Na_2CO_3), 0.1 M sodium chloride (NaCl), 10 percent sodium hydroxide (NaOH), dilute (3 M) sulfuric acid (H_2SO_4), and 0.1 M zinc nitrate [$Zn(NO_3)_2$].

DISCUSSION

Double displacement reactions are among the most common of the simple chemical reactions and are comparatively easy to study.

In each part of this experiment two water solutions, each containing positive and negative ions, will be mixed in a test tube. Consider the hypothetical reaction.

$$AB + CD \longrightarrow AD + CB$$

where AB exists as A^+ and B^- ions in solution and CD exists as C^+ and D^- ions in solution. As the ions come in contact with each other, there are six possible combinations that might conceivably cause chemical reaction. Two of these combinations are the meeting of ions of like charge; that is, $A^+ + C^+$ and $B^- + D^-$. But since like charges repel, no reaction will occur. Two other possible combinations are those of the original two compounds; that is, $A^+ + B^-$ and $C^+ + D^-$. Since we originally had a solution containing each of these pairs of ions, they can mutually exist in the same solution; therefore they do not recombine. Thus the two possibilities for chemical reaction are the combination of each of the positive ions with the negative ion of the other compound; that is, $A^+ + D^-$ and $C^+ + B^-$. Let us look at some examples.

Example 1. When solutions of sodium chloride and potassium nitrate are mixed, the equation for the double displacement reaction (hypothetical) is

$$NaCl(aq) + KNO_3(aq) \longrightarrow KCl(aq) + NaNO_3(aq)$$

We get the hypothetical products by simply combining each positive ion with the other negative ion. But has there been a reaction? When we do the experiment, we see no evidence of reaction. There is no precipitate formed, no gas evolved, and no obvious temperature change. Thus we must conclude that no reaction occurred. Both hypothetical products are soluble salts, so the ions are still present in solution. We can say that we simply have a solution of four kinds of ions, Na^+, Cl^-, K^+, and NO_3^-.

The situation is best expressed by changing the equation to

$$NaCl(aq) + KNO_3(aq) \longrightarrow \text{No reaction}$$

– 85 –

Example 2. When solutions of sodium chloride and silver nitrate are mixed, the equation for the double displacement reaction (hypothetical) is

$$NaCl + AgNO_3 \longrightarrow NaNO_3 + AgCl$$

A white precipitate is produced when these solutions are mixed. This precipitate is definite evidence of a chemical reaction. One of the two products, sodium nitrate ($NaNO_3$) or silver chloride ($AgCl$), is insoluble. Although the precipitate can be identified by further chemical testing, we can instead look at the **Solubility Table in Appendix 6** to find that sodium nitrate is soluble but silver chloride is insoluble. We may then conclude that the precipitate is silver chloride and indicate this in the equation with an (s). Thus

$$NaCl(aq) + AgNO_3(aq) \longrightarrow NaNO_3(aq) + AgCl(s)$$

Example 3. When solutions of sodium carbonate and hydrochloric acid are mixed, the equation for the double displacement reaction (hypothetical) is

$$Na_2CO_3(aq) + 2\ HCl(aq) \longrightarrow 2\ NaCl(aq) + H_2CO_3(aq)$$

Bubbles of a colorless gas are evolved when these solutions are mixed. Although this gas is evidence of a chemical reaction, neither of the indicated products is a gas. But carbonic acid, H_2CO_3, is an unstable compound and readily decomposes into carbon dioxide and water.

$$H_2CO_3(aq) \longrightarrow H_2O(l) + CO_2(g)$$

Therefore, CO_2 and H_2O are the products that should be written in the equation. The original equation then becomes

$$Na_2CO_3(aq) + 2\ HCl(ag) \longrightarrow 2\ NaCl(aq) + H_2O(l) + CO_2(g)$$

The evolution of a gas is indicated by a (g).

Examples of other substances that decompose to form gases are sulfurous acid (H_2SO_3) and ammonium hydroxide (NH_4OH):

$$H_2SO_3(aq) \longrightarrow H_2O(l) + SO_2(g)$$
$$NH_4OH(aq) \longrightarrow H_2O(l) + NH_3(g)$$

Example 4. When solutions of sodium hydroxide and hydrochloric acid are mixed, the equation for the double displacement reaction (hypothetical) is

$$NaOH(aq) + HCl(aq) \longrightarrow NaCl(aq) + H_2O(l)$$

The mixture of these solutions produces no visible evidence of reaction, but on touching the test tube we notice that it feels warm. The evolution of heat is evidence of a chemical reaction. This example and **Example 1** appear similar because there is no visible evidence of reaction. However, the difference is very important. In **Example 1** all four ions are still uncombined. In the present example the hydrogen ions (H^+) and hydroxide ions (OH^-) are no longer free in solution but have combined to form water. The reaction of H^+ (an acid) and OH^- (a base) is called **neutralization.** The formation of the slightly ionized compound (water) caused the reaction to occur and was the source of the heat liberated.

Water is the most common slightly ionized substance formed in double displacement reactions; other examples are acetic acid ($HC_2H_3O_2$), oxalic acid ($H_2C_2O_4$), and phosphoric acid (H_3PO_4).

From the four examples cited we see that a double displacement reaction will occur if at least one of the following classes of substances is formed by the reaction:

1. A precipitate

2. A gas

3. A slightly ionized compound, usually water

PROCEDURE

Wear protective glasses.

Each part of the experiment (except No. 12) consists of mixing equal volumes of two solutions in a test tube. Use about a **3 mL sample** of each solution (about 1.5 cm of liquid in a standard test tube). It is not necessary to measure each volume accurately. Record your observations at the time of mixing. Where there is no visible evidence of reaction, feel each tube, or check with a thermometer, to determine if heat is evolved (exothermic reaction). In each case where a reaction has occurred, complete and balance the equation, properly indicating precipitates and gases. When there is no evidence of reaction, write the words "No reaction" as the right-hand side of the equation.

1. Mix 0.1 M sodium chloride and 0.1 M potassium nitrate solutions.

2. Mix 0.1 M sodium chloride and 0.1 M silver nitrate solutions.

3. Mix 0.1 M sodium carbonate and conc. (12 M) hydrochloric acid solutions.

4. Mix 10 percent sodium hydroxide and dil. (6 M) hydrochloric acid solutions.

5. Mix 0.1 M barium chloride and dil. (3 M) sulfuric acid solutions.

6. Mix **dilute** (6 M) ammonium hydroxide and **dilute** (3 M) sulfuric acid solutions.

7. Mix 0.1 M copper(II) sulfate and 0.1 M zinc nitrate solutions.

8. Mix 0.1 M sodium carbonate and 0.1 M calcium chloride solutions.

9. Mix 0.1 M copper(II) sulfate and 0.1 M ammonium chloride solutions.

10. Mix 10 percent sodium hydroxide and dil. (6 M) nitric acid solutions.

11. Mix 0.1 M iron(III) chloride and dil. (6 M) ammonium hydroxide solutions.

 12. **Do this part in the fume hood.** Add 1 g of solid sodium sulfite to 3 mL of water and shake to dissolve. Add about 1 mL of conc. (12 M) hydrochloric acid solution, dropwise, using a medicine dropper.

 Dispose of mixtures from reactions 2, 5, 7, 9 in the "heavy metal waste" container. Dispose of the contents of reaction 12 in the sink under the hood. Dispose of the contents of all other tubes in the sink and flush with water.

REPORT FOR EXPERIMENT 10

Double Displacement Reactions

Directions for completing table below:

1. Record your observations (Evidence of Reaction) of each experiment. Use the following terminology: (a) "Precipitate formed" (include the color), (b) "Gas evolved," (c) "Heat evolved," or (d) "No reaction observed."

2. Complete and balance the equation for each case in which a reaction occurred. First write the correct formulas for the products, taking into account the charges (oxidation numbers) of the ions involved. Then balance the equation by placing a whole number in front of each formula (as needed) to adjust the number of atoms of each element so that they are the same on both sides of the equation. Use (g) or (s) to indicate gases and precipitates. Where no evidence of reaction was observed, write the words "No reaction" as the right-hand side of the equation.

	Evidence of Reaction	Equation
1.		$NaCl + KNO_3 \longrightarrow$
2.		$NaCl + AgNO_3 \longrightarrow$
3.		$Na_2CO_3 + HCl \longrightarrow$
4.		$NaOH + HCl \longrightarrow$
5.		$BaCl_2 + H_2SO_4 \longrightarrow$
6.		$NH_4OH + H_2SO_4 \longrightarrow$
7.		$CuSO_4 + Zn(NO_3)_2 \longrightarrow$
8.		$Na_2CO_3 + CaCl_2 \longrightarrow$
9.		$CuSO_4 + NH_4Cl \longrightarrow$
10.		$NaOH + HNO_3 \longrightarrow$
11.		$FeCl_3 + NH_4OH \longrightarrow$
12.		$Na_2SO_3 + HCl \longrightarrow$

QUESTIONS AND PROBLEMS

1. The formation of what three classes of substances caused double displacement reactions to occur in this experiment?

   (a)

   (b)

   (c)

2. Write the equation for the decomposition of sulfurous acid.

3. Using three criteria for double displacement reactions, together with the Solubility Table in Appendix 6, predict whether a double displacement reaction will occur in each example below. If reaction will occur, complete and balance the equation, properly indicating gases and precipitates. If you believe no reaction will occur, write "no reaction" as the right-hand side of the equation. All reactants are in aqueous solution.

   (a) K_2S + $CuSO_4$ $\longrightarrow$

   (b) NH_4OH + $H_2C_2O_4$ $\longrightarrow$

   (c) KOH + NH_4Cl $\xrightarrow{\Delta}$

   (d) $NaC_2H_3O_2$ + HCl $\longrightarrow$

   (e) Na_2CrO_4 + $Pb(C_2H_3O_2)_2$ $\longrightarrow$

   (f) $(NH_4)_2SO_4$ + $NaCl$ $\longrightarrow$

   (g) $BiCl_3$ + $NaOH$ $\longrightarrow$

   (h) $KC_2H_3O_2$ + $CoSO_4$ $\longrightarrow$

   (i) Na_2CO_3 + HNO_3 $\longrightarrow$

   (j) $ZnBr_2$ + K_3PO_4 $\longrightarrow$

EXPERIMENT 11

Single Displacement Reactions

MATERIALS AND EQUIPMENT

Solids: strips of sheet copper, lead, and zinc measuring about 1 × 2 cm; and sandpaper or emery cloth. **Solutions:** 0.1 M copper(II) nitrate [$Cu(NO_3)_2$], 0.1 M lead(II) nitrate [$Pb(NO_3)_2$], 0.1 M magnesium sulfate ($MgSO_4$), 0.1 M silver nitrate ($AgNO_3$), and dilute (3 M) sulfuric acid (H_2SO_4).

DISCUSSION

The chemical reactivity of elements varies over an immense range. Some, like sodium and fluorine, are so reactive that they are never found in the free or uncombined state in nature. Others, like xenon and platinum, are nearly inert and can be made to react with other elements only under special conditions.

The **reactivity** of an element is related to its tendency to lose or gain electrons; that is, to be oxidized or reduced. In principle it is possible to arrange nearly all the elements into a single series in order of their reactivities. A series of this kind indicates which free elements are capable of displacing other elements from their compounds. Such a list is known as an **activity** or **electromotive series**. To illustrate the preparation of an activity series, we will experiment with a small group of selected elements and their compounds.

A generalized single displacement reaction is represented by the equation

$$A + BC \longrightarrow B + AC$$

Element A is the more active element and replaces element B from the compound BC. But if element B is more active than element A, no reaction will occur.

Let us consider two specific examples, using copper and mercury.

Example 1. A few drops of mercury metal are added to a solution of copper(II) chloride ($CuCl_2$).

Example 2. A strip of metallic copper is immersed in a solution of mercury(II) chloride ($HgCl_2$).

In Example 1 no change is observed even after the solution has been standing for a prolonged time, and we conclude that there is no reaction. In Example 2 the copper strip is soon coated with metallic mercury, and the solution becomes pale green. From this evidence we conclude that mercury will not displace copper in copper compounds but copper will displace mercury in mercury compounds. Therefore copper is a more reactive metal than mercury and is above mercury in the activity series. In terms of chemical equations these facts may be represented as

Example 1. $Hg(l) + CuCl_2(aq) \longrightarrow$ No reaction

Example 2. $Cu(s) + HgCl_2(aq) \longrightarrow Hg(l) + CuCl_2(aq)$

The second equation shows that, in terms of oxidation numbers (or charges), the chloride ion remained unchanged, mercury changed from +2 to 0, and copper changed from 0 to +2. The +2 oxidation state of copper is the one normally formed in solution.

Expressed another way, the actual reaction that occurred was the displacement of a mercury ion by a copper atom. This can be expressed more simply in equation form:

$$Cu^0(s) + Hg^{2+}(aq) \longrightarrow Hg^0(l) + Cu^{2+}(aq)$$

In contrast to double displacement reactions, single displacement reactions involve changes in oxidation numbers and therefore are also classified as **oxidation-reduction reactions.**

PROCEDURE

Wear protective glasses.

1. Place six clean test tubes in a rack and number them 1–6. To each, add about 4 mL of the solutions listed below.

2. Obtain three pieces of sheet zinc, two of copper, and one of lead. Clean the metal pieces with fine sandpaper or emery cloth to expose fresh metal surfaces. Add the metals to the test tubes with the solutions as listed.

> Tube 1: silver nitrate + copper strip
>
> Tube 2: copper(II) nitrate + lead strip
>
> Tube 3: lead(II) nitrate + zinc strip
>
> Tube 4: magnesium sulfate + zinc strip
>
> Tube 5: dilute (3M) sulfuric acid + copper strip
>
> Tube 6: dilute (3M) sulfuric acid + zinc strip

3. Observe the contents of each tube carefully and record any evidence of chemical reaction.

> Evidence of reaction will be either evolution of a gas (bubbles) or appearance of a metallic deposit on the surface of the metal strip. Metals deposited from a solution are often black or gray (in the case of copper, very dark reddish brown) and bear little resemblance to commercially prepared metals.
>
> With some of the combinations used in these experiments, the reactions may be slow or difficult to detect. If you see no immediate evidence of reaction, set the tube aside and allow it to stand for about 10 minutes, then reexamine it.

4. Pour the solutions in each test tube into the "heavy metals waste" container. Rinse the metals in tap water and dispose of the strips in the trash. Do not allow the metal strips to go into the sink or into the waste bottle.

NAME ____________________

SECTION ______ DATE ____________

REPORT FOR EXPERIMENT 11

INSTRUCTOR ____________________

Single Displacement Reactions

Evidence of Reaction Describe any evidence of reaction; if no reaction was observed, write "None".	Equation (to be completed) Write "No reaction", if no reaction was observed.
1.	$Cu + AgNO_3(aq) \longrightarrow$
2.	$Pb + Cu(NO_3)_2(aq) \longrightarrow$
3.	$Zn + Pb(NO_3)_2(aq) \longrightarrow$
4.	$Zn + MgSO_4(aq) \longrightarrow$
5.	$Cu + H_2SO_4(aq) \longrightarrow$
6.	$Zn + H_2SO_4(aq) \longrightarrow$

QUESTIONS AND PROBLEMS

1. Complete the following table by writing the symbols of the two elements whose reactivities are being compared in each test:

	Tube Number					
	1	2	3	4	5	6
Greater activity						
Lesser activity						

2. Arrange Pb, Mg, and Zn in order of their activities, listing the most active first.

(1) ____________

(2) ____________

(3) ____________

3. Arrange Cu, Ag, and Zn in order of their activities, listing the most active first.

    (1) _____

    (2) _____

    (3) _____

4. Arrange Mg, H, and Ag in order of their activities, listing the most active first.

    (1) _____

    (2) _____

    (3) _____

5. Arrange all five of the metals (excluding hydrogen) in an activity series, listing the most active first.

    (1) _____

    (2) _____

    (3) _____

    (4) _____

    (5) _____

6. On the basis of the reactions observed in the six test tubes, explain why the position of hydrogen cannot be fixed exactly with respect to all of the other elements listed in the activity series in Question 5.

7. What additional test(s) would be needed to establish the exact position of hydrogen in the activity series of the elements listed in Question 5?

8. On the basis of the evidence developed in this experiment:

    (a) Would silver react with dilute sulfuric acid? Why or why not?

    (b) Would magnesium react with dilute sulfuric acid? Why or why not?

EXPERIMENT 12

Ionization—Acids, Bases, and Salts

MATERIALS AND EQUIPMENT

Demonstration. **Solids:** sodium chloride (NaCl) and sugar ($C_{12}H_{22}O_{11}$). **Liquid:** glacial acetic acid ($HC_2H_3O_2$). **Solutions:** 0.1 M ammonium chloride (NH_4Cl), 1 M ammonium hydroxide (NH_4OH), 1 M acetic acid ($HC_2H_3O_2$), saturated barium hydroxide [$Ba(OH)_2$], 0.1 M copper(II) sulfate ($CuSO_4$), 1 M hydrochloric acid (HCl), 0.1 M nickel(II) nitrate [$Ni(NO_3)_2$], 0.1 M sodium bromide (NaBr), 1 M sodium hydroxide (NaOH), 0.1 M sodium nitrate ($NaNO_3$), and dilute (3 M) sulfuric acid (H_2SO_4). Conductivity apparatus; magnetic stirrer and stirring bar.

Solids: calcium hydroxide [$Ca(OH)_2$], calcium oxide (CaO), iron wire (paper clips), magnesium ribbon (Mg), magnesium oxide (MgO), marble chips ($CaCO_3$), sodium bicarbonate ($NaHCO_3$), sulfur (S), and wood splints. **Solutions:** dilute (6 M) acetic acid ($HC_2H_3O_2$), concentrated (15 M) ammonium hydroxide (NH_4OH), dilute (6 M) hydrochloric acid (HCl), dilute (6 M) nitric acid (HNO_3), phenolphthalein, 10 percent sodium hydroxide (NaOH), and dilute (3 M) sulfuric acid (H_2SO_4). 0.001 M HCl, 0.01 M HCl, 0.1 M HCl for pH measurements, pH meter.

DISCUSSION

A. Acids and Bases

Acids can be described as substances that yield hydrogen ions (H^+) when dissolved in water. This definition was first proposed by the Swedish chemist Arrhenius (over 100 years ago) for solutions that conduct electricity and share common properties such as sour taste and the ability to change the color of the plant dye, litmus. The corresponding definition for **bases** describes them as substances that yield hydroxide ions (OH^-) in water solutions. These definitions are the simplest way to think of acids and bases and still apply. However, they are limited because they apply only to aqueous solutions.

The more inclusive Brønsted-Lowry acid-base theory defines an acid as a substance that will liberate or give up a proton (a hydrogen ion, H^+) and a base as a substance that will combine with or accept a proton. In short, an **acid** is a **proton donor** and a **base** is a **proton acceptor.**

Thus, water behaves as both an acid and a base, as illustrated by the equation

$$H_2O + H_2O \longrightarrow H_3O^+ + OH^-$$
$$\text{acid} \quad \text{base} \qquad \text{acid} \quad \text{base}$$

One water molecule has donated a proton (acted as an acid) and another water molecule has accepted a proton (acted as a base). The hydronium ion, H_3O^+, is a hydrated hydrogen ion (H^+H_2O). To simplify writing equations, the formula of the hydronium ion is often abbreviated H^+. However, free hydrogen ions do not actually exist in solution.

Many compounds may be recognized as acids from their written formulas. The ionizable hydrogen atoms, which are responsible for the acidity, are written first, followed by the symbols of the other elements in the formula. Examples are

HCl	Hydrochloric acid	H_2CO_3	Carbonic acid
HNO_3	Nitric acid	HNO_2	Nitrous acid
H_2SO_4	Sulfuric acid	H_2SO_3	Sulfurous acid
$HC_2H_3O_2$	Acetic acid	$H_2C_2O_4$	Oxalic acid
H_3PO_4	Phosphoric acid		

In a like manner common bases may be recognized by their formulas as a hydroxide ion (OH^-) combined with a metal or other positive ion. Examples are

NaOH	Sodium hydroxide
KOH	Potassium hydroxide
$Ca(OH)_2$	Calcium hydroxide
$Mg(OH)_2$	Magnesium hydroxide
NH_4OH	Ammonium hydroxide

An aqueous solution will be either acidic, basic, or neutral, depending on the relative concentrations of H^+ and OH^-. In acidic solutions the concentration of the H^+ ions is greater than that of the OH^- ions. In basic solutions the concentration of the OH^- ions is greater than that of the H^+ ions. The terms **alkali** and **alkaline solutions** are used synonymously with base and basic solutions. If the concentrations of H^+ and OH^- are equal (as in water), the solution is neutral.

Nonmetallic oxides that react with water to form acids are called **acid anhydrides**. For example:

$$CO_2(g) + 2 H_2O(l) \longrightarrow 2H_2CO_3(aq)$$

Metal oxides that react with, water to form bases are **basic anhydrides**. For example:

$$CaO(s) + H_2O(l) \longrightarrow Ca(OH)_2(aq)$$

There are two general methods for determining the relative concentrations of H^+ and OH^- and thus whether a solution is acid, alkaline, or neutral. The older and simpler method is the use of **indicators**. These are organic compounds that change color at a particular hydrogen or hydroxide ion concentration. For example, litmus, a vegetable dye, shows a pink color in acidic solutions and a blue color in alkaline solutions. Another common indicator is phenolphthalein; it is colorless in acid solutions and pink in basic solutions. An indicator can only determine the relative concentrations of H^+ and OH^- within the range of its color changes.

A more precise method for measuring the H^+ in solutions is with an instrument called a pH meter. A pH meter is designed so that it measures the H^+ directly and is used when an accurate measurement of the concentration of H^+ is needed. The pH meter is described and explained more fully in section C of this discussion.

The reaction of an acid and a base to form water and a salt is known as **neutralization**. For example,

$$HCl(aq) + NaOH(aq) \longrightarrow H_2O(l) + NaCl(aq)$$

An indicator can be used to determine when the acid in a solution has been neutralized. For instance, when sodium hydroxide solution is added to a hydrochloric acid solution containing phenolphthalein, the solution turns faintly pink when the acid is neutralized and one more drop of the base is added.

B. Salts

Salts consist of a positively charged ion (H^+ excluded) and a negatively charged ion (O^{2-} and OH^- excluded). Salts may be formed by the reaction of acids and bases, or by replacing the hydrogen atoms in an acid with a metal, or by the interaction of two other salts. There are many more salts than acids and bases. For example, from a single acid such as HCl we may produce many chloride salts (e.g., $NaCl$, KCl, $RbCl$, $CaCl_2$, NH_4Cl, $FeCl_3$, etc.).

C. The Importance and Measurement of H^+ Ion Concentration

The H^+ ion has a great effect on many chemical reactions, including biological processes that sustain life. For example, the H^+ of human blood is regulated to very close tolerances. The concentration of this important ion is expressed as pH rather than other concentration expressions such as molarity. The pH is defined by this formula:

$$pH = -\log [H^+]$$

The logarithm (log) of a number is simply the power to which 10 must be raised to give that number. Thus, the log of 0.001 is 3 ($0.001 = 10^{-3}$). The H^+ written in brackets, $[H^+]$, means the concentration in moles/liter.

The pH of pure water is 7.0 at 25°C and is said to be neutral, that is, it is neither acidic nor basic because $[H^+]$ and $[OH^-]$ are equal (10^{-7} moles/liter). Solutions that are acidic have pH values less than 7.0. Solution that are basic have pH values greater than 7.0.

pH < 7.0	acid solutions
pH = 7.0	neutral solutions
pH > 7.0	basic (alkaline) solutions

A pH meter is a delicate instrument that comes in many versions and sizes which share some of these common features.

1. pH meters use a glass electrode that is immersed in the solution being tested. The electrode converts the H^+ concentration into an electrical potential which is read by a voltmeter calibrated in pH units.

2. pH meters are calibrated against standard solutions of known pH. After calibration, the electrode is immersed in the test solutions and the pH meter provides a value relative to the standard.

3. The operation of pH meters varies with different models. Your instructor will demonstrate the use of the pH meter available to you.

D. Electrolytes

Pure water will not conduct an electric current. However, when acids, bases, and salts interact with water the resulting aqueous solutions will conduct electricity. Therefore, acids, bases, and salts are called **electrolytes.** Other substances such as sugar and alcohol dissolve in water but are nonconductors and are called **nonelectrolytes.**

When aqueous solutions conduct electricity it is because they contain ions which are free to move in the solution. The electrical current through the solution is the movement of these ions to the positive and negative electrodes. The ions are the result of the dissociation of ionic substances such as the base NaOH and the salt NaCl:

$$NaOH(s) \longrightarrow Na^+(aq) + OH^-(aq)$$

$$NaCl(s) \longrightarrow Na^+(aq) + Cl^-(aq)$$

or the ionization of compounds which react with water to form hydrated ions. For example:

$$HCl(g) + H_2O(l) \longrightarrow H_3O^+(aq) + Cl^-(aq)$$

The necessity for water in this ionization process is illustrated by the fact that, when hydrogen chloride is dissolved in benzene, no ions are formed and the solution is a nonconductor. The formula HCl is used to represent both hydrogen chloride and hydrochloric acid.

Electrolytes are classified as strong or weak depending on the extent to which they exist as ions in solution. **Strong electrolytes** are essentially 100 percent ionized; that is, they exist totally as ions in solution. **Weak electrolytes** are considerably less ionized; only a small amount of the dissolved substance exists as ions, the remainder being in the un-ionized or molecular form. Most salts are strong electrolytes; acids and bases occur as both strong and weak electrolytes. Examples are as follows:

Strong Electrolytes	**Weak Electrolytes**
Most salts	$HC_2H_3O_3$
HCl	H_2SO_3
H_2SO_4	HNO_2
HNO_3	H_2CO_3
NaOH	H_2S
KOH	$H_2C_2O_4$
$Ba(OH)_2$	H_3PO_4
$Ca(OH)_2$	NH_4OH

In the first part of this experiment, the conductivity of many aqueous solutions will be demonstrated.

PROCEDURE

Wear protective glasses.

A. Conductivity of Solutions—Instructor Demonstration

All of the following tests (except number 8) are performed in 18 × 150 mm test tubes, using the conductivity apparatus shown in Figure 12.1 or other suitable conductivity apparatus. The electrodes should be rinsed thoroughly with distilled water between the testing of different solutions.

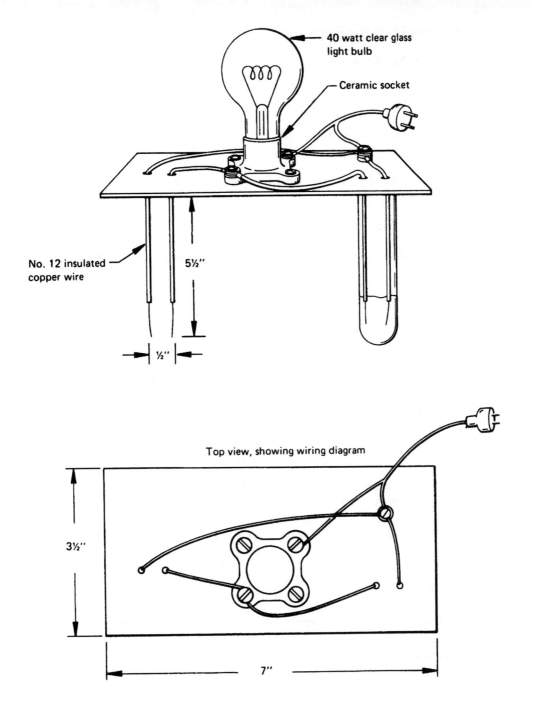

Figure 12.1 Conductivity apparatus.

Each test is performed by filling a test tube about half full of the liquid to be tested, then raising the test tube up around a pair of electrodes. When a measurable number of ions are in solution, the solution will conduct the electric current and the light will glow. A dimly glowing light indicates a relatively small number of ions in solution; a brightly glowing light indicates a relatively large number of ions in solution.

Note: The student should complete the data table in the report form at the time the demonstration is performed.

1. Test the conductivity of distilled water.

2. Test the conductivity of tap water.

3. Add a small amount of sugar to a test tube that is half full of distilled water. Dissolve the sugar and test the solution for conductivity.

4. Add a small amount of sodium chloride to a test tube that is half full of distilled water. Dissolve the salt and test the solution for conductivity.

5. Remove the plug from the electrical outlet, clean and dry the electrodes, and reconnect the plug.

    (a) Test the conductivity of glacial acetic acid.

    (b) Pour out half of the acid, replace with distilled water, mix, and test the solution for conductivity.

    (c) Pour out half of the solution in 5(b), replace with distilled water, mix, and test the solution for conductivity.

6. Strong and weak acids and bases. Test the following 1 molar solutions for conductivity: (a) acetic acid, (b) hydrochloric acid, (c) ammonium hydroxide, (d) sodium hydroxide. If the conductivity apparatus has two sets of electrodes, as shown in Figure 12.1, the relative conductivity of the strong and weak acids or bases may be compared by alternately raising a tube of each solution around the electrodes.

7. Test the following 0.1 M salt solutions for conductivity: (a) sodium nitrate, (b) sodium bromide, (c) nickel(II) nitrate, (d) copper(II) sulfate, and (e) ammonium chloride.

8. Clean the electrodes well. Place about 25 mL of distilled water and 1 drop of dil. (3 M) sulfuric acid in a 150 mL beaker. Place the beaker on a magnetic stirrer and dip one set of electrodes into the solution. With the stirrer turning, add saturated barium hydroxide solution dropwise until the light goes out completely. Add a few more milliliters of barium hydroxide solution.

B. Properties of Acids

Dispose of all solutions in the sink and flush with water. Take care to make sure that solids such as metal strips, splints, and unreacted marble chips do not go into the sink. They should be put into the wastebasket.

1. Reaction with a Metal

    (a) Into four consecutive test tubes place about 5 mL of dil. (6 M) hydrochloric, (3 M) sulfuric, (6 M) nitric, and (6 M) acetic acids.

    (b) Place a small strip of magnesium ribbon into each tube, one at a time, and test the gas evolved for hydrogen by bringing a burning splint to the mouth of the tube. If the liberation of gas is slow, stopper the test tube loosely for a minute or two before testing for hydrogen.

2. Measurement of Acidity and pH

    (a) Test dilute solutions of hydrochloric acid, acetic acid, and sulfuric acid by placing a drop of each acid from a stirring rod onto a strip of red and onto a strip of blue litmus paper. Note any color changes.

(b) Add 2 drops of phenolphthalein solution to about 5 mL of distilled water. Add several drops of dilute hydrochloric acid, mix, and note any color change.

(c) Use the pH meter to measure the pH of three dilutions of hydrochloric acid: 0.001 M HCl, 0.01 M HCl, and 0.1 M HCl. *Rinse the electrodes thoroughly with distilled water when done.*

3. Reaction with Carbonates and Bicarbonates

(a) Cover the bottom of a 150 mL beaker with a small quantity of sodium bicarbonate powder. Now add about 4 to 5 mL of dil. (6 M) hydrochloric acid to the beaker and cover with a glass plate. After about 30 seconds lower a burning splint into the beaker and observe the results. Dispose of the reaction mixture in the sink.

(b) Repeat the above experiment, using a few granules of marble chips (calcium carbonate) instead of sodium bicarbonate. Allow the reaction to proceed for 2 minutes before testing with the burning splint. Dispose of unreacted marble chips in the wastebasket, not the sink.

4. Reaction with Bases—Neutralization. To about 25 mL of water in a beaker, add 3 drops of phenolphthalein solution and 5 drops of dil. (6 M) hydrochloric acid. Using a medicine dropper, add 10 percent sodium hydroxide solution dropwise, stirring after each drop, until the indicator in the solution changes color. Then add dilute hydrochloric acid, drop by drop, stirring after each drop, until the indicator becomes colorless again. Repeat the additions of base and acid one or two more times.

5. Nonmetal Oxide plus Water

(a) **Do this part in the fume hood.** Place a small lump of sulfur in a deflagrating spoon and start it burning by heating in the burner flame. Lower the burning sulfur into a wide-mouth bottle containing 15 mL of distilled water and let the sulfur burn for 2 minutes. Remove the deflagrating spoon and quench the excess burning sulfur in a beaker of water. Cover the bottle with a glass plate and shake the bottle back and forth to dissolve the sulfur dioxide gas. Test the solution with blue litmus paper.

(b) As shown in Figure 12.2, fit a test tube with a one-hole stopper containing a glass delivery tube long enough to extend to the bottom of another test tube. Place several pieces of marble chips and a few milliliters of dil. (6 M) hydrochloric acid into the tube and insert the stopper. Bubble the liberated carbon dioxide into another test tube containing 10 mL water, 2 drops of 10 percent sodium hydroxide solution, and 2 drops of phenolphthalein solution. Record the results.

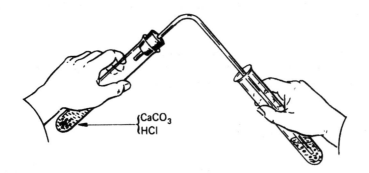

$\begin{cases} CaCO_3 \\ HCl \end{cases}$

Figure 12.2 Generator for carbon dioxide.

C. Properties of Bases

1. "Feel" Test. Make very dilute solutions by adding 3 drops of concentrated ammonium hydroxide to 10 mL of water in a test tube and 3 drops of 10 percent sodium hydroxide solution to 10 mL of water in another test tube. Rub a small amount of each very dilute solution between your fingers to obtain the characteristic "feel" of a hydroxide (base) solution. Wash your hands thoroughly immediately after making the "feel" test. Save the very dilute base solutions for the measurement of pH in the next section, C2(c).

2. **Measurement of Alkalinity**

   (a) Test the two base (alkaline) solutions prepared in C.1 with both red and blue litmus paper. Note any color changes.

   (b) Add 2 drops of phenolphthalein solution to each of the two alkaline solutions prepared in C.1. Note any color changes.

   (c) Pour the dilute ammonium hydroxide and sodium hydroxide that were prepared in the previous step into separate small beakers. Use the pH meter to measure the pH of these alkaline solutions. *Dip the electrode in dilute acetic acid and rinse thoroughly with distilled water when done.*

3. **Metal Oxides plus Water**

   (a) Place 10 mL of water and 2 drops of phenolphthalein solution in each of 3 test tubes. Add a pinch of calcium oxide to the first, magnesium oxide to the second, and calcium hydroxide to the third tube. Note and record the results.

   (b) Wind the end of a 5 cm piece of iron wire (or paper clip) around a small marble chip. Grasp the wire with tongs and heat the marble chip in the hottest part of the burner flame for about 2 minutes—the edges of the chip should become white hot while being heated. Allow the chip to cool; then drop it into a beaker containing 15 mL of water and 2 drops of phenolphthalein solution. For comparison, repeat this part of the experiment with a marble chip which has not been heated. Note the results. **Return the iron wire to the reagent shelf.**

4. **Reaction with Acids—Neutralization**

Review Part B.4.

REPORT FOR EXPERIMENT 12

Ionization—Acids, Bases, and Salts

A. Conductivity of Solutions—Instructor Demonstration

Complete the table for each of the substances tested in the ionization demonstration. Place an "X" in the column where the property of the substance tested fits the column description.

	Nonelectrolyte	Strong Electrolyte	Weak Electrolyte
1. Distilled Water			
2. Tap water			
3. Sugar			
4. NaCl			
5. a. $HC_2H_3O_2$ (glacial)			
b. 1st dilution			
c. 2nd dilution			
6. a. 1 M $HC_2H_3O_2$			
b. 1 M HCl			
c. 1 M NH_4OH			
d. 1 M NaOH			
7. a. $NaNO_3$			
b. NaBr			
c. $Ni(NO_3)_2$			
d. $CuSO_4$			
e. NH_4Cl			

8. (a) Write an equation for the chemical reaction that occurred between sulfuric acid and barium hydroxide.

   (b) Explain in terms of the properties of the products formed why the light went out when barium hydroxide was added to sulfuric acid solution, even though both of these reactants are electrolytes.

   (c) Explain why the light came on again when additional barium hydroxide was added.

9. In the conductivity tests, what controlled the brightness of the light?

10. Write an equation to show how acetic acid reacts with water to produce ions in solution.

11. What classes of compounds tested are electrolytes?

B. Properties of Acids

  1. **Reaction with a Metal**

   (a) Write the formulas of the acids which liberated hydrogen gas when reacting with magnesium metal.

   _____

   (b) Write equations to represent the reactions in which hydrogen gas was formed.

2. **Measurement of Acidity and pH**

   (a) What is the effect of acids on the color of red litmus?

   (b) What is the effect of acids on the color of blue litmus?

   (c) What color is phenolphthalein in an acid solution?   _____

   (d) What was the pH of the hydrochloric acids tested?

   0.001 M _____   0.01M _____   0.1M _____

   (e) Which pH measured has the highest number of H^+ in solution?   _____

   (f) What is the H^+ concentration in an acid with of pH 4.6?
        Express your answer as a power of 10   _____

        Refer to Study Aid 4 if you need help with using your calculator to convert the pH into
        H^+ concentration using the antilog function.

3. **Reaction with Carbonates and Bicarbonates**

   (a) What gas is formed in these reactions?

        Name _____   Formula _____

   (b) What happened to the burning splint when it was thrust into the beaker?

   (c) What do you conclude about one of the properties of the gas in the beaker, based on
        the behavior of the burning splint?

   (d) Complete and balance the equations representing the reactions:

        $NaHCO_3$ +   HCl $\longrightarrow$

        $CaCO_3$ +   HCl $\longrightarrow$

4. Reaction with Bases—Neutralization

(a) Write an equation for the neutralization reaction of HCl and NaOH.

(b) How did you know when all the acid was neutralized?

5. Nonmetal Oxide plus Water

(a) Write an equation for the combustion of sulfur in air.

(b) What acid is formed when the product of the sulfur combustion reacts with water?

Name _____   Formula _____

Write the equation for its formation.

(c) What evidence in this experiment leads you to believe that carbon dioxide in water has acidic properties?

(d) What acid is formed when carbon dioxide reacts with water?

Name _____   Formula _____

C. Properties of Bases

1. **"Feel" Test.** What is the characteristic feel of basic solutions?

2. **Measurement of Alkalinity**

(a) What is the effect of bases on the color of red litmus?

(b) What is the effect of bases of the color of blue litmus?

(c) What color is phenolphthalein in a basic solution?   _____

(d) What was the pH for each dilute base tested?

$NH_4OH(aq)$ _____ $NaOH(aq)$ _____

(e) Which base tested has the highest number of H^+ in solution?   _____

(f) What is the H^+ concentration in the strongest base tested?
   Express your answer as a power of 10.   _____

   Refer to Study Aid 4 if you need help using your calculator to convert the pH into H^+ concentration using the antilog function.

3. **Metal Oxides plus Water**

   (a.1) Color (if any) produced by phenolphthalein.

   Color with CaO in water   _____

   Color with MgO in water   _____

   Color with $Ca(OH)_2$ in water   _____

   (a.2) Complete and balance these equations:

   $$CaO + \quad H_2O \longrightarrow$$

   $$MgO + \quad H_2O \longrightarrow$$

   (b.1) The formula for marble is $CaCO_3$. What compounds are formed when it is heated strongly?

   (b.2) Write the equation representing this decomposition:

   $$CaCO_3(s) \xrightarrow{\Delta}$$

   (b.3) What evidence led you to formulate the composition of the solid residue after heating the marble chip?

ADDITIONAL QUESTIONS AND PROBLEMS

1. State whether each of the formulas below represents an **acid**, a **base**, a **salt**, or **none** of these types of compounds:

CuF_2	_____	$CaSO_4$	_____
$Ba(OH)_2$	_____	C_2H_4	_____
$LiOH$	_____	$C_{12}H_{22}O_{11}$	_____
$HBrO_3$	_____	HI	_____
$RaCO_3$	_____	P_2O_5	_____
KNO_2	_____	HCN	_____
$H_2C_2O_4$	_____	ZnS	_____

2. Complete and balance the following equations and name the product formed. (Only one product is formed in each case.)

   (a) $K_2O + H_2O \longrightarrow$ _____

   (b) $SrO + H_2O \longrightarrow$ _____

   (c) $SO_3 + H_2O \longrightarrow$ _____

   (d) $N_2O_5 + H_2O \longrightarrow$ _____

EXPERIMENT 13

Identification of Selected Anions

MATERIALS AND EQUIPMENT

Liquids: Decane ($C_{10}H_{22}$). **Solutions:** 0.1 M barium chloride ($BaCl_2$), freshly prepared chlorine water (Cl_2), dilute (6 M) hydrochloric acid (HCl), dilute (6 M) nitric acid (HNO_3), 0.1 M silver nitrate ($AgNO_3$), 0.1 M sodium arsenate (Na_3AsO_4), 0.1 M sodium bromide (NaBr), 0.1 M sodium carbonate (Na_2CO_3), 0.1 M sodium chloride (NaCl), 0.1 M sodium iodide (NaI), 0.1 M sodium phosphate (Na_3PO_4), 0.1 M sodium sulfate (Na_2SO_4), and unknown solutions.

DISCUSSION

The examination of a sample of inorganic material to identify the ions that are present is called **qualitative analysis.** To introduce qualitative analysis, we will analyze for seven anions (negatively charged ions). The ions selected for identification are chloride (Cl^-), bromide (Br^-), iodide (I^-), sulfate (SO_4^{2-}), phosphate (PO_4^{3-}), carbonate (CO_3^{2-}), and arsenate (AsO_4^{3-}).

Qualitative analysis is based on the fact that no two ions behave identically in all of their chemical reactions. Identification depends on appropriate chemical tests coupled with careful observation of such characteristics as solution color, formation and color of precipitates, evolution of gases, etc. Test reactions are selected to identify the ions in the fewest steps possible. In this experiment only one anion is assumed to be present in each sample. If two or more anions must be detected in a single solution, the scheme of analysis can be considerably more complex.

Silver Nitrate Test

When solutions of the sodium salts of the seven anions are reacted with silver nitrate solution, the following precipitates are formed: AgCl, AgBr, AgI, Ag_3PO_4, Ag_2CO_3, and Ag_3AsO_4. Ag_2SO_4 is moderately soluble and does not precipitate at the concentrations used in these solutions. When dilute nitric acid is added, the precipitates Ag_3PO_4, Ag_2CO_3, and Ag_3AsO_4 dissolve; AgCl, AgBr, and AgI remain undissolved.

In some cases a tentative identification of an anion may be made from the silver nitrate test. This identification is based on the color of the precipitate and on whether or not the precipitate is soluble in nitric acid. However, since two or more anions may give similar results, second or third confirmatory tests are necessary for positive identification.

Barium Chloride Test

When barium chloride solution is added to solutions of the sodium salts of the seven anions, precipitates of $BaSO_4$, $Ba_3(PO_4)_2$, $BaCO_3$, and $Ba_3(AsO_4)_2$ are obtained. No precipitate is obtained with Cl^-, Br^-, or I^-.

When dilute hydrochloric acid is added, the precipitates $Ba_3(PO_4)_2$, $BaCO_3$, and $Ba_3(AsO_4)_2$ dissolve; $BaSO_4$ does not dissolve.

Organic Solvent Test

The silver nitrate test can prove the presence of a halide ion (Cl^-, Br^-, or I^-) because the silver precipitates of the other four anions dissolve in nitric acid. But the colors of the three silver halides do not differ sufficiently to establish which halide ion is present.

Adding chlorine water (Cl_2 dissolved in water) to halide salts in solution will oxidize bromide ion to free bromine (Br_2) and iodide ion to free iodine (I_2). The free halogen may be extracted from the water solution by adding an immiscible organic solvent such as decane and shaking vigorously. The colors of the three halogens in organic solvents are quite different. Cl_2 is pale yellow, Br_2 is yellow-orange to reddish-brown, and I_2 is pink to violet. After adding chlorine water and shaking, a yellow-orange to reddish-brown color in the decane layer indicates that Br^- was present in the original solution; a pink to violet color in the decane layer indicates that I^- was present. However, a pale yellow color does not indicate Cl^-, since Cl_2 was added as a reagent. But if the silver nitrate test gives a white precipitate that is insoluble in nitric acid, and the organic solvent test shows no Br^- or I^-, then you can conclude that Cl^- was present.

Though we have described many of the expected results of these tests, it is necessary to test known solutions to actually see the results of the tests and to develop satisfactory experimental techniques. During this experiment, you will perform these tests on seven known anions.

Then, two "unknown" solutions, each containing one of the seven anions, will be analyzed. When an unknown is analyzed, the results should agree in all respects with one of the known anions. If the results do not fully agree with one of the seven known ions, either the testing has been poorly done or the unknown does not contain any of the specified ions.

Three different kinds of equations may be used to express the behavior of ions in solution. For example, the reaction of the chloride ion (from sodium chloride) may be written.

(1) $NaCl(aq) + AgNO_3(aq) \longrightarrow AgCl(s) + NaNO_3(aq)$

(2) $Na^+ + Cl^- + Ag^+ + NO_3^- \longrightarrow AgCl(s) + Na^+ + NO_3^-$

(3) $Cl^-(aq) + Ag^+(aq) \longrightarrow AgCl(s)$

Equation (1) is the **un-ionized equation;** it shows the formulas of the substances in the equation as they are normally written. Equation (2) is the **total ionic equation;** it shows the substances as they occur in solution. Strong electrolytes are written as ions; weak electrolytes, precipitates, and gases are written in their un-ionized or molecular form. Equation (3) is the **net ionic equation;** it includes only those substances or ions in Equation (2) that have undergone a chemical change. Thus Na^+ and NO_3^- (sometimes called the "spectator" ions) have not changed and do not appear in the net ionic equation. In both the total ionic equation or net equation, the atoms and charges must be balanced.

PROCEDURE

Wear protective glasses.

1. Clean nine test tubes and rinse each twice with 5 mL of distilled water. The first seven test tubes are for the known solutions that will be tested to demonstrate the expected reactions

with each anion. Use a marker to label these tubes as follows: NaCl, NaBr, NaI, Na$_2$SO$_4$, Na$_3$PO$_4$, Na$_2$CO$_3$, and Na$_3$AsO$_4$. The last two tubes are for your unknowns and should be left blank for now. Arrange these test tubes in order in your test tube rack.

> ⚠ Arsenic and its compounds are highly toxic even in very small quantities. Although you will use only small quantities, be very cautious and avoid contact with any known or unknown that contains arsenate. All waste that contains arsenate ions must be poured into the "arsenic waste" container.

2. Clean and rinse two more test tubes and take them to your instructor for your unknown solutions and their identification code. Label them with the code numbers immediately. To avoid possible confusion with the empty unknown test tubes in the rack, put these coded tubes aside in a beaker. Record the code of these unknowns in the top right-hand columns of your report form and label each of the blank tubes in the rack with one of these unknown code numbers. Pour 2 mL (no more) of each of the seven known solutions—one solution per tube—and 2 mL of the corresponding unknown into each unknown tube. Save the remaining portions of the unknown solutions for tests B and C.

You can save considerable time by measuring out 2 mL into the first test tube and using the height of this liquid in the test tube as a guide for measuring out the others.

 Dispose of solutions containing decane in the container marked "Waste solvents." Dispose of solutions containing silver, and barium, in the "heavy metals waste" container. Dispose of the contents of the tubes containing arsenic compounds in the container marked "arsenic waste."

> For each of the following tests that will be performed on known and unknown solutions, there is a corresponding block on the report form where observations should be recorded. If a precipitate forms, record "ppt formed" and include its color. If no precipitate forms, record "no ppt." When dissolving precipitates, record "ppt dissolved" or "ppt did not dissolve." For the decane solubility test, indicate the color of the decane layer.

A. Silver Nitrate Test

⚠ Silver nitrate will stain your skin black. If any silver nitrate gets on your hands, wash it off immediately to avoid these stains.

Add about 1 mL of 0.1 M silver nitrate solution to each test tube. Record the results. Now add about 3 mL of dilute (6 M) nitric acid to each test tube; stopper and shake well. Record the results.

B. Barium Chloride Test

Wash all nine test tubes and rinse each tube twice with distilled water. Again put about 2 mL of the specified solution into each of the nine test tubes. Add about 2 mL of 0.1 M barium chloride solution to each test tube and mix. Record the results. Now add 3 mL of dilute hydrochloric acid to each tube; stopper and shake well. Record the results.

C. Organic Solvent Test

Again wash and rinse all nine test tubes. Again put about 2 mL of the specified solution into each of the nine test tubes. Now add about 2 mL of decane and about 2 mL of chlorine water to each test tube; stopper and shake well. Record the results.

After completing the three tests, compare the results of the known solutions with your observations for your unknown solutions. Record the formula of the anion present in each solution on the report form (Part D).

REPORT FOR EXPERIMENT 13

Identification of Selected Anions

	NaCl	NaBr	NaI	Na_2SO_4	Na_3PO_4	Na_2CO_3	Na_3AsO_4	Unknown No. ___	Unknown No. ___
A. AgNO₃ Test Addition of AgNO₃ sol.									
Addition of dil. HNO₃									
B. BaCl₂ Test Addition of BaCl₂ sol.									
Addition of dil. HCl									
C. Organic Solvent Test Color of decane layer									
D. Formula of anion present in the solution tested.									

QUESTIONS AND PROBLEMS

1. The following three solutions were analyzed according to the scheme used in this experiment. Which one, if any, of the seven ions tested, is present in each solution? If the data indicate that none of the seven is present, write the word "None" as your answer.

    (a) **Silver Nitrate Test.**   Reddish-brown precipitate formed, which dissolved in dilute nitric acid.

    Barium Chloride Test.   White precipitate formed, which dissolved in dilute hydrochloric acid.

    Organic Solvent Test.   The decane layer remained almost colorless after treatment with chlorine water.

    Anion present _____

    (b) **Silver Nitrate Test.**   Red precipitate formed, which dissolved in dilute nitric acid to give an orange solution.

    Barium Chloride Test.   Yellow precipitate formed, which dissolved in dilute hydrochloric acid to give an orange solution.

    Organic Solvent Test.   The decane layer remained almost colorless after treatment with chlorine water.

    Anion present _____

    (c) **Silver Nitrate Test.**   Yellow precipitate formed, which dissolved in dilute nitric acid.

    Barium Chloride Test.   White precipitate formed, which dissolved in dilute hydrochloric acid.

    Organic Solvent Test.   The decane layer remained almost colorless after treatment with chlorine water.

    Anion present _____

2. Write un-ionized, total ionic, and net ionic equations for the following reactions:

(a) Sodium bromide and silver nitrate.

(b) Sodium carbonate and silver nitrate.

(c) Sodium arsenate and barium chloride.

3. Write net ionic equations for the following reactions. Assume that a precipitate is formed in each case.

(a) Sodium iodide and silver nitrate.

(b) Sodium arsenate and silver nitrate.

(c) Sodium phosphate and barium chloride.

(d) Sodium sulfate and barium chloride.

EXPERIMENT 14

Properties of Lead(II), Silver, and Mercury(I) Ions

MATERIALS AND EQUIPMENT

Solutions: concentrated (15 M) ammonium hydroxide (NH_4OH), dilute (6 M) hydrochloric acid (HCl), 0.1 M lead(II) nitrate [$Pb(NO_3)_2$], 0.1 M mercury(I) nitrate [$Hg_2(NO_3)_2$], dilute (6 M) nitric acid (HNO_3), 0.1 M potassium chromate (K_2CrO_4), and 0.1 M silver nitrate ($AgNO_3$); "known" solution containing Pb^{2+}, Ag^+, and Hg_2^{2+} ions; and "unknown" solutions to be analyzed.

DISCUSSION

The object of this experiment is to investigate some of the chemical properties of lead(II), silver, and mercury(I) cations and to identify these ions in solution. In the broader scheme of qualitative analysis these three cations are known as the silver group. They are grouped together because of the common property of forming water-insoluble chlorides, a property that enables then to be separated from most other cations.

First we will run some selected chemical reactions for each of these ions. Then a scheme of analysis based on these reactions will be used to separate and identify these cations in a "known" solution containing all three ions and in an "unknown" solution containing one or more of the ions.

If necessary, see Experiment 13 for an explanation of un-ionized, total ionic, and net ionic equations.

The net ionic equations representing the reactions to be observed are:

Lead(II) [$Pb(NO_3)_2$ solution]

$$Pb^{2+}(aq) + 2\ Cl^-(aq) \longrightarrow PbCl_2(s) \qquad \text{(white ppt forms)}$$

$$PbCl_2(s) \xrightarrow[\text{H}_2\text{O}]{\text{hot}} Pb^{2+}(aq) +\ 2Cl^-(aq) \qquad \text{(ppt dissolves)}$$

$$Pb^{2+}(aq) + CrO_4{}^{2-}(aq) \longrightarrow PbCrO_4(s) \qquad \text{(bright yellow ppt forms)}$$

Silver [$AgNO_3$ solution]

$$Ag^+(aq) + Cl^-(aq) \longrightarrow AgCl(s) \qquad\qquad\qquad \text{(white ppt forms)}$$

$$AgCl(s) + 2\ NH_4OH(aq) \longrightarrow Ag(NH_3)_2{}^+(aq) + Cl^-(aq) + 2\ H_2O(l) \quad \text{(ppt dissolves)}$$

$$Ag(NH_3)_2{}^+(aq) + Cl^-(aq) + 2\ H^+(aq) \longrightarrow AgCl(s) + 2\ NH_4{}^+(aq) \qquad \text{(white ppt forms)}$$

Mercury(I) [Hg(NO₃)₂ solution]

$$Hg_2^{2+} + 2\ Cl^- \longrightarrow Hg_2Cl_2(s) \quad \text{(white ppt forms)}$$

$$Hg_2Cl_2 + 2\ NH_4OH \longrightarrow Hg(l) + HgNH_2Cl(s) + 2\ H_2O + NH_4^+ + Cl^- \quad \text{(black ppt forms)}$$

PROCEDURE

Wear protective glasses.

General Instructions

1. Use **distilled water** throughout this experiment.

2. Since hot water is frequently needed in the procedure, fill a 400 mL beaker half full of water and start heating it to boiling before beginning to work with your cation solutions.

3. One mL quantities are used frequently throughout the procedure. You can save considerable time by determining how many drops are needed to deliver 1 mL from your medicine dropper or pipet and using this number of drops whenever 1 mL of reagent is required.

4. Submit a clean, labeled test tube to your instructor for your unknown solution.

5. Precipitates are washed to remove soluble ions. Washing is accomplished by adding the specified amount of solvent (usually water), agitating the mixture by gently shaking back and forth, and pouring off (decanting) the washing solvent. Shaking back and forth, rather than up and down, prevents the accumulation of large amounts of precipitate on the walls of the tube.

6. **Immediately** after making your observations, record them on the report form.

 Several toxic heavy metals and their solutions are used in this experiment. Dispose of all solutions and precipitates of Pb, Ag, Hg, and chromate compounds in the "heavy metals" waste bottle. This includes liquids decanted off a precipitate during washing.

A symbol will be used to remind you to dispose of these wastes properly.

A. Tests for Lead(II) Ion

1. To 2 mL (no more) of 0.1 M lead(II) nitrate solution in a test tube add 1 mL of dil. (6 M) hydrochloric acid.

 2. Allow the precipitate to settle and then separate it from the liquid by decanting the liquid. (Decanting means to pour the liquid off carefully, leaving the solid behind.) A loss of a small amount of precipitate at this point has no effect on the experimental results.

 3. Wash the precipitate with 2 mL of cold water, again decanting the liquid after the precipitate settles.

4. Add 5 mL of water to the precipitate and place the tube in the beaker of boiling water. Heat the mixture for about 2 minutes, shaking frequently. All the precipitate should dissolve.

5. Remove the tube from the beaker and add a few drops of 0.1 M potassium chromate solution. A yellow precipitate of lead(II) chromate confirms the presence of lead(II) ions.

B. Tests for Silver Ion

1. To 2 mL of 0.1 M silver nitrate solution in a test tube add 1 mL of dilute (6 M) HCl.

2. Allow the precipitate to settle and then decant the liquid.

3. Wash the precipitate with 2 mL of cold water, again decanting the liquid after the precipitate settles.

4. Add concentrated ammonium hydroxide dropwise to the precipitate until it all dissolves. (This should take less than 1 mL.)

5. Now add dil. (6 M) nitric acid dropwise until the solution is acidic (test with blue litmus paper). A white precipitate of silver chloride confirms the presence of silver ions.

C. Tests for Mercury(I) Ion

1. To 2 mL of 0.1 M mercury(I) nitrate solution in a test tube add 1 mL of dilute (6 M) HCl.

2. Allow the precipitate to settle and then decant the liquid.

3. Wash the precipitate with 2 mL of cold water, again decanting the liquid after the precipitate settles.

4. Add about 1 mL of concentrated (15 M) ammonium hydroxide to the precipitate. The formation of a black precipitate confirms the presence of mercury(I) ions. The precipitate is composed of finely divided black mercury (Hg) and white mercury(II) amido chloride ($HgNH_2Cl$) and appears black overall.

D. Analysis of a Known and an Unknown Solution

The sequence of steps that follows is to be performed on a known solution containing all three silver group cations and on an unknown solution obtained from your instructor. With the known solution you should see evidence of all reactions described for each cation. The unknown may contain one, two, or three cations; you will not see any evidence of reaction for a cation that is not present. Step 4 of the analysis requires the use of a funnel for each solution tested. If you have two funnels, you can run the known and unknown solutions simultaneously. If you have only one funnel, run the known solution first. A flow chart of the silver group analysis is shown in Figure 14.1.

1. To 2 mL of the sample in a test tube add 1 mL of dil. (6 M) HCl.

2. Mix by shaking, allow the precipitate to settle, and decant the liquid.

3. Wash the precipitate with 2 mL of cold water and decant the liquid. Wash the precipitate again with 2 mL of cold water and decant the liquid.

4. Add 5 mL of water to the precipitate and place the tube in the beaker of boiling water for about 3 minutes, shaking frequently. Put about 10 mL of water into each of two additional tubes and heat them in the boiling water. While the liquids are heating prepare a filter cone, place it in a funnel, wet the paper, and place the funnel in a test tube in the rack.

5. Mix and quickly pour the hot solution and the precipitate from D.4 into the funnel, catching the filtrate in the test tube. (Save this filtrate for part D.7.)

 6. Transfer the funnel to another test tube. Wash the precipitate **twice** by pouring 5 mL portions of hot water over the precipitate in the funnel. Use the hot water from one of the tubes that you have been heating in the boiling water. Discard this filtrate.

 7. Test the filtrate from D.5 for lead(II) ions by adding a few drops of 0.1 M K_2CrO_4 solution to it. A yellow precipitate confirms the presence of lead(II) ion.

8. Transfer the funnel to another test tube and add 1 mL of concentrated NH_4OH to the precipitate in the funnel. The formation of a black residue confirms the presence of mercury(I) ion.

 9. Add dil. (6 M) HNO_3 to the filtrate from D.8 until the solution is acidic to litmus or a permanent white precipitate is formed. The formation of this precipitate confirms the presence of silver ion. (A slightly cloudy filtrate from D.8 is due to incomplete removal of the lead(II) chloride but does not interfere with this test.)

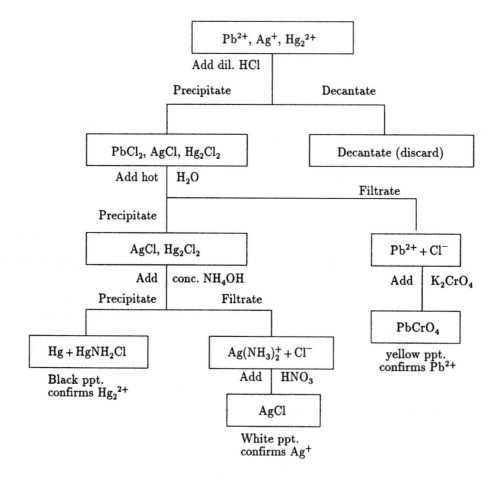

Figure 14.1 Flow Chart of the Silver Group Analysis

NAME ______________________

SECTION ______ DATE ______________

INSTRUCTOR ______________________

REPORT FOR EXPERIMENT 14

Properties of Lead(II), Silver and Mercury(I) Ions

A. Test for Lead(II) Ion

1. Record your observations for

   (a) Part A.1.

   (b) Part A.4.

   (c) Part A.5.

2. Write the name, formula, and color of the precipitate formed in Part A.1.

   ____

   ____

   ____

3. Write the name, formula, and color of the precipitate formed in the confirmatory test for lead(II) ion (Part A.5).

   ____

   ____

   ____

4. What is accomplished by washing a precipitate?

B.   Test for Silver Ion

   1.   Record your observations for

      (a) Part B.1.

      (b) Part B.4.

      (c) Part B.5.

   2.   Write the name, formula, and color of the precipitate formed in Part B.1.

                                                    _____

                                                    _____

                                                    _____

   3.   Write the name, formula, and color of the precipitate formed in the confirmatory test for silver ion (Part B.5.).

                                                    _____

                                                    _____

                                                    _____

C.   Tests for Mercury(I) Ion

   1.   Record your observations for

      (a) Part C.1.

      (b) Part C.4.

2. Write the name, formula, and color of the precipitate formed in Part C.1.

    _____

    _____

    _____

3. Write the name, formula, and color of the two substances formed in the confirmatory test for mercury(I) ion (Part C.4).

    _____   _____

    _____   _____

    _____   _____

D. Analysis of a Known and an Unknown Solution

1. The following questions pertain to the **known solution**.

    (a) Write the formulas of the substances precipitated when HCl was added (Part D.1).

        _____

    (b) What occurred when this precipitate was treated with hot water? What is the evidence for your conclusion (Part D.4)?

    (c) After filtering (Part D.5), why was the precipitate washed with more hot water?

    (d) What two reactions occurred simultaneously to the precipitate in Part D.8 when concentrated NH_4OH was added after the hot water wash?

    (e) What did you observe when you added dilute HNO_3 to the filtrate in Part D.9?

2. Unknown No. _____ ;   Cations present _____

QUESTIONS AND PROBLEMS

1. Suggest another reagent that could be used in place of hydrochloric acid to precipitate the cations of the silver group and still allow a sample to be analyzed by the scheme used in this experiment.

_____

2. For each of the following pairs of chlorides, select a reagent that will dissolve one of them and thus allow the separation of the two compounds.

$AgCl - PbCl_2$ _____

$AgCl - Hg_2Cl_2$ _____

$PbCl_2 - Hg_2Cl_2$ _____

3. What conclusions can be drawn about the cation(s) present in silver group unknowns that showed the following characteristics? Use formulas for the ions present.

(a) A white chloride precipitate was partially soluble in hot water and turned black when concentrated ammonium hydroxide was added to it.

Cation(s) Present   _____

(b) A white precipitate was formed on addition of hydrochloric acid. The precipitate was insoluble in hot water and soluble in concentrated ammonium hydroxide.

Cation(s) Present   _____

(c) A white precipitate was formed on addition of hydrochloric acid and it dissolved when the solution was heated.

Cation(s) Present   _____

EXPERIMENT 15

Quantitative Precipitation of Chromate Ion

MATERIALS AND EQUIPMENT

Solid: potassium chromate (K_2CrO_4). **Solution:** 0.50 M lead(II) nitrate [$Pb(NO_3)_2$].

DISCUSSION

In this experiment you will examine and verify the mole and mass relationships involved in the quantitative precipitation of chromate ion as lead(II) chromate. Potassium chromate is the source of the chromate ion, and lead(II) ion from lead(II) nitrate is the precipitating agent. The chemistry is expressed in these equations.

$$K_2CrO_4(aq) + Pb(NO_3)_2(aq) \longrightarrow PbCrO_4(s) + 2\ KNO_3(aq) \qquad \text{(Un-ionized equation)}$$

$$CrO_4^{2-}(aq) +\ Pb^{2+}(aq) \longrightarrow PbCrO_4(s) \qquad \text{(Net ionic equation)}$$

The first equation above shows that potassium chromate and lead(II) nitrate react with each other in a 1-to-1 mole ratio. Furthermore, for every mole of chromate ions present, 1 mole of lead(II) chromate is formed. From these molar relationships we can calculate the amount of lead(II) chromate that is theoretically obtainable from any specified amount of potassium chromate in the reaction. The theoretical value can then be compared to the experimental value.

To conduct the experiment quantitatively, we need to precipitate all the chromate ion from a known amount of potassium chromate in solution and to isolate the precipitate in as pure a form as feasible. To ensure complete precipitation of the chromate, an excess of lead(II) nitrate is used. Lead(II) chromate is only slightly soluble (2×10^{-7} mol/L), and the excess lead(II) nitrate reduces the solubility even further.

Use the following relationships in your calculations:

1. Moles K_2CrO_4 reacted = moles $Pb(NO_3)_2$ reacted = moles $PbCrO_4$ produced

2. $\text{Moles} = \dfrac{\text{g of solute}}{\text{molar mass of solute}}$

3. $\text{Molarity} = \dfrac{\text{moles}}{\text{liter}} = \dfrac{\text{g of solute}}{\text{molar mass of solute} \times \text{liters of solution}}$

Note that molarity is an expression of concentration, the units of which are moles of solute per liter of solution. Thus a 1.00 molar (1.00 M) solution contains 1.00 mole of solute in 1 liter of solution.

4. $\text{Percentage error} = \dfrac{(\text{experimental value}) - (\text{theoretical value})}{\text{theoretical value}} \times 100$

PROCEDURE

Wear protective glasses.

> 1. Make all weighings to the highest precision possible with the balance available.
>
> 2. Use the same balance for all weighings.
>
> 3. Record all data directly on the report form as they are obtained.

The sequence of major experimental steps in this experiment is as follows:

1. Weigh beaker.
2. Weigh potassium chromate into beaker.
3. Dissolve potassium chromate in distilled water.
4. Precipitate lead(II) chromate by adding lead(II) nitrate solution.
5. Weigh filter paper.
6. Filter and wash precipitated lead(II) chromate.
7. Heat and dry lead(II) chromate.
8. Determine the mass of lead(II) chromate precipitated.

> ⚠ Caution: the chromate ion (CrO_4^{2-}) and its compounds are possible carcinogens. Avoid contact with the solutions and precipitates in this experiment and dispose of all substances and washes in the "heavy metals" waste. The symbol 🗑 will be found throughout this procedure to remind you of this disposal warning.

Weigh a clean, **dry** 150 mL beaker. Add between 0.7 and 0.8 gram of K_2CrO_4 to the beaker; weigh again. Add 40 mL distilled water and dissolve the potassium chromate. Heat this solution to almost boiling. Obtain 10.0 mL of 0.50 M $Pb(NO_3)_2$ solution in a graduated cylinder and slowly add, while stirring, to the hot K_2CrO_4 solution. (To avoid losses of solution and precipitate, keep the stirring rod in the beaker.) Again heat the mixture almost to the boiling point while stirring.

Accurately weigh a piece of filter paper. Fold it into a cone, place it into a funnel, wet the paper with distilled water and press the top of the paper cone against the funnel. Separate the liquid from the precipitate by carefully decanting the liquid into the funnel, **keeping as much** **of the precipitate as possible in the beaker**. The decantation is best accomplished without losses by pouring the liquid down a stirring rod (see Figure 1.9 in Experiment 1). Dispose of the filtrate (not the precipitate).

The precipitate $PbCrO_4$, must now be washed free of excess $Pb(NO_3)_2$. Wash the precipitate by adding 15 mL of distilled water to the beaker and stirring. Allow the precipitate to settle for about a minute. Decant the wash water into the funnel, again keeping as much of the precipitate as possible in the beaker. Repeat the washing once more with another 15 mL of distilled water.

 After the last washing, spread the precipitate evenly on the bottom of the beaker with the stirring rod.

Using forceps, remove the filter paper from the funnel and use it to wipe off any $PbCrO_4$ adhering to the stirring rod. Place the filter cone down in the beaker as shown in Figure 15.1.

The following method of drying must be strictly followed to avoid spattering and scorching the precipitate and the filter paper. Adjust the burner so you have a non-luminous, 10 to 15 cm (4 to 6 in.) flame without a distinct inner cone. Place the beaker on a wire gauze on a ring stand so that the wire gauze is 10 to 15 cm (4 to 6 in.) above the tip of the flame. Heat the beaker and contents for 60 minutes. If spattering of the precipitate occurs remove the burner and either lower the flame or raise the beaker before continuing heating. Make a similar adjustment if you see evidence that the paper is scorching. **Do not leave the drying setup unattended.**

While the precipitate is drying start working on the report form.

Cool the beaker, weigh, and reheat for an additional 10 minutes. Cool again and reweigh. If the second weighing is within 0.05 g of the first, the contents of the beaker may be considered dry. If the second weighing has decreased more than 0.05 g, a third heating and weighing is necessary. The experiment is complete after obtaining constant weight (within 0.05 g).

Use forceps to remove the filter paper cone and put it into the solid "heavy metals" waste jar. Using a stirring rod or spatula, loosen the dry $PbCrO_4$ precipitate from the bottom of the beaker and add it to the waste jar. Add a small amount of water to remove the remaining precipitate.

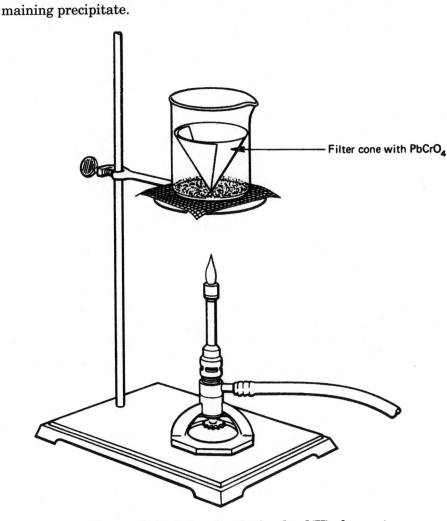

Filter cone with $PbCrO_4$

Figure 15.1   Setup for drying lead(II) chromate

REPORT FOR EXPERIMENT 15

Quantitative Precipitation of Chromate Ion

Data Table

	Mass
Mass of beaker	
Mass of beaker and K_2CrO_4	
Mass of filter paper	
Mass of beaker, $PbCrO_4$, and filter paper:	
After first heating	
After second heating	
After third heating (if needed)	

CALCULATIONS

Show calculation setups and answers. Remember to use the proper number of significant figures in all calculations. (The number 0.004 has only one significant figure!)

1. Using the data table, determine:

    (a) Mass of potassium chromate used. _____ g

    (b) Moles of potassium chromate used. _____ mol

    (c) Mass of dry lead(II) chromate obtained. _____ g

    (d) Moles of lead(II) chromate obtained. _____ mol

2. Calculate the number of moles and the mass of $PbCrO_4$ that can be theoretically produced from the mass of K_2CrO_4 that you used.

_____ mol

_____ g

3. Calculate the percentage error in your experimental mass of $PbCrO_4$ based on the theoretical mass calculated in 2.

_____

QUESTIONS AND PROBLEMS

1. Why is the mass of the filter paper needed?

2. Why is the mass of $PbCrO_4$ recovered greater than the starting mass of K_2CrO_4?

3. Calculate the moles and grams of $Pb(NO_3)_2$ present in the 10.0 mL of 0.50 M $Pb(NO_3)_2$ solution you used.

_____ mol

_____ g

4. Would the 10.0 mL of 0.50 M $Pb(NO_3)_2$ be sufficient to precipitate the chromate in 0.850 g of Na_2CrO_4? Show supporting calculations and explanation.

5. Calculate the moles of chromate ion in the K_2CrO_4 you used and in the $PbCrO_4$ obtained. Theoretically, why should these two values be the same?

moles $CrO_4{}^{2-}$ in K_2CrO_4 _____

moles $CrO_4{}^{2-}$ in $PbCrO_4$ _____

EXPERIMENT 16

Electromagnetic Energy and Spectroscopy

MATERIALS AND EQUIPMENT

Solutions: Nickel nitrate, $Ni(NO_3)_2$, 0.1M; potassium permanganate, $KMnO_4$, 0.002M.
Special equipment: Hand-held spectroscopes; 1.75 m long springs for simulating wave motion (CENCO #84740G) ; meter sticks; spectrophotometer and cuvettes; vapor lamps (hydrogen, neon) with power supplies; incandescent 60-100 watt bulbs, fluorescent light, spectrum chart, colored pencils, stopwatch.

DISCUSSION

Background: Electron Arrangements and Electromagnetic Energy

All substances will emit or absorb electromagnetic (EM) energy in a unique manner. You are most familiar with a range of EM energy known as visible light. A common example of emitted visible light energy is the red glow of a neon sign. A common example of the effects of absorbed visible light is the green color of the plant pigment, chlorophyll. The patterns of emitted and absorbed light by substances is the basis on which Neils Bohr and others built the model of atomic structure which is accepted today. In this model, the neutrons (no charge) and protons (positive charge) are packed tightly into a dense nucleus. The electrons are arranged in energy levels with the lowest energy electrons closest to the nucleus and the most energetic electrons farthest away. This can be shown in a diagram similar to the one below for sulfur.

increasing energy level $\longrightarrow$

(16P 17N)   n= 1   2   3
            2 e⁻   8 e⁻   6 e⁻

The hydrogen atom has only one proton in its nucleus (no neutrons) and one electron which will be in the first energy level when the atom is in its most stable configuration, called the ground state. When the atom absorbs energy this electron becomes excited and moves farther away from the nucleus into a higher energy level. Eventually, the excited electron loses energy and returns to a lower energy level, emitting energy during the transition. The absorption of certain wavelengths of emitted energy by the bonds of pigment molecules in the retina enable us to perceive light and color.

For the hydrogen electron, some of the possible transitions for an excited electron moving closer to its ground state are shown in Figure 16.1. If the transition of the electron is into energy level 2, the emitted energy fits the absorption spectrum of the vision pigments. For hydrogen, there are 4 such transitions. If the transition is into energy level 1 from a higher energy level, the emitted energy is not absorbed by the electrons of vision molecules and, although energetic and significant, the energy is not visible. If the transition is into energy level 3, the emitted energy is also not absorbed by vision pigments.

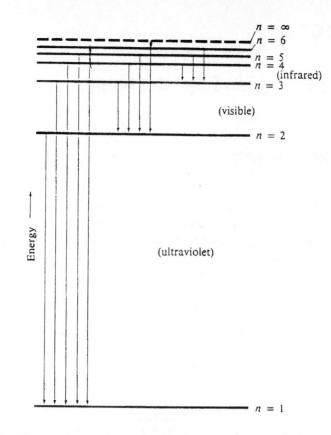

Figure 16.1   **Electron transitions in the H atom**

A. Wave Properties of Electromagnetic Energy

Because only certain values of EM energy can be absorbed for any particular atom, the absorption is said to be quantized. Because of this, any change in the electronic energy level of an atom involves the absorption or emission of a definite amount or quantum of energy. These packets or quanta of energy are called photons. They are emitted or absorbed by electrons in an atom as they change energy levels; and as they travel through space, they are referred to as electromagnetic radiation. All photons of light, regardless of their energy content, exhibit wave-like behavior and travel at the same speed in a vacuum (3.00×10^8 m/s). Therefore, photons are often described by their wave properties, specifically, their wavelength (lambda, λ) and their frequency (nu, v).

As photons move through space they do not move in a smooth straight line. Instead, their straight-line path is displaced slightly from its position (depending on its energy content) and once displaced, it tends to correct itself, returning to its original position, and then overcorrect itself. In this process, a wave pattern in generated. Light waves are described as transverse and have particle displacement perpendicular to the motion of the wave. The resultant wave is usually drawn as shown in Figure 16.2.

The wavelength of these waves is shown by λ and can be measured from peak to peak or trough to trough. Frequency tells how many waves pass a particular point per second. Depending on the energy content of the photon, the wave properties of a given particle of energy will vary greatly. Thus, there are photons with very short wavelengths (10^{-12} m) and very long wavelengths (10^4 m), or with very high frequencies (10^{20} waves/s) or with very low frequencies (10^9 waves/s). The relationship between wavelength, energy, and frequency will be demonstrated in the first activity of this experiment.

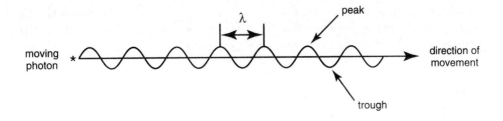

Figure 16.2   Transverse wave pattern of EM energy

B. Emission Spectra

Photons of energy are released or emitted by excited electrons as they fall back to lower energy levels. The range of photons characteristic of a given atom as its excited electrons fall closer to the nucleus is called its emission spectrum. Photons with a wavelength between 350 and 650 nm are absorbed by electrons in the retinal molecules of the eye which initiates a chain of events leading to their perception in the brain as visible light. Thus, these photons are called the visible light spectrum. A special filter or prism, called a spectroscope, can be used to separate photons of various wavelengths emitted by a given source so they can be identified.

C. Absorption Spectra

Just as atoms emit characteristic photons of energy when their excited electrons fall to lower energy levels, so all substances absorb characteristic photons when ground state electrons jump to higher energy levels. Thus, the range of photons absorbed by a given element or compound is unique for that element and is called its absorption spectrum. The instrument often used to identify the photons absorbed by a given substance is a spectrophotometer. Photons which are not absorbed are said to be transmitted.

The Spectronic 20 is the instrument commonly used in academic laboratories. The light source is a tungsten lamp which produces light over a specific range of wavelengths. A "tuner" or "filter" (usually a grating or a prism) selects a narrow band of wavelengths produced by the light source and sends it at a given intensity through the sample. A sample of a substance in a special glass sample holder called a cuvette, will absorb some of the light and allow some of the light to pass through (be transmitted). A detector (usually a phototube) measures the light beam which is transmitted and converts it to an electric current which will then move a needle on the dial or generate a digital readout. The dial or readout can be calibrated as absorbance (range from 0 to 1.5) or percent transmittance (range from 0 to 100%).

PROCEDURE

Wear protective glasses.

 Dispose of all solutions in waste containers provided.

A. Using a spring to generate and observe a transverse wave and its properties

Find a space on the floor where there is plenty of room to stretch a long (about 1.75 m) spring to about 4 m. A long hallway works very well for this. Two students should sit or kneel on the floor with the spring flat on the floor between them, each holding opposite ends securely. If the spring has a ring on each end, do **not** hold the spring by these rings which sometimes slip off and allow a sudden recoil of the spring. Place a short strip of masking tape on the floor at each end of the spring as shown in Figure 16.3. This facilitates keeping both ends of the spring at the same positions on the floor as the waves are generated.

strip of masking tape

strip of masking tape

stretched spring flat on floor
(about 4 m between tape strips)

Figure 16.3

One of the two students should now briskly displace the spring (Figure 16.4) using a flip of the wrist while holding their end on the floor in the taped position. The other end of the spring does not move at all.

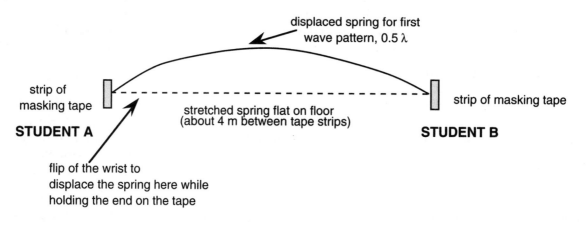

displaced spring for first
wave pattern, 0.5 λ

strip of
masking tape

strip of masking tape

stretched spring flat on floor
(about 4 m between tape strips)

STUDENT A

STUDENT B

flip of the wrist to
displace the spring here while
holding the end on the tape

Figure 16.4

Patiently experiment with manipulating the spring to generate the first wave pattern (shown in Figure 16.4) on the floor and then practice generating the second, third, and fourth wave patterns shown on the report form. Several students standing around the stretched spring can evaluate the patterns from a higher vantage and assist student A to know when the displacement and its frequency will best achieve the desired patterns. The first pattern (with a long wavelength) requires the least energy from the wrist of student A. As the wavelengths shorten, the energy required from the wrist of student A increases. Once the team understands what and how the spring manipulations are to be done, proceed as follows.

1. Measure and record the distance between the two pieces of tape in centimeters.

2. Students A and B should begin to generate the first wave form with a wavelength of 0.5 λ. When the wave form is correct, a third student will start counting the vibrations and timing (a stopwatch is best but a watch or clock with a second hand will also work). After 50 vibrations, note and record the elapsed time.

3. Repeat this procedure for waves with 1 λ, 1.5 λ, and 2 λ as shown in the first column on the report form. Remove the tape from the floor.

4. Calculate and record the frequency of each wave pattern on the report form. To find frequency, v:

$$v = \frac{\text{no. of waves}}{\text{elapsed time}} = \frac{50 \text{ waves}}{20 \text{ seconds}} = 2.5 \text{ waves / s}$$

5. Calculate and record the wavelength, λ, for each wave generated. For example, with 400 cm between tapes, the calculation for the wave pattern shown in Figure 16.5 is:

$$\lambda = \frac{400 \text{ cm}}{1.0 \text{ wave}} = 400 \text{ cm / wave}$$

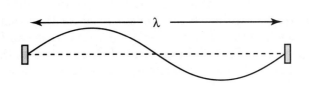

Figure 16.5

6. Use the graph paper provided in the report form to plot frequency vs. wavelength for the waves generated by the spring. In this experiment, the wavelength was the variable being determined by the experimenters. Follow the guidelines in Study Aid 3 for the completion of this graph.

B. Emission Spectra

1. **Examining the continuous spectrum:** Take a hand-held spectroscope and aim the end with the slit at an incandescent light. Put the viewing end (with the circular eyepiece that has a diffraction grating over it) up to your eye. To adjust the spectroscope, rotate the entire unit until the spectrum appears vertically on the side walls of the tube. Then, turn the slit portion until the widest spectral band is attained. The bands of color, merging smoothly into each other, are called a continuous spectrum. Make a sketch of the continuous spectrum on your report form using colored pencils.

2. **Examining bright-line spectra:** A bright-line spectrum is produced by hot gases of low density when electrons, excited by an electrical current, fall back to lower energy levels and emit photons. The spectrum which results has bright lines which correspond to the energy level transitions separated by dark spaces. The bright lines are determined by the kinds of atoms present in the gases and the amount of energy supplied. Each gas emits its own unique bright-line spectrum. Use the hand-held spectroscope to examine the bright line spectrum of a fluorescent light. Make a sketch of the bright-line spectrum for the fluorescent light in the space provided on your report form.

⚠️ Set up in the laboratory are spectrum tubes for hydrogen and neon. Do NOT try to adjust these tubes without unplugging the power supply. Observe the bright line spectrum for each of these gases using your spectroscope. Sometimes it is necessary to decrease the room light to see the lines from the spectrum tubes clearly. Sketch the bright-line spectrum for each gas on your report form.

C. Absorption Spectrum

1. The Spectrophotometer

Plug in the power line and switch the instrument on by turning the control knob clockwise past the click. Allow the instrument to warm-up for twenty minutes before making any measurements.

2. Measurement of the Absorption Spectra for Colored Solutions

The absorption spectrum for two aqueous solutions with visible color (referred to as the samples) will be measured using a Spectronic 20 instrument. The glass sample holders (cuvettes) in which the solutions are placed for these measurements must be clean and unscratched. Fill one cuvette half full with the solvent used for these solutions (distilled water). Fill the other two cuvettes half full with the nickel nitrate (green) and potassium permanganate (purple) solutions provided. This makes a total of three cuvettes that will be used. Several students can work together to measure the absorption spectrum for these solutions as follows:

a. Be sure the machine has been warmed up for about twenty minutes and that you understand the operation of this instrument before starting. This may require a brief demonstration by the instructor.

b. Select the desired wavelength using the wavelength knob. You will start with a wavelength of 350 nm. Insert the cuvette with distilled water into the sample holder and close the cover. Adjust the light control knob so 100% transmittance is read on the scale. This step is known as **calibrating** the instrument. You have adjusted the instrument so that 100% of the light of this wavelength (350 nm) has passed through the sample.

c. Remove the cuvette of distilled water from the sample holder, replace it with the green nickel nitrate solution and close the cover. Read and record the percent transmittance on the report form.

d. Remove the cuvette with the nickel nitrate solution, replace it with the purple potassium permanganate solution and close the cover. Read and record the percent transmittance as before.

e. Repeat steps b–d using the next wavelength shown on the data table (375 nm). It is necessary to calibrate the instrument with distilled water each time you change the wavelength, so repeat step b at each wavelength.

f. Continue until you have measured the percent transmittance for both solutions at wavelengths increasing by 25 nm up to 700 nm. If the available instrument cannot provide the full range of 350–700 nm wavelengths, ask the instructor how to proceed.

3. Graphing the Absorption Spectra Data

Plot the transmittance vs. wavelength data on the graph paper provided. If available, a computer can be used to plot and print this graph as described in Study Aid 3. If the computer option is used, attach a print-out of the resulting graph to the report form.

a. For this graph, the independent variable is the wavelength; the dependent variable is the percent transmittance.

b. Plotting this data will result in two lines, one for each solution. Design your graph with a figure legend (key) using lines and symbols that make it clear which line corresponds to which solution. It is very helpful to use colored pencils to make or cover the lines.

REPORT FOR EXPERIMENT 16

Electromagnetic Energy and Spectroscopy

A. Wave Properties

1. Length of stretched spring _____ cm

2. Complete the table below for each of the waves generated by your group.

Form of Wave	λ/cycle	No. of cycles	Time (s)	Frequency, cycles/s	Wavelength, cm/wave
	0.5 λ	50			
	1 λ	50			
	1.5 λ	50			
	2 λ	50			

3. What does the spring have to do with electromagnetic energy?

4. Plot a graph of frequency vs. wavelength for the data produced by the spring. Label the graph with a suitable title, determine an appropriate scale for each axis, and label with units that match the data. See Study Aid 3 for additional help if necessary.

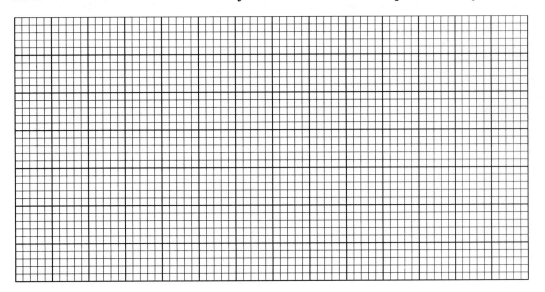

B. Emission Spectra

1. Use colored pencils and sketch the spectrum observed with the spectroscope for each of the light sources observed.

Light Source	Emission Spectrum Observed
Incandescent Bulb	
Fluorescent Bulb	
Hydrogen Gas	
Neon Gas	

2. a. Why must the hydrogen vapor lamp be turned on before it gives off light?

   b. Why is the spectroscope necessary to observe the hydrogen spectrum?

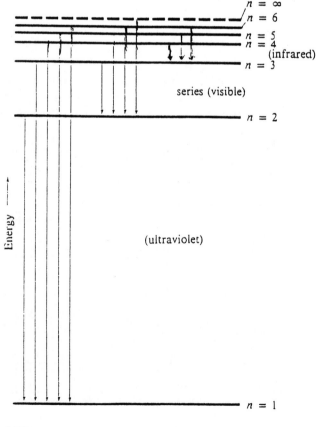

3. Use the colored pencils to color the arrows on the electron transition diagram for H_2 so they correspond to the colors of the visible H_2 spectral lines. Refer to the atomic spectrum chart to match the color to corresponding wavelength then decide the corresponding energy content. Remember that *the length of the arrow is a function of the energy content of the photon released and NOT its wavelength.*

4. Why can we see only 3 or 4 lines in the spectroscope when there are many more arrows in the hydrogen electron transition diagram?

C. Absorption Spectra for colored solutions

1. Record the percent transmittance data measured from the spectrophotometer for each of the solutions shown on the table below.

Percent transmittance

Wavelength, nm	$Ni(NO_3)_2$ (green)	$KMnO_4$ (purple)
350		
375		
400		
425		
450		
475		
500		
525		
550		
575		
600		
625		
650		
675		
700		

2. Graph these data as described in the procedure using either the graph paper provided or a computer if available.

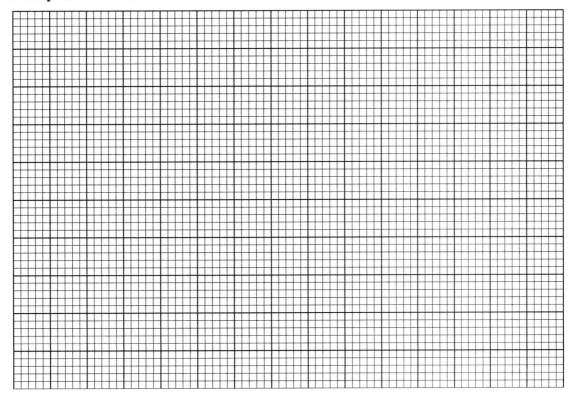

QUESTIONS AND PROBLEMS

1. Draw a diagram on the line below which shows 2.5 transverse waves. Measure the line and calculate the wavelength of a single wave in centimeters.

    _____

2. If your diagram represents a wave being generated with a spring like the one used in the experiment, and it took 25 seconds to generate 60 of these wave forms, what is the frequency of the wave? Show calculations.

3. What is the difference between an **emission spectrum** and an absorption spectrum?

4. What is the relationship **between** percent transmittance and absorption?

5. Where do the photons that are absorbed go when they are absorbed by the solution?

EXPERIMENT 17

Lewis Structures and Molecular Models

MATERIALS AND EQUIPMENT

Special equipment: Ball-and-stick molecular model sets

DISCUSSION

Molecules are stable groups of covalently bonded atoms, usually nonmetallic atoms. Chemists study models of molecules to learn more about their bonds, the spatial relationships between atoms and the shapes of molecules. Using models helps us to predict molecular structure.

A. Valence Electrons

Every atom has a nucleus surrounded by electrons which are held within a region of space by the attractive force of the positive protons in the nucleus. The electrons in the outermost energy level of an atom are called valence electrons. The valence electrons are involved in bonding atoms together to form compounds. For the representative elements, the number of valence electrons in the outermost energy level is the same as their group number in the periodic table (Groups IA–VIIA). For example, sulfur in Group VIA has six valence electrons and potassium in Group IA has one valence electron.

B. Lewis Structures

Lewis electron dot structures are a useful device for keeping track of valence electrons for the representative elements. In this notation, the nucleus and core electrons are represented by the atomic symbol and the valence electrons are represented by dots around the symbol. Although there are exceptions, Lewis structures emphasize an octet of electrons arranged in the noble gas configuration, $ns^2 np^6$. Lewis structures can be drawn for individual atoms, monatomic ions, molecules, and polyatomic ions.

1. Atoms and Monatomic Ions: A Lewis structure for an atom shows its symbol surrounded by dots to represent its valence electrons. Monatomic ions form when an atom loses or gains electrons to achieve a noble gas electron configuration. The Lewis structure for a monatomic ion is enclosed by brackets with the charge of the ion shown. The symbol is surrounded by the valence electrons with the number adjusted for the electrons lost or gained when the ion formed. This is the basis of ionic bond formation which is not included in this experiment.

Examples:

	sulfur atom	sulfide ion	potassium atom	potassium ion
	$:\overset{\cdot\cdot}{\underset{\cdot}{S}}\cdot$	$[:\overset{\cdot\cdot}{\underset{\cdot\cdot}{S}}:]^{2-}$	$K\cdot$	$[K]^{+}$

2. Molecules and Polyatomic Ions: Lewis structures for molecules and polyatomic ions emphasize the principle that atoms in covalently bonded groups achieve the noble gas configuration, ns^2np^6. Since all noble gases except helium have eight valence electrons, this is often called the octet rule. Although many molecules and ions have structures which support the octet rule, it is only a guideline. There are many exceptions. One major exception is the hydrogen atom which can covalently bond with only one atom and share a total of two electrons to form a noble gas configuration like helium. All of the examples in this experiment follow the octet rule except hydrogen.

A Lewis structure for covalently bonded atoms is a two-dimensional model in which one pair of shared electrons between two atoms is a single covalent bond represented by a short line; unshared or lone pairs of electrons are shown as dots. Sometimes two pairs of electrons are shared between two atoms forming a double bond and are represented by two short lines. It is even possible for two atoms to share three pairs of electrons forming a triple bond, represented by three short lines. For a polyatomic ion, the rules are the same except that the group of atoms is enclosed in brackets and the overall charge of the ion is shown. For example:

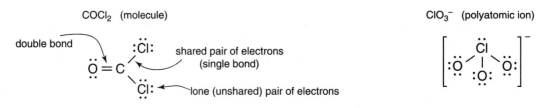

The rules for writing Lewis structures for molecules and polyatomic ions will be provided in the procedure section so you can use your Lewis structures to build three-dimensional models.

C. Molecular Model Building

The three-dimensional structure of a molecule is difficult to visualize from a two-dimensional Lewis structure. Therefore, in this experiment, a ball-and-stick model kit (molecular "tinker toys") is used to build models so the common geometric patterns into which atoms are arranged can be seen. Each model that is constructed must be checked by the instructor and described by its geometry and its bond angles on the report form.

D. Molecular Geometry

Atoms in a molecule or polyatomic ion are arranged into geometric patterns that allow their electron pairs to get as far away from each other as possible (which minimizes the repulsive forces between them). The theory underlying this molecular model is known as the valence shell electron pair repulsion (VSEPR) theory. All of the geometric structures in this experiment fall into the following patterns:

1. **Tetrahedral:** four pairs of shared electrons (no pairs of lone (unshared) electrons) around a central atom.

2. **Trigonal pyramidal:** three pairs of shared electrons and one pair of unshared electrons around a central atom.

3. **Trigonal planar:** three groups of shared electrons around a central atom. Two of these groups are single bonds and one group is a double bond made up of two pairs of shared electrons. There are no unshared electrons around the central atoms.

4. **Bent:** two groups of shared electrons (in single or double bonds) and one or two pairs of unshared electrons around a central atom.

5. **Linear:** two groups of shared electrons, usually double bonds with two shared electron pairs between two atoms, and no unshared electrons around a central atom. When there are only two atoms in a molecule or ion, and there is no central atom (HBr, for example), the geometry is also linear. These patterns are described more extensively in Section E, which follows.

> **Note:** There are other electron arrangements and molecular geometries. Since they do not follow the octet rule, they are not included in this experiment.

E. Bond Angles

Bond angles always refer to the angle formed between two end atoms with respect to the central atom.

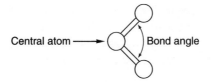

Central atom ⟶ ⬡ Bond angle

The size of the angle depends mainly on the repulsive forces of the electrons around the central atom. The molecular model kits are designed so that these angles can be determined when sticks representing electron pairs are inserted into pre-drilled holes.

1. Bond angles for atoms bonded to a central atom **without** unshared electrons on the central atom.

    a. For four pairs of shared electrons around a central atom (tetrahedral geometry) the angle between the bonds is approximately **109.5°**.

$$:\overset{\cdot\cdot}{\underset{\cdot\cdot}{Cl}}:$$
$$:\overset{\cdot\cdot}{\underset{\cdot\cdot}{Cl}} - C - \overset{\cdot\cdot}{\underset{\cdot\cdot}{Cl}}: \quad 109.5°$$
$$:\overset{\cdot\cdot}{\underset{\cdot\cdot}{Cl}}: \quad 109.5°$$

    b. For three atoms bonded to a central atom, (trigonal planar) the angle is **120°**. The shared electron pairs can be arranged in single or double bonds.

H   120°
120° C=Ö
H   120°

    c. For two atoms bonded to a central atom (linear) the angle is **180°**. The shared electrons are usually arranged in double bonds.

180°
Ö=C=Ö

    d. Linear diatomic molecules or ions with no central atom do not have a bond angle.

H—Cl:
[H—O:]⁻

2. Bond angles for atoms bonded to a central atom **with** unshared electrons on the central atom.

When some of the valence electrons around a central atom are unshared, the VSEPR theory can be used to predict changes in spatial arrangements. An unshared pair of electrons on the central atom has a strong influence on the shape of the molecule. It reduces the angle of bonding pairs by squeezing them toward each other.

For example:

tetrahedral	trigonal pyramidal	bent
No unshared electrons	1 unshared electron pair	2 unshared lone pairs
4 pairs shared electrons	3 pairs shared electrons	2 pairs shared electrons
CH_4	NH_3	H_2O

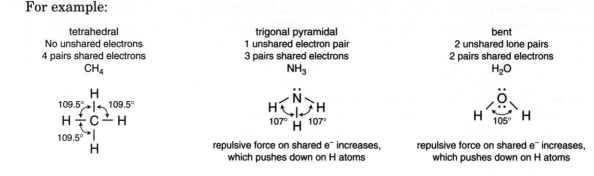

repulsive force on shared e^- increases, which pushes down on H atoms

repulsive force on shared e^- increases, which pushes down on H atoms

F. Bond Polarity

Electrons shared by two atoms are influenced by the positive attractive forces of both atomic nuclei. For like atoms, these forces are equal. For example, in diatomic molecules such as H_2 or Cl_2 the bonded atoms have exactly the same electronegativity (affinity for the bonding electrons). Electronegativity values for most of the elements have been assigned.

$$H — H$$

$$:\ddot{C}l — \ddot{C}l:$$

Electronegativity Table

1 H 2.1																	2 He
3 Li 1.0	4 Be 1.5											5 B 2.0	6 C 2.5	7 N 3.0	8 O 3.5	9 F 4.0	10 Ne
11 Na 0.9	12 Mg 1.2											13 Al 1.5	14 Si 1.8	15 P 2.1	16 S 2.5	17 Cl 3.0	18 Ar
19 K 0.8	20 Ca 1.0	21 Sc 1.3	22 Ti 1.4	23 V 1.6	24 Cr 1.6	25 Mn 1.5	26 Fe 1.8	27 Co 1.8	28 Ni 1.8	29 Cu 1.9	30 Zn 1.6	31 Ga 1.6	32 Ge 1.8	33 As 2.0	34 Se 2.4	35 Br 2.8	36 Kr
37 Rb 0.8	38 Sr 1.0	39 Y 1.2	40 Zr 1.4	41 Nb 1.6	42 Mo 1.8	43 Tc 1.9	44 Ru 2.2	45 Rh 2.2	46 Pd 2.2	47 Ag 1.9	48 Cd 1.7	49 In 1.7	50 Sn 1.8	51 Sb 1.9	52 Te 2.1	53 I 2.5	54 Xe
55 Cs 0.7	56 Ba 0.9	57–71 La–Lu 1.1–1.2	72 Hf 1.3	73 Ta 1.5	74 W 1.7	75 Re 1.9	76 Os 2.2	77 Ir 2.2	78 Pt 2.2	79 Au 2.4	80 Hg 1.9	81 Tl 1.8	82 Pb 1.8	83 Bi 1.9	84 Po 2.0	85 At 2.2	86 Rn
87 Fr 0.7	88 Ra 0.9	89–103 Ac–Lr 1.1–1.7	104 Rf —	105 Db —	106 Sg —	107 Bh —	108 Hs —	109 Mt —									

Key: 9 — Atomic number; F — Symbol; 4.0 — Electronegativity

* The electronegativity value is given below the symbol of each element.

In general, electronegativity increases as we move across a period and up a group on the periodic table. Identical atoms with identical attractions for their shared electron pairs form **nonpolar covalent bonds.** Unlike atoms exert unequal attractions for their shared electrons and form **polar covalent bonds.**

Electronegativity is used to determine the direction of bond polarity which can be indicated in the Lewis structure by replacing the short line for the bond with a modified arrow (+———➞) pointed towards the more electronegative atom. For example, nitrogen and hydrogen have electronegativity values of 3.0 and 2.1, respectively. The N—H bond is thus represented as

N ⟷ H with the arrow directed toward the more electronegative nitrogen atom. Then, the Lewis structure can be redrawn with arrows replacing the dashes as shown for NH_4^+ and NH_3.

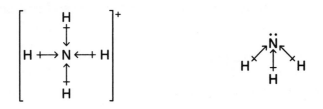

G. Molecular Dipoles

When there are several polar covalent bonds within a molecule as in NH_4^+ and NH_3 the polar effect of these bonds around a central atom can be cancelled if they are arranged symmetrically as shown in CCl_4 below. On the other hand, if the arrangement of the polar bonds is asymmetrical, as in the bent water molecule, H_2O, the resulting molecule has a definite positive end and oppositely charged negative end, and the molecule is called a dipole. In water, the H atoms have a partial positive charge, δ^+, and the O atom has a partial negative charge, δ^-. The symmetry, or lack of symmetry of molecules and polyatomic ions, can generally be seen in the three-dimensional model.

PROCEDURE

Follow steps **A–G** for each of the molecules or polyatomic ions listed on the report form. Refer back to the previous discussion, organized into corresponding sections A–G, for help with each step if necessary.

A. Number of Valence Electrons in a Molecule or Polyatomic Ion

Use a periodic table to determine the number of valence electrons for each group of atoms in the first column of the report form.

    example: SiF_4 Si is in Group IVA, it has 4 valence electron
                                  F is in Group VIIA, it has 7 valence electrons

    Total valence electrons is 4 + 4(7) = 32 electrons

If the group is a polyatomic ion, total the electrons as above, then add one electron for each negative charge or subtract one electron for each positive charge.

    example: CO_3^{2-} C is in group IVA, it has 4 valence electrons
                                  O is in Group VIA, it has 6 valence electrons
                                  Ion has a –2 charge, add 2 electrons

    Total valence electrons is 4 + 3(6) +2 = 24 electrons

B. Lewis Structures for Molecules and Polyatomic Ions

Use the following rules to show the two-dimensional Lewis structure for each molecule or polyatomic ion. Put your structure in the space provided. Use a **sharp** pencil and be as neat as possible.

1. Write down the skeletal arrangement of the atoms and connect them with a single covalent bond (a short line). We want to keep the rules at a minimum for this step, but we also want to avoid arrangements which will later prove incorrect. Useful guidelines are

   a. carbon is usually a central atom or forms bonds with itself; if carbon is absent, the central atom is usually the least electronegative atom in the group;

   b. hydrogen, which has only one valence electron, can form only one covalent bond and is never a central atom;

   c. oxygen atoms are not normally bonded to each other except in peroxides, and oxygen atoms normally have a maximum of two covalent bonds (two single bonds or one double bond).

Using these guidelines, skeletal arrangements for SiF_4 and CO_3^{2-} are

2. Subtract two electrons from the total valence electrons for each single bond used in the skeletal arrangement. This calculation gives the net number of electrons available for completing the structure. In the examples above, there are 4 and 3 single bonds, respectively. With 2 e^- per bond the calculation is

SiF_4: 32 e^- – 4(2e^-) = 24 e^- left to be assigned to the molecule

CO_3^{2-}: 24 e^- – 3(2e^-) = 18 e^- left to be assigned to the polyatomic ion

3. Distribute these remaining electrons as pairs of dots around each atom (except hydrogen) to give each atom a total of eight electrons around it. If there are not enough electrons available, move on to step 4.

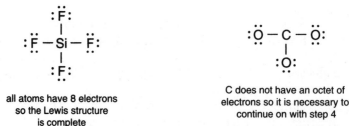

all atoms have 8 electrons
so the Lewis structure
is complete

C does not have an octet of
electrons so it is necessary to
continue on with step 4

4. Check each Lewis structure to determine if every atom except hydrogen has an octet of electrons. If there are not enough electrons to give each of these atoms eight electrons, change

single bonds between atoms to double or triple bonds by shifting unshared pairs of electrons as needed. A double bond counts as 4 e$^-$ for each atom to which it is bonded.

For CO_3^{2-}, shift 2e$^-$ from one of the O atoms and place it between C and that O.

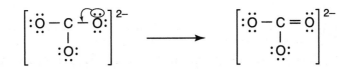

Now, all the atoms have 8 e– around them. (Don't forget the $^{2-}$)

C. Model Building

1. Use the balls and sticks from the kit provided to build a 3-dimensional model of the molecule or polyatomic ion for each Lewis structure in the report form.

    a. Use a ball with 4 holes for the central atom.
    b. Use inflexible sticks for single bonds.
    c. Use flexible connectors for double or triple bonds.
    d. Use inflexible sticks for lone pairs *around the central atom* only.

2. **Leave the model together until it is checked by the instructor.** If you have to wait for someone to check your model, start building the next model on the list. If you complete each structure so fast that you run out of components before someone checks your models, work on other parts of the experiment. If the models are not checked, you will not get credit for them.

D. Molecular Geometry

Look at your model from all angles and compare its structure to the description in the discussion (Section D). Then identify its molecular geometry from the following list and write the name of the geometric pattern on the report form in column D.

    1. tetrahedral
    2. trigonal pyramidal
    3. trigonal planar
    4. bent
    5. linear

E. Central Bond Angles

Fill in column E with the bond angles between the central atom and all atoms attached to it. Review the discussion (Section E) to find the value of the angles associated with each geometric form. For molecules with more than one central atom, give bond angles for each. For molecules without a central atom and hence no bond angle, write *no central atom.*

F. Bond Polarity

Bond polarity can be determined by looking up the electronegativity values for both atoms in the Electronegativity table. In the F column of the report form, draw the symbols for both atoms involved in a bond and connect them with an arrow pointing toward the more electronegative atom. If there are several identical bonds it is only necessary to draw one. Use the following as examples.

$$N \longleftarrow\!\!\!\!+ H \qquad\qquad\qquad S +\!\!\!\longrightarrow O$$

G. Molecular Dipoles

Look at the model and evaluate its symmetry. Decide if the polar bonds within it cancel each other around the central atom resulting in a nonpolar molecule or if they do not cancel one another and result in a dipole. Some examples:

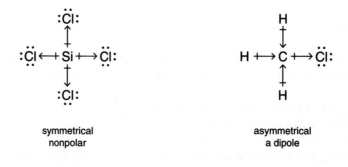

symmetrical nonpolar	asymmetrical a dipole

Remember, it is also possible for all the polar bonds within a polyatomic ion to cancel each other so the resultant effect is nonpolar even though the group as a whole has a net charge.

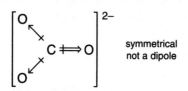

symmetrical
not a dipole

NAME ⎯⎯⎯⎯⎯⎯

SECTION ⎯⎯⎯ DATE ⎯⎯⎯

INSTRUCTOR ⎯⎯⎯

REPORT FOR EXPERIMENT 17

Lewis Structures and Molecular Models

For each of the following molecules or polyatomic ions, fill out columns A through G using the instructions provided in the procedure section. These instructions are summarized briefly below.

A. Calculate the total number of valence electrons in each formula.
B. Draw a Lewis structure for the molecule or ion which satisfies the rules provided in the procedure.
C. Build a model of the molecule and have it checked by the instructor.
D. Use your model to determine the molecular geometry for this molecule (don't try to guess the geometry without the model):

    tetrahedral, trigonal pyramidal, trigonal planar, bent, linear

E. Determine the bond angle between the central atom and the atoms bonded to it. If there are only two atoms write "no central atom" in the space provided.
F. Use the electronegativity table to determine the electronegativity of the bonded atoms.
    If the bonds are polar, indicate this with a modified arrow (+⟶) pointing to the more electronegative element.
    If the bonds are nonpolar, indicate this with a short line (—).
    If there are two or more different atoms bonded to the central atom, include each bond.
G. Use your model and your knowledge of the bond polarity to determine if the molecule as a whole is nonpolar or a dipole. If it is polar, write *dipole* in G. If it is not, write *nonpolar*.

	A	B	C	D	E	F	G
Molecule or Polyatomic Ion	No. of Valence Electrons	Lewis Structure		Molecular Geometry	Bond Angles	Bond Polarity	Molecular Dipole or Nonpolar
CH₄							
CS₂							

– 151 –

Molecule or Polyatomic Ion	A No. of Valence Electrons	B Lewis Structure	C	D Molecular Geometry	E Bond Angles	F Bond Polarity	G Molecular Dipole or Nonpolar
H_2S							
N_2							
SO_4^{2-}							
H_3O^+							
CH_3Cl							
C_2H_6							
C_2H_4							

Molecule or Polyatomic Ion	No. of Valence Electrons	Lewis Structure		Molecular Geometry	Bond Angles	Bond Polarity	Molecular Dipole or Nonpolar
	A	B	C	D	E	F	G
$C_2H_2Cl_2$		*					
SO_3^{2-}							
CH_2O							
OF_2							
PO_3^{3-}							
O_2							
NO_3^{-}		**					

*More than one possible Lewis structure can be drawn. See questions 1, 2.
**More than one possible Lewis structure can be drawn. See question 3.

QUESTIONS

1. There are three acceptable Lewis structures for $C_2H_2Cl_2$ (*) and you have drawn one of them on the report form. Draw the other two structures and indicate whether each one is nonpolar or a dipole.

2. Explain why one of the three structures for $C_2H_2Cl_2$ is nonpolar and the other two are molecular dipoles.

3. There are three Lewis structures for $[NO_3]^-$ (**). Draw the two structures which are not on the report form. Compare the molecular polarity of the three structures.

EXPERIMENT 18

Boyle's Law

MATERIALS AND EQUIPMENT

Special equipment: Elasticity of Gases Kit (or Simple Form Boyle's Law Apparatus); tube of silicone grease, barometer, slotted 0.5 and 1.0 kg masses, assorted bricks, and a vernier caliper.

DISCUSSION

Matter in the gaseous state has neither a definite volume nor a definite shape. Therefore, a confined sample of gas will take the shape of its container and depending on conditions, its volume can increase or decrease. In this experiment, air is the gas and its container is a large plastic syringe as shown in Figure 18.1. There are two conditions that can change the volume of this confined gas sample: pressure and temperature. We will keep the temperature constant (at room conditions) and examine the quantitative relationship between the volume and pressure by adding masses to the platform on the syringe. As these masses are added, the barrel will move relative to the piston and the volume of the confined gas sample will decrease as pressure, P_{gas}, increases.

This relationship was first recognized by the British scientist Robert Boyle in 1662 and is known as Boyle's Law: *at constant temperature, the volume of a sample of gas varies inversely with the pressure*. The statement may be symbolized as follows:

$$V \propto \frac{1}{P} \qquad \text{(constant T)} \qquad \qquad (1)$$

$$V = \mathbf{k} \times \frac{1}{P} \qquad \text{(constant T)} \qquad \qquad (2)$$

$$PV = \mathbf{k} \qquad \text{(constant T)} \qquad \qquad (3)$$

where $\mathbf{k}$ is a constant that depends on the mass and the temperature of the gas.

Equation (2) emphasizes the inverse relationship between pressure and volume which can be stated in simpler language: As the pressure on a gas is increased, the volume is decreased and vice versa.

Equation (3) is obtained by rearranging equation (2) and states that, at constant temperature, the product of the pressure and volume of a given mass of gas is constant. From equation (3) it follows that

$$P_1V_1 = P_2V_2 \qquad \text{(constant T)} \qquad \qquad (4)$$

where P_1V_1 is the pressure-volume product at one set of conditions and P_2V_2 is the product at a second set of conditions. Solving equation (4) for V_2 gives

$$V_2 = \frac{V_1P_1}{P_2} \qquad \text{(constant T)} \qquad \qquad (5)$$

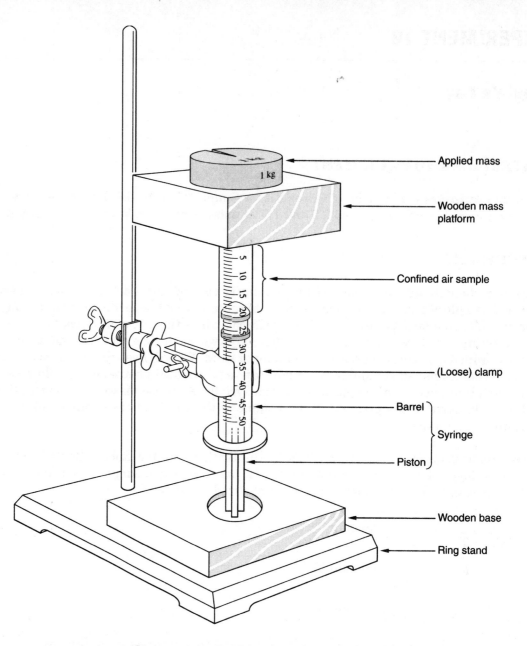

Figure 18.1   Boyle's Law Apparatus

This equation is commonly used in Boyle's law calculations. To calculate a new volume (V_2), the initial volume is multiplied by the ratio of the initial pressure over the final pressure (P_1/P_2).

In this experiment the pressure and volume of a gas are measured. The data are plotted as volume vs. pressure to obtain a curve typical of an inverse relationship. The pressure-volume product (PV) is calculated for each set of data. The constancy of these PV products proves the validity of equation (3) and thus the validity of Boyle's law.

The apparatus is essentially a large syringe mounted on a wooden base. The barrel of the syringe is the container for the gas with a pressure P_{gas}. This pressure is equal to the pressure of

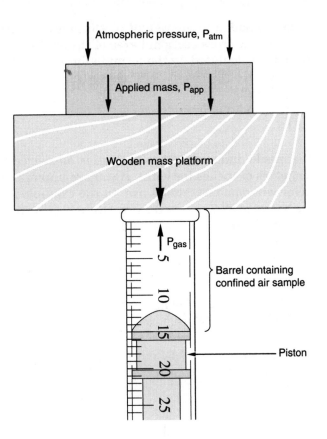

Figure 18.2 P_{atm}, P_{app}, P_{gas}

the atmosphere, P_{atm}, pushing down on the top of the barrel and any applied pressure added to the platform, P_{app}. This is illustrated in Figure 18.2 and summarized as

$$P_{gas} = P_{atm} + P_{app} \tag{6}$$

Gas pressure can be measured in various units. Atmospheric pressure P_{atm} is usually measured with a barometer calibrated in mm Hg or in. Hg. The applied pressure (P_{app}) is measured in units of mass per surface area.

$$P_{app} = \frac{mass}{surface\ area} \tag{7}$$

One commonly used pressure unit is lbs/in.^{2} (read as "pounds per square inch" and sometimes abbreviated as "psi"). It is measured by calculating the weight of the masses in pounds and dividing by the cross-sectional area of the barrel in square inches. Different equivalent units of pressure are related as follows:

1 atm = 760 mmHg = 760 torr = 14.7 lb/in.^{2} (psi)

For a barometer reading of 747 mmHg, the P_{atm} in lb/in.^{2} is calculated as follows:

$$(747\ mmHg)\left(\frac{14.7\ lb\,/\,in.^2}{760\ mmHg}\right) = 14.4\ lb\,/\,in.^2$$

In the procedure, the masses added to the weight platform are kilogram slotted masses and bricks. Their force is applied to the cross-sectional area of the syringe. The cross-sectional area of the syringe is circular and is calculated by the formula for area (A) of a circle:

$$A = \pi\, r^2 \qquad \text{(where } r = d/2)$$

(8)

$$r = \text{radius and } d = \text{diameter}$$
$$\pi = 3.14$$

To calculate the area, the inside diameter of the syringe is measured using a vernier caliper calibrated in either centimeters or inches (or both). If you are unfamiliar with the use of a vernier caliper, your instructor will demonstrate its use. In the example illustrated (Figure 18.3), the inside diameter of the barrel is 32.7 mm or 3.27 cm. Substituting into equation 8, the cross-sectional area of the barrel illustrated is calculated as

$$A = (3.14)\left(\frac{3.27 \text{ cm}}{2}\right)^2 = 8.39 \text{ cm}^2$$

With a 2.5 kg mass on the weight platform, the applied pressure, P_{app} is 2.5 kg/8.39 cm² which, when converted to lbs/in.², is 4.2 lb/in.²

$$P_{app} = \frac{\text{mass}}{A} = \left(\frac{2.5 \text{ kg}}{8.39 \text{ cm}^2}\right)\left(\frac{2.2 \text{ lb}}{1 \text{ kg}}\right)\left(\frac{2.54 \text{ cm}}{1 \text{ in.}}\right)^2 = 4.2 \,\frac{\text{lb}}{\text{in.}^2}$$

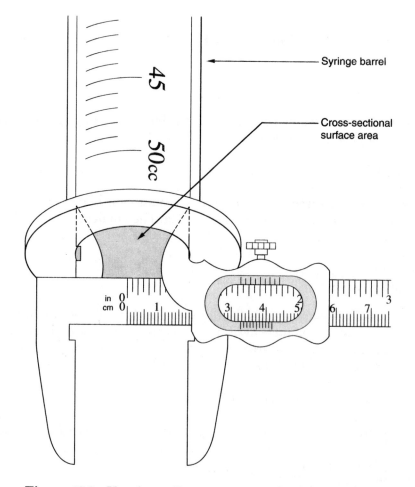

Figure 18.3   Vernier caliper measuring inside diameter

Therefore, the pressure on the gas in the syringe at atmospheric pressure with a 2.5 kg mass on the mass platform is:

$$P_{gas} = P_{atm} + P_{app}$$

$$P_{gas} = 14.4 \text{ lb/in.}^2 + 4.2 \text{ lb/in.}^2 = 18.6 \text{ lb/in.}^2$$

PROCEDURE

Wear protective glasses.

WASTE
DISPOSE OF
PROPERLY   **No waste for disposal in this experiment.**

> Record all measurements on the data tables of the report form and perform calculations as indicated. Every calculated number entered on the report form must have a setup shown clearly in the space provided.

1. Read the barometer and record the atmospheric pressure in mm Hg.

2. Remove the red cover from the nozzle of the syringe and pull the piston all the way out. Measure the inside diameter of the barrel with the vernier caliper. Be careful not to lose the red cap.

3. Lubricate the side wall of the black rubber gasket on the bottom end of the piston with silicone grease.

4. Seat the top end of the piston firmly in the hole in the thin wooden block, the base.

5. Place the barrel over the gasket on the piston and move it to the 35.0 cm^3 (cc) mark on the barrel. Put the red cap back on the nozzle of the barrel.

6. Place the yellow plastic cap over the top of the barrel with the cap passing through the center hole (some kits may not have this yellow cover). Put the wooden base of this whole assembly on the ring stand and place a clamp around the barrel but do not tighten it. Turn the barrel so the calibrations are visible with the clamp in place.

7. Put the wooden mass platform (the thick wooden block) over the top of the barrel so the red cap on the nozzle is below the surface of the surrounding wood.

8. With no applied mass on the mass platform, record the volume of the air inside the cylinder in the appropriate space on Data Table 2 of the report form (top line of the table, applied mass = 0). The only pressure now exerted on the gas inside the syringe is atmospheric pressure and that of the wooden platform.

9. Weigh one of the 0.5 or 1 kg masses to three significant figures and place it carefully on the mass platform so the syringe does not bend and touch the stabilizing clamp. One person should hold the syringe steady while another person carefully adds the mass. A third person can read and record the volume. As the masses are added, the syringe has a tendency to become unsteady and the masses can slide off or the syringe can break. The ring stand and clamp are supposed to prevent this kind of disaster but teamwork is necessary to make sure the

clamp doesn't create more problems by interfering with the movement of the barrel on the piston. Read and record the volume.

10. Repeat Step 9 by adding various combinations of weighed masses and bricks.

> When complete, there should be five values recorded in the first and fifth columns of Data Table 2, starting with an applied mass of 0 in the top row.

11. On your report form, complete Data Table 1, 5–10 for the first applied mass. Show the setups for every calculated number and pay attention to significant figures. Complete the calculations for the remaining applied masses.

12. Graph the data using the following guidelines.

a. The variables in this experiment are the pressure of the confined gas, P_{gas}, and the volume of the gas under each pressure condition. The independent variable is the variable which the experimenter controlled. This variable is plotted on the x-axis. The dependent variable changes in response to changes made in the independent variable. It is plotted on the y-axis.

b. If necessary, review the instructions in Study Aid 3 for determining scale values, choosing suitable starting values, numbering the major increments, and writing axes labels and a title.

c. Plot each PV point and draw a smooth line that best fits the 5 points. This line should be slightly curved and need not pass through each point.

NAME ______________________

SECTION ______ DATE ______________

REPORT FOR EXPERIMENT 18

INSTRUCTOR ______________________

Boyle's Law

Data Table 1 Show the setup used for every calculated number. Include all units in dimensional analysis setups.

1. Atmospheric pressure __________ mm Hg __________ lbs/in.^{2}
2. Inside diameter of syringe __________ cm __________ in.
3. Inside radius __________ in.
4. Area of syringe (cross-section) __________ in.^{2}

Record measurements and show your calculations for the first applied mass in Data Table 2.

5. First applied mass __________ kg
6. Applied mass __________ lb
7. Applied pressure __________ lb/in.^{2}
8. Total pressure __________ lb/in.^{2}
9. Volume of air __________ cm^3
10. Pressure-Volume Product, PV __________ lb cm^3/in.^{2}

Data Table 2 Complete the table below for 0.0 applied mass and 4 additional masses. Calculations for the first applied mass should be shown above in 5–10.

Applied Mass, kg	Applied Mass, lb	Applied Pressure, P_{app} (lb/in.^{2})	Total Pressure $P_{gas} = P_{atm} + P_{app}$ (lb/in.^{2})	Volume of Air, V (cm^3)	Pressure-Volume Product, PV (lb cm^3/ in.^{2})
0	0	0			

Average PV __________

QUESTIONS AND PROBLEMS

1   What is the independent variable in this experiment? Explain your choice.

2.   Why must the temperature be constant during this experiment?

3.   What part of the tabulations in the data table proves Boyle's Law? How?

4.   Plot Total Pressure vs. Volume of Air for the five values in Data Table 2. Use the graph paper provided or attach a computer-generated graph to your report form. If necessary, refer to Study Aid 3 for instructions on how to complete either type of graph.

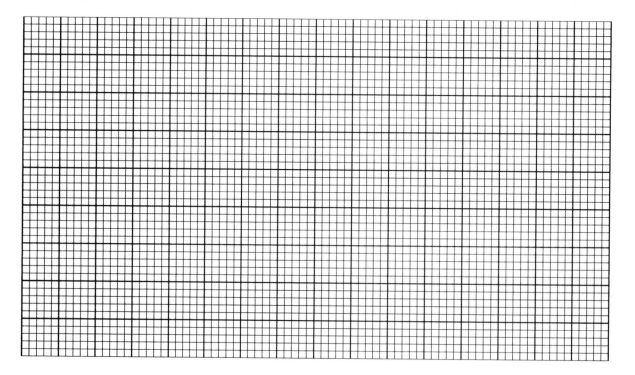

5. What is the relationship between gas pressure and volume shown by the data in your graph?

6. If you repeated this experiment at a lower temperature (for example in a walk-in refrigerator), how would the P vs. V curve obtained differ from the curve on your graph?

7. Given the following data: volume of air without any applied pressure, 25.0 cm^3; inside diameter of the syringe, 3.20 cm; barometric pressure, 630 mmHg. Show setups and answers for each of the following problems based on this data.

   a. What is the cross-sectional area of the syringe in in.^{2}?

   b. What is the barometric pressure in lb/in.^{2}?

   c. A brick with a mass of 6.00 lb is placed on the barrel of the syringe. What is the total pressure, P_{gas}, of the gas in the syringe after adding the brick?

   d. What is the change in volume after adding the brick?

EXPERIMENT 19

Charles' Law

MATERIALS AND EQUIPMENT

125 mL Erlenmeyer flask, one-hole rubber stopper, glass and rubber tubing, pneumatic trough, thermometer, screw clamp.

DISCUSSION

The quantitative relationship between the volume and the absolute temperature of a gas is summarized in Charles' law. This law states: at constant pressure, the volume of a particular sample of gas is directly proportional to the absolute temperature.

Charles' law may be expressed mathematically:

$$V \propto T \text{ (constant pressure)} \tag{1}$$

$$V = kT \text{ or } \frac{V}{T} = k \text{ (constant pressure)} \tag{2}$$

where V is volume, T is Kelvin temperature, and k is a proportionality constant dependent on the number of moles and the pressure of the gas.

If the volume of the same sample of gas is measured at two temperatures, $V_1/T_1 = k$ and $V_2/T_2 = k$, and we may say that

$$\frac{V_1}{T_1} = \frac{V_2}{T_2} \text{ or } V_2 = (V_1)\left(\frac{T_2}{T_1}\right) \text{ (constant pressure)} \tag{3}$$

where V_1 and T_1 represent one set of conditions and V_2 and T_2 a different set of conditions, with pressure the same at both conditions.

Experimental Verification of Charles' Law

This experiment measures the volume of an air sample at two temperatures, a high temperature, T_H, and a low temperature, T_L. The volume of the air sample at the high temperature, (V_H), decreases when the sample is cooled to the low temperature and becomes V_L. All of these measurements are made directly. The experimental data is then used to verify Charles' law by two methods:

1. The experimental volume (V_{exp}) measured at the low temperature is compared to the V_L predicted by Charles' law where

$$V_L(\textit{theoretical}) = (V_H)\left(\frac{T_L}{T_H}\right)$$

2. The V/T ratios for the air sample measured at both the high and the low temperatures are compared. Charles' law predicts that these ratios will be equal.

$$\frac{V_H}{T_H} = \frac{V_L}{T_L}$$

Pressure Considerations

The relationship between temperature and volume defined by Charles' law is valid only if the pressure is the same when the volume is measured at each temperature. That is not the case in this experiment.

1. The volume, V_H, of air at the higher temperature, T_H, is measured at atmospheric pressure, P_{atm} in a dry Erlenmeyer flask. The air is assumed to be dry and the pressure is obtained from a barometer.

2. The experimental air volume, (V_{exp}) at the lower temperature, T_L, is measured over water. This volume is saturated with water vapor that contributes to the total pressure in the flask. Therefore, the experimental volume must be corrected to the volume of dry air at atmospheric pressure. This is done using Boyle's law as follows:

   a. The partial pressure of the dry air, P_{DA}, is calculated by subtracting the vapor pressure of water from atmospheric pressure:

   $$P_{atm} - P_{H_2O} = P_{DA}$$

   b. The volume that this dry air would occupy at P_{atm} is then calculated using the Boyle's law equation:

   $$(V_{DA})(P_{atm}) = (V_{exp})(P_{DA})$$

   $$(V_{DA}) = \frac{(V_{exp})(P_{DA})}{(P_{atm})}$$

PROCEDURE

Wear protective glasses.

WASTE
DISPOSE OF
PROPERLY
   No waste for disposal in this experiment.

> **NOTE:** It is essential that the Erlenmeyer flask and rubber stopper assembly be as dry as possible in order to obtain reproducible results.

Dry a 125 mL Erlenmeyer flask by gently heating the entire outer surface with a burner flame. Care must be used in heating to avoid breaking the flask. If the flask is wet, first wipe the inner and outer surfaces with a towel to remove nearly all the water. Then, holding the flask with a test tube holder, gently heat the entire flask. Avoid placing the flask directly in the flame. Allow to cool.

While the flask is cooling select a 1-hole rubber stopper to fit the flask and insert a 5 cm piece of glass tubing into the stopper so that the end of the tubing is flush with the bottom of the stopper. Attach a 3 cm piece of rubber tubing to the glass tubing (see Figure 19.1). Insert the stopper into the flask and mark (wax pencil) the distance that it is inserted. Clamp the flask so that it is submerged as far as possible in water contained in a 400 mL beaker (without the flask touching the bottom of the beaker) (see Figure 19.2).

Heat the water to boiling. Keep the flask in the gently boiling water for at least 8 minutes to allow the air in the flask to attain the temperature of the boiling water. Add water as needed to maintain the water level in the beaker. Read and record the temperature of the boiling water.

While the flask is still in the boiling water, seal it by clamping the rubber tubing tightly with a screw clamp. Remove the flask from the hot water and submerge it in a pan of cold water, keeping the top down at all times to avoid losing air (see Figure 19.3). Remove the screw clamp, letting the cold water flow into the flask. Keep the flask totally submerged for about 6 minutes to allow the flask and contents to attain the temperature of the water. Read and record the temperature of the water in the pan.

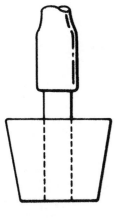

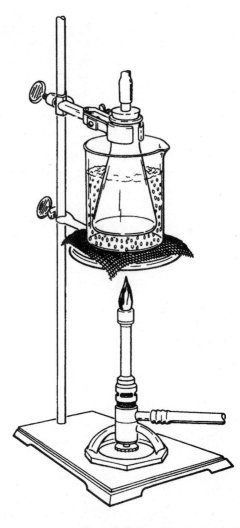

Figure 19. 1 Rubber stopper assembly

Figure 19.2 Heating the flask (and air) in boiling water

In order to equalize the pressure inside the flask with that of the atmosphere, bring the water level in the flask to the same level as the water in the pan by raising or lowering the flask (see Figure 19.3). With the water levels equal, pinch the rubber tubing to close the flask. Remove the flask from the water and set it down on the laboratory bench.

Using a graduated cylinder carefully measure and record the volume of liquid in the flask.

Repeat the entire experiment. Use the same flask and flame dry again; make sure that the rubber stopper assembly is thoroughly dried inside and outside.

After the second trial fill the flask to the brim with water and insert the stopper assembly to the mark, letting the glass and rubber fill to the top and overflow. Measure the volume of water in the flask. Since this volume is the total volume of the flask, record it as the volume of air at the higher temperature. Because the same flask is used in both trials, it is necessary to make this measurement only once.

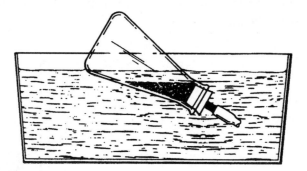

Figure 19.3   Equalizing the pressure in the flask. The water level inside the flask is adjusted to the level of the water in the pan by raising or lowering the flask.

REPORT FOR EXPERIMENT 19

Charles' Law

Data Table

	Trial 1	Trial 2
Temperature of boiling water, T_H	_____ °C, _____ K	_____ °C, _____ K
Temperature of cold water, T_L	_____ °C, _____ K	_____ °C, _____ K
Volume of water collected in flask (decrease in volume due to cooling)		
Volume of air at higher temperature, V_H (volume of flask measured only after Trial 2)		
Volume of wet air at lower temperature (volume of flask less volume of water collected), V_{exp}		
Atmosphere pressure, P_{atm} (barometer reading)		
Vapor pressure of water at lower temperature, P_{H_2O} (see Appendix 6)		

CALCULATIONS: In the spaces below, show calculation setups for Trial 1 only. Show answers for both trials in the boxes

	Trial 1	Trial 2

1. Corrected experimental volume of dry air at the lower temperature calculated from data obtained at the lower temperature.

    (a) Pressure of dry air (P_{DA})

    $$P_{DA} = P_{Atm} - P_{H_2O}$$

    (b) Corrected experimental volume of dry air (lower temperature).

    $$V_{DA} = \left(V_{exp}\right)\left(\frac{P_{DA}}{P_{Atm}}\right) =$$

2. Predicted volume of dry air at lower temperature V_L calculated by Charles' law from volume at higher temperature (V_H).

    $$V_L = \left(V_H\right)\left(\frac{T_L}{T_H}\right)$$

3. Percentage error in verification of Charles' law.

    $$\% \text{ error } = \frac{V_{DA} - V_L}{V_L} \times 100 =$$

4. Comparison of experimental V/T ratios. (Use dry volumes and absolute temperatures.)

    (a) $\dfrac{V_H}{T_H} =$

    (b) $\dfrac{V_{DA}}{T_L} =$

5. On the graph paper provided, plot the volume- temperature values used in Calculation 4. Temperature data **must be in** °C. Draw a straight line between the two plotted points and extrapolate (extend) the line so that it crosses the temperature axis.

QUESTIONS AND PROBLEMS

1. (a) In the experiment, why are the water levels inside and outside the flask equalized before removing the flask from the cold water?

    (b) When the water level is higher inside than outside the flask, is the gas pressure in the flask higher than, lower than, or the same as, the atmospheric pressure? (specify which)

    _____

2. A 125 mL sample of dry air at 230°C is cooled to 100°C at constant pressure. What volume will the dry air occupy at 100°C?

    _____ mL

3. A 250 mL container of a gas is at 150°C. At what temperature will the gas occupy a volume of 125 mL, the pressure remaining constant?

    _____ °C

4. (a) An open flask of air is cooled. Answer the following:

    1. Under which conditions, before or after cooling, does the flask contain more gas molecules?

        _____

    2. Is the pressure in the flask at the lower temperature the same as, greater than, or less than the pressure in the flask before it was cooled?

        _____

(b) An open flask of air is heated, stoppered in the heated condition, and then allowed to cool back to room temperature. Answer the following:

1. Does the flask contain the same, more, or fewer gas molecules now compared to before it was heated?

_____

2. Is the volume occupied by the gas in the flask approximately the same, greater, or less than before it was heated?

_____

3. Is the pressure in the flask the same, greater, or less than before the flask was heated?

_____

4. Do any of the above conditions explain why water rushed into the flask at the lower temperature in the experiment? Amplify your answer.

5. On the graph you plotted,

(a) At what temperature does the extrapolated line intersect the x-axis?

_____ °C

(b) At what temperature does Charles' law predict that the extrapolated line should intersect the x-axis?

_____ °C

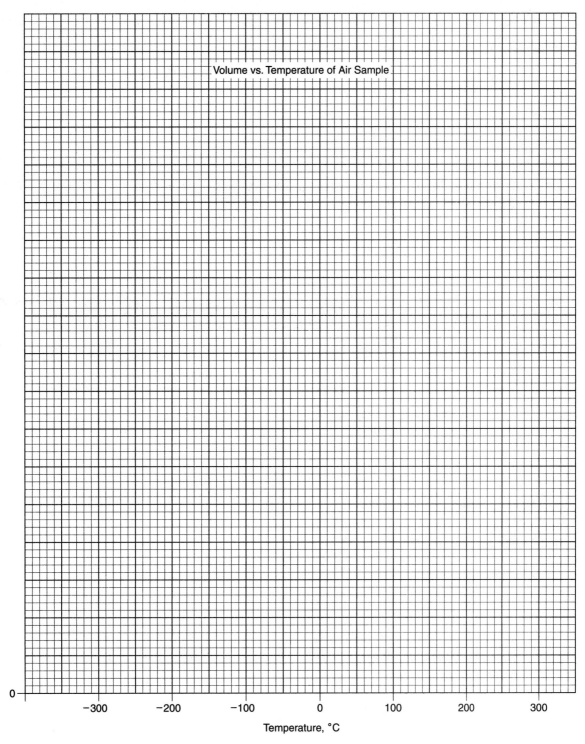

Volume vs. Temperature of Air Sample

Volume, mL

Temperature, °C

EXPERIMENT 20

Molar Volume of a Gas

MATERIALS AND EQUIPMENT

Solution: Hydrogen peroxide (3.0%, commercial preparation). **Solid**: manganese dioxide (powder). **Special equipment**: Graduated cylinder, 50.0 ml; disposable syringe, 3.0 cc or 5.0 cc; stopper assembly with needle for attaching the syringe; battery jar or large beaker (1 or 2 L); large test tube.

DISCUSSION

Often, gases must be handled and measured in the same experiment with solids and liquids. The amount of solid used or produced can be determined by measuring the mass of the material on a balance but it is difficult to measure the mass of a gas. It is much easier to measure gas volume and use this volume to calculate moles or mass. It is possible to do this because of Avogadro's law and the other gas laws. This experiment will illustrate the chemical significance of these important principles.

The term **molar volume** is used to describe the volume occupied by exactly one mole of any gas at a given temperature and pressure. Because a mole contains 6.022×10^{23} molecules (Avogadro's number), and Avogadro determined that equal volumes of different gases at the same temperature and pressure contain the same number of molecules (Avogadro's law) a mole of different gases will have exactly the same volume at the same temperature and pressure. The temperature and pressure at which molar volumes of gases are usually compared is standard temperature (0°C or 273 K) and standard pressure (1 atm). The molar volume of 22.4 L has been experimentally determined and verified over and over again for many gases in many laboratories. In this experiment, we will try to verify this value of 22.4 L for the molar volume of oxygen generated by the decomposition of hydrogen peroxide. The rate of this decomposition is greatly increased by the addition of MnO_2, a catalyst which remains unchanged by the reaction.

$$2\ H_2O_2 \xrightarrow{\ MnO_2\ } 2\ H_2O + O_2$$

If we think of molar volume as the liters per mole of gas at STP, then we need two measurements: the volume of oxygen at STP and the number of moles of oxygen that occupy that volume.

Using dimensional analysis and stoichiometry, the number of moles of oxygen gas generated can be calculated from the concentration and volume of H_2O_2 . For example, if 2.0 ml of 5.0% hydrogen peroxide is decomposed in the reaction above, the number of moles of oxygen generated is calculated as follows:

$$\text{mol } O_2 = \left(2.0 \text{ mL } H_2O_2(aq)\right)\left(\frac{5.0 \text{ g } H_2O_2}{100. \text{ mL } H_2O_2(aq)}\right)\left(\frac{1 \text{ mol } H_2O_2}{34.02 \text{ g } H_2O_2}\right)\left(\frac{1 \text{ mol } O_2}{2 \text{ mol } H_2O_2}\right)$$

$$= 0.0015 \text{ mol } O_2$$

The oxygen will be collected by the downward displacement of water in an inverted graduated cylinder very much the way oxygen was collected in Experiment 3. The H_2O_2 solution is added to the generator with a syringe so there will be no loss of gas while putting the stopper in the tube after adding the solution. Figure 20.1 illustrates the setup for the generation and collection of oxygen in this experiment.

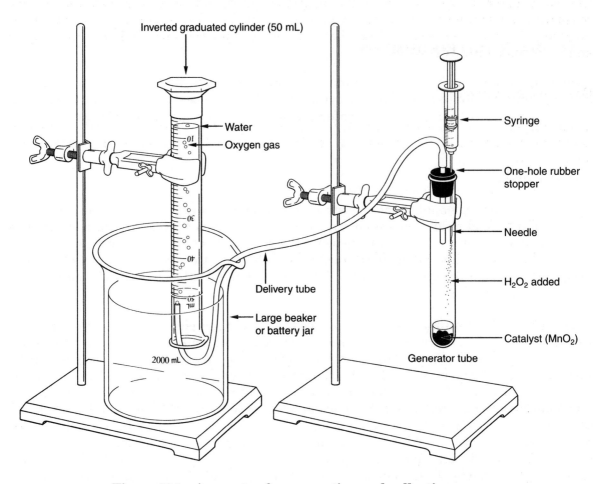

Figure 20.1   Apparatus for generating and collecting oxygen

The gas collected will contain some air because the oxygen generated will push air out of the generator and delivery tube and into the graduated cylinder. The mixture of air/oxygen/water vapor in the graduated cylinder will also include the air pushed out of the system by the addition of the hydrogen peroxide (2.0 mL in this example) so we must subtract the volume of peroxide added from the volume of gas collected to obtain the net volume of oxygen generated. Finally, before we can calculate the experimental molar volume the net volume of oxygen generated at laboratory conditions must be corrected to STP. This correction is done using the combined gas law:

$$\frac{V_1 P_1}{T_1} = \frac{V_2 P_2}{T_2} \qquad V_2 = \left(\frac{V_1 P_1}{T_1}\right)\left(\frac{T_2}{P_2}\right)$$

Therefore, we must be able to measure V_1, T_1, and P_1 so we can solve for V_2.

V_1 is the volume of the gas collected after subtracting the volume of the H_2O_2 added. T_1 is the temperature in the laboratory changed to Kelvin. The pressure of the gas in the graduated cylinder is equal to the atmospheric pressure when the water levels inside and outside the graduate are the same. However, the gas collected also contains water vapor.

$$P_{O_2} + P_{H_2O} = P_{total} \text{ (atmospheric pressure)}$$

Therefore, to obtain the pressure of the oxygen collected (P_1) we need to subtract the vapor pressure of water from the atmospheric pressure. V_2 is the volume of oxygen at STP; it is calculated using the combined gas laws where P_2 is standard pressure (760 mm Hg) and T_2 is standard temperature (273 K).

Sample Calculation

When 2.0 mL of 5.0% H_2O_2 was reacted, 38.6 mL of gas was collected at 19.0°C. The vapor pressure of water at 19.0°C is 16.5 mm Hg (See Appendix 6). The barometric pressure was 743.5 mm Hg.

$$V_1 = 38.6 \text{ mL} - 2.0 \text{ mL} = 36.6 \text{ mL}$$

$$T_1 = 19.0 + 273 = 292 \text{ K}$$

$$P_1 = 743.5 \text{ mm Hg} - 16.5 \text{ mm Hg} = 727.0 \text{ mm Hg}$$

$$T_2 = 273 \text{ K}$$

$$P_2 = 760 \text{ mm Hg}$$

$$V_2 = \frac{V_1 P_1 T_2}{T_1 P_2} = \frac{(36.6 \text{ mL})(727.0 \text{ mmHg})(273 \text{ K})}{(292 \text{ K})(760 \text{ mmHg})} = 32.7 \text{ mL}$$

We now have all the information to determine the molar volume of O_2, 32.7 mL and 0.0015 mole O_2.

$$\frac{L}{mol} = \left(\frac{32.7 \text{ mL}}{0.0015 \text{ mol}}\right)\left(\frac{1 \text{ L}}{1000 \text{ mL}}\right) = 22 \frac{L}{mol} \text{ (molar volume)}$$

If this experimental value is close to the theoretical value of 22.4 L/mol, we have verified Avogadro's law.

The percent error for our molar volume is determined as we have done in previous experiments:

$$\frac{\text{Theoretical value} - \text{Actual value}}{\text{Theoretical value}} \times 100 = \frac{22.4 \text{ L} / \text{mol} - 22 \text{ L} / \text{mol}}{22.4 \text{ L} / \text{mol}} \times 100 = 1.8\%$$

PROCEDURE

Wear protective glasses.

1. Set up the apparatus as shown in Figure 20.1. Weigh about 2 g of MnO_2 and add it to the dry generator tube. It is not necessary to record this mass on the report form. Cover the MnO_2 with about 2 ml of distilled water and replace the stopper/syringe assembly. The needle should

NOT be removed from the stopper at any time during this experiment. Make sure the stopper is firmly in the generator tube.

2. Fill the large beaker with water to about 1 inch from the top. The diameter of this container must accomodate a hand holding the inverted graduated cylinder so it must be fairly large. A 2 L beaker or battery jar works best. Fill the graduated cylinder to overflowing with water and cup your hand over the top being careful to keep all the water in the cylinder while you invert and submerge its top in the large beaker of water. Try to do this so there is no air at the "top" of the inverted cylinder when you are done. This may require several attempts and there will probably be one air bubble at the top of the graduate despite your best efforts. When successful, attach the inverted graduate to the ring stand as shown in Figure 20.1. Note that the clamp is attached near the top of the inverted cylinder. Do not put the delivery tube into the cylinder yet.

3. Determine the temperature of the gas to be collected by measuring the temperature in the laboratory. Then, read the mercury barometer to determine the atmospheric pressure in mm Hg. If a different type barometer is available, convert the units if necessary. Record these measurements on the report form.

4. Remove the syringe from the needle by twisting. It may be necessary to hold the needle with pliers when you twist the barrel. Pour about 10 mL of the peroxide solution into a small (50 mL) beaker and fill the syringe barrel to the 3.0 ml mark. Do not replace the barrel on the needle yet. Accurately read and record the volume in the syringe. If you are unsure about reading the volume in the syringe, ask the instructor. Every significant digit is important here. Record the concentration of the peroxide solution on the report form. A solution which is 3.0% H_2O_2 (m/v) in water has 3.0 g of H_2O_2/100 ml of solution.

5. Carefully reattach the syringe to the needle. Hold the syringe by the barrel and be very careful to avoid moving the plunger prematurely. Put the gas delivery tube into the graduated cylinder. If a few air bubbles go into the cylinder at this point, make a note on your report form and proceed anyway. Inject the H_2O_2 into the generator by pushing firmly down on the syringe plunger. You will notice that air immediately bubbles up into the cylinder as you push the peroxide in. Then, as the peroxide and catalyst meet, the reaction begins immediately and proceeds slowly. Remove the entire generator tube from the ring stand and shake gently. It will take about two minutes to react all the H_2O_2. Be careful that the rubber stopper stays tightly in the top of the test tube and that the delivery tube stays in the graduated cylinder.

6. When the reaction is over, remove the delivery tube and carefully loosen the clamp holding the graduate to the ring stand. Move it up or down on the pole to equalize the levels of water inside and outside the graduate. Retighten the clamp to the ring stand. Read the volume of gas collected paying close attention to the meniscus. Although the gas in not pure oxygen, its volume is equal to the volume of oxygen generated plus the volume of air pushed out of the generator by the addition of the peroxide solution.

7. If you have any excess hydrogen peroxide solution, do not return it to the stock bottle. Pour it into the sink and flush generously with water.

8. Pour the contents of the generator tube into the sink and flush generously with water. Return the stopper with needle and syringe to your instructor.

REPORT FOR EXPERIMENT 20

Molar Volume of a Gas

Measurements

1. Concentration of H_2O_2 _____ %

2. Volume of H_2O_2 added to the generator   _____ mL

3. Barometric pressure   _____ mmHg

4. Water temperature   _____ °C   _____ K

5. Gas volume collected in the graduated cylinder   _____ mL

Calculations: *Show the setup for every calculation.*

1. Oxygen volume generated by the reaction   _____ mL

2. Moles of oxygen generated by the decomposition of H_2O_2 _____ mol

     a. balanced equation for the reaction

     b. stoichiometric setup using continuous calculation method

3. Pressure of dry gas in the graduated cylinder

     a. vapor pressure of water   _____ mmHg

     b. pressure of dry gas   _____ mmHg

4. Experimental volume of dry oxygen converted to STP   _____ mL

5. Experimental molar volume   _____ L/mol

6. Theoretical molar volume (L/mol)   _____ L/mol

7. Percent error   _____ %

QUESTIONS AND PROBLEMS

1. The curved surface of an aqueous solution is called _____ .
   If the top of this surface were used to measure the volume of the gas rather than the
   bottom of the surface, the volume of the gas above the liquid would be

   _____ .
                                                          (too large or too small)

2. Why was it necessary to subtract the volume of the H_2O_2 solution injected into the generator from the volume of gas collected in the graduated cylinder?

3. If a student did this experiment using 5.0 mL of 10% hydrogen peroxide, how many moles
   of O_2 would be generated?

4. Why was it not necessary to measure the MnO_2 precisely when we have to be so careful
   about measuring the volume of H_2O_2 and the volume of oxgyen collected?

5. At 20.0°C, a student collects H_2 gas in a gas collecting tube. The barometric pressure is
   755.2 mm Hg and the water levels inside and outside the tube are exactly equal.

   a. What is the total gas pressure in the gas collecting tube?

   b. What is the pressure of the water vapor in the gas collecting tube?

   c. What is the pressure of the dry hydrogen in the gas collecting tube?

6. What would be the effect on the molar volume if hydrogen instead of oxygen gas had been
   collected during this experiment? Explain your answer.

7. After correction to standard conditions, the volume of a gas collected was 43.8 mL. If this
   volume represents 0.00184 mol, what is the percent error for the experimental molar
   volume?

EXPERIMENT 21

Neutralization — Titration I

MATERIALS AND EQUIPMENT

Solid: potassium acid phthalate, abbreviated KHP ($KHC_8H_4O_4$). **Liquids:** phenolphthalein indicator, unknown base solution (NaOH). One buret (25 mL or 50 mL) and buret clamp.

DISCUSSION

The reaction of an acid and a base to form a salt and water is known as **neutralization.** In this experiment potassium acid phthalate (abbreviated KHP) is used as the acid. Potassium acid phthalate is an organic substance having the formula $HKC_8H_4O_4$, and like HCl, has only one acid hydrogen atom per molecule. Because of its complex formula, potassium acid phthalate is commonly called KHP. Despite its complex formula we see that the reaction of KHP with sodium hydroxide is similar to that of HCl. One mole of KHP reacts with one mole of NaOH.

$$HKC_8H_4O_4 + NaOH \longrightarrow NaKC_8H_4O_4 + H_2O$$

$$HCl + NaOH \longrightarrow NaCl + H_2O$$

Titration is the process of measuring the volume of one reagent required to react with a measured volume or mass of another reagent. In this experiment we will determine the molarity of a base (NaOH) solution from data obtained by titrating KHP with the base solution. The base solution is added from a buret to a flask containing a weighed sample of KHP dissolved in water. From the mass of KHP used we calculate the moles of KHP. Exactly the same number of moles of base is needed to neutralize this number of moles of KHP since one mole of NaOH reacts with one mole of KHP. We then calculate the molarity of the base solution from the titration volume and the number of moles of NaOH in that volume.

In the titration, the point of neutralization, called the **end-point,** is observed when an indicator, placed in the solution being titrated, changes color. The indicator selected is one that changes color when the stoichiometric quantity of base (according to the chemical equation) has been added to the acid. A solution of phenolphthalein, an organic acid, is used as the indicator in this experiment. Phenolphthalein is colorless in acid solution but changes to pink when the solution becomes slightly alkaline. When the number of moles of sodium hydroxide added is equal to the number of moles of KHP originally present, the reaction is complete. The next drop of sodium hydroxide added changes the indicator from colorless to pink.

Use the following relationships in your calculations:

1. According to the equation for the reaction,

   Moles of KHP reacted = Moles of NaOH reacted

2. Moles $= \dfrac{\text{g of solute}}{\text{molar mass of solute}}$

3. Molarity is an expression of concentration, the units of which are moles of solute per liter of solution:

$$\text{Molarity} = \frac{\text{moles}}{\text{liter}}$$

Thus, a 1.00 molar (1.00 M) solution contains 1.00 mole of solute in 1 liter of solution. A 0.100 M solution, then, contains 0.100 mole of solute in 1 liter of solution.

4. The number of moles of solute present in a known volume of solution of known concentration can be calculated by multiplying the volume of the solution (in liters) by the molarity of the solution:

$$\text{Moles} = \text{liters} \times \text{molarity} = \text{liters} \times \frac{\text{moles}}{\text{liter}}$$

PROCEDURE

Wear protective glasses.

 Dispose of all solutions in the sink.

Obtain some solid KHP in a test tube or vial. Weigh two samples of KHP into two 125 mL Erlenmeyer flasks, numbered for identification. (The flasks should be rinsed with distilled water, but need not be dry on the inside.) First weigh the flask to the highest precision of the balance. Add KHP to the flask by tapping the test tube or vial until 1.000 to 1.200 g has been added (see Figure 21.1). Determine the mass of the flask and the KHP. In a similar manner weigh another sample of KHP into the second flask. To each flask add approximately 30 mL of distilled water. If some KHP is sticking to the walls of the flask, rinse it down with water from a wash bottle. Warm the flasks slightly and swirl them until all the KHP is dissolved.

Figure 21.1   Method of adding KHP from a vial to a weighed Erlenmeyer flask

Obtain one buret and clean it. See "Use of the Buret," on the following page for instructions on cleaning and using the buret. Read and record all buret volumes to the nearest 0.01 mL.

Obtain about 250 mL of a base (NaOH) of unknown molarity in a clean, dry 250 mL Erlenmeyer flask as directed by your instructor. Record the number of this unknown.

1. Keep your base solution stoppered when not in use.

2. The 250 mL sample of base is intended to be used in both this experiment and Experiment 22. Be sure to label and save it.

Rinse the buret with two 5 to 10 mL portions of the base, running the second rinsing through the buret tip. Discard the rinsing in the sink. Fill the buret with the base, making sure that the tip is completely filled and contains no air bubbles. Adjust the level of the liquid in the buret so that the bottom of the meniscus is at exactly 0.00 mL. Record the initial buret reading (0.00 mL) in the space provided on the report form.

Add 3 drops of phenolphthalein solution to each 125 mL flask containing KHP and water. Place the first (Sample 1) on a piece of white paper under the buret extending the tip of the buret into the flask (see Figure 21.2).

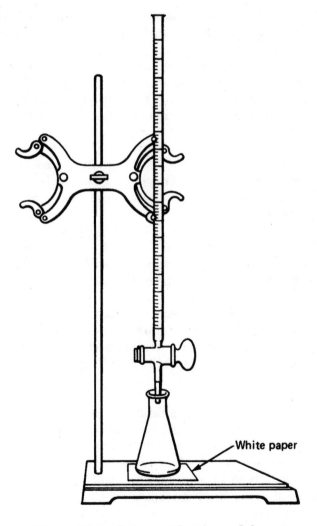

Figure 21.2   Setup with stopcock buret

Titrate the KHP by adding base until the end-point is reached. The titration is conducted by swirling the solution in the flask with the right hand (if you are right handed) while manipulating the stopcock with the left. As base is added you will observe a pink color caused by localized high base concentration. Toward the end-point the color flashes throughout the solution, remaining for a longer time. When this occurs, add the base drop by drop until the end-point is reached, as indicated by the first drop of base which causes a faint pink color to remain in the entire solution for at least 30 seconds. Read and record the final buret reading (see Figure 21.5). Refill the buret to the zero mark and repeat the titration with Sample 2. Then, calculate the molarity of the base in each sample. If these molarities differ by more than 0.004, titrate a third sample.

When you are finished with the titrations, empty and rinse the buret at least twice (including the tip) with tap water and once with distilled water. Return the vial with the unused KHP.

Use of the Buret

A buret is a volumetric instrument that is calibrated to deliver a measured volume of solution. The 50 mL buret is calibrated from 0 to 50 mL in 0.1 mL increments and is read to the nearest 0.01 mL. All volumes delivered from the buret should be between the calibration marks. (Do not estimate above the 0 mL mark or below the 50 mL mark.)

1. **Cleaning the Buret.** The buret must be clean in order to deliver the calibrated volume. Drops of liquid clinging to the sides as the buret is drained are evidence of a dirty buret.

To clean the buret, first rinse it a couple of times with tap water, pouring the water from a beaker. Then scrub it with a detergent solution, using a long-handled buret brush. Rinse the buret several times with tap water and finally with distilled water. Check for cleanliness by draining the distilled water through the tip and observe whether droplets of water remain on the inner walls of the buret.

2. **Using the Buret.** After draining the distilled water, rinse the buret with two 5 to 10 mL portions of the solution to be used in it. This rinsing is done by holding the buret in a horizontal position and rolling the solution around to wet the entire inner surface. Allow the final rinsing to drain through the tip.

Fill the buret with the solution to slightly above the 0 mL mark and adjust it to 0.00 mL, or some other volume below this mark, by draining the solution through the tip. The buret tip must be completely filled to deliver the volume measured.

To deliver the solution from the buret, turn the stopcock with the forefinger and the thumb of your left hand (if you are right handed) to allow the solution to enter the flask. (See Figure 21.3). This procedure leaves your right hand free to swirl the solution in the flask during the titration. With a little practice you can control the flow so that increments as small as 1 drop of solution can be delivered.

3. **Reading the Buret.** The smallest calibration mark of a 50 mL buret is 0.1 mL. However, the buret is read to the nearest 0.01 mL by estimating between the calibration marks. When reading the buret be sure your line of sight is level with the bottom of the meniscus in order to avoid parallax errors (see Figure 21.4). The exact bottom of the meniscus may be made more prominent and easier to read by allowing the meniscus to pick up the reflection from a heavy dark line on a piece of paper (see Figure 21.5).

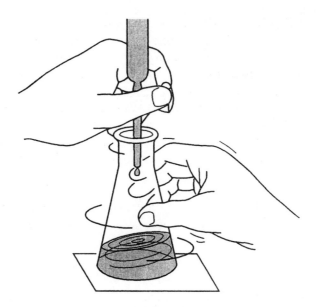

Figure 21.3   Titration technique

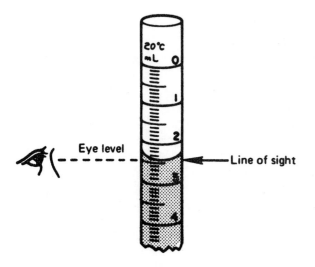

Figure 21.4   Reading the buret. The line of sight must be level with the bottom of the meniscus to avoid parallax.

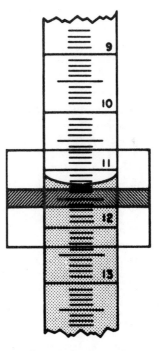

Figure 21.5   Reading the meniscus. A heavy dark line brought to within one division of the meniscus will make the meniscus more prominent and easier to read. The volume reading is 11.28 mL.

REPORT FOR EXPERIMENT 21

Neutralization—Titration I

Data Table

	Sample 1	Sample 2	Sample 3 (if needed)
Mass of flask and KHP			
Mass of empty flask			
Mass of KHP			
Final buret reading			
Initial buret reading			
Volume of base used			

CALCULATIONS: In the spaces below show calculation setups for Sample 1 only. Show answers for both samples in the boxes. Remember to use the proper number of significant figures in all calculations. (The number 0.005 has only one significant figure.)

	Sample 1	Sample 2	Sample 3 (if needed)
1. Moles of acid (KHP, Molar mass = 204.2)			
2. Moles of base used to neutralize (react with) the above number of moles of acid			
3. Molarity of base (NaOH)			

4. Average molarity of base   _____

5. Unknown base number   _____

QUESTIONS AND PROBLEMS

1. If you had added 50 mL of water to a sample of KHP instead of 30 mL, would the titration of that sample then have required more, less, or the same amount of base? Explain.

2. A student weighed out 1.106 g of KHP. How many moles was that?

_____ mol

3. A titration required 18.38 mL of 0.1574 M NaOH solution. How many moles of NaOH were in this volume?

_____ mol

4. A student weighed a sample of KHP and found it weighed 1.276 g. Titration of this KHP required 19.84 mL of base (NaOH). Calculate the molarity of the base.

_____ M

5. Forgetful Freddy weighed his KHP sample, but forgot to bring his report sheet along, so he recorded his masses on a paper towel. During his titration, which required 18.46 mL of base, he spilled some base on his hands. He remembered to wash his hands, but forgot about the data on the towel, and used it to dry his hands. When he went to calculate the molarity of his base, Freddy discovered that he didn't have the mass of his KHP. His kindhearted instructor told Freddy that his base was 0.2987 M. Calculate the mass of Freddy's KHP sample.

_____ g

6. What mass of solid NaOH would be needed to make 645 mL of Freddy's NaOH solution?

_____ g

EXPERIMENT 22

Neutralization–Titration II

MATERIALS AND EQUIPMENT

Solutions: Acid of unknown molarity, standard base solution (NaOH), vinegar, phenolphthalein indicator. Suction bulb, buret, buret clamp, 10 mL volumetric pipet.

DISCUSSION

This experiment may follow Experiment 21 or it may be completed independently of Experiment 21. In either case the discussion section of Experiment 21 supplements the following discussion.

The reaction of an acid and a base to form water and a salt is known as neutralization. Hydrochloric acid and sodium hydroxide, for example, react to form water and sodium chloride.

$$HCl(aq) + NaOH(aq) \longrightarrow H_2O(l) + NaCl(aq)$$

The ionic reaction in neutralizations of this type is that of hydrogen (or hydronium) ion reacting with hydroxide ion to form water.

$$H^+ + OH^- \longrightarrow H_2O \quad \text{or} \quad H_3O^+ + OH^- \longrightarrow 2\ H_2O$$

A monoprotic acid—i.e., an acid having one ionizable hydrogen atom per molecule—reacts with sodium hydroxide (or any other monohydroxy base) on a 1:1 mole basis. This fact is often utilized in determining the concentrations of solutions of acids by titration.

Titration is the process of measuring the volume of one reagent to react with a measured volume or mass of another reagent. In this experiment an acid solution of unknown concentration is titrated with a base solution of known concentration, Phenolphthalein is used as an indicator. This substance is colorless in acid solution, but changes to pink when the solution becomes slightly basic or alkaline. The change of color, caused by a single drop of the base solution in excess over that required to neutralize the acid, marks the end-point of the titration.

Molarity (M) is the concentration of a solution expressed in terms of moles of solute per liter of solution.

$$\text{Molarity} = \frac{\text{moles}}{\text{liter}}$$

Thus a solution containing 1.00 mole of solute in 1.00 liter of solution is 1.00 molar (1.00 M). If only 0.155 mole is present in 1.00 liter of solution, it is 0.155 M, etc. To determine the molarity of any quantity it is only necessary to divide the total number of moles of solute present in the solution by the volume (in liters).

To determine the number of moles of solute present in a known volume of solution, multiply the volume in liters by the molarity.

$$\text{Moles} = \text{liters} \times \text{molarity} = \text{liters} \times \frac{\text{moles}}{\text{liter}}$$

For titrations involving monoprotic acids and monohydroxy bases (one hydroxide ion per formula unit), the number of moles of acid is identical to the number of moles of base required to neutralize the acid. In this experiment we measure the volume of base of known molarity required to neutralize a measured volume of acid of unknown molarity. The molarity of the acid can then be calculated.

$$\text{Moles base} = \text{liters} \times \text{molarity} = \text{liters base} \times \frac{\text{moles base}}{\text{liters}}$$

$$\text{Moles acid} = \text{moles base} \times \frac{1 \text{ mole acid}}{1 \text{ mole base}}$$

$$\text{Molarity of acid} = \frac{\text{moles acid}}{\text{liters acid}}$$

In order to determine the molarity of an acid solution, it is not actually necessary to know what the acid is—only whether it is monoprotic, diprotic, or triprotic. The calculations in this experiment are based on the assumption that the acid in the unknown is monoprotic.

If the molarity and the formula of the solute are known, the concentration in grams of solute per liter of the solution may be calculated by multiplying by the molar mass.

$$\text{Molarity} \times \text{molar mass} = \frac{\text{moles}}{\text{liter}} \times \frac{\text{grams}}{\text{mole}} = \frac{\text{grams}}{\text{liter}}$$

In determining the acid content of commercial vinegar, it is customary to treat the vinegar as a dilute solution of acetic acid, $HC_2H_3O_2$. The acetic acid concentration of the vinegar may be calculated as grams of acetic acid per liter or as percent acid by mass. If the acetic acid content is to be expressed on a mass percent basis, the density of the vinegar must also be known.

PROCEDURE

Wear protective glasses.

 Do not pipet by mouth.

 Dispose of all solutions in the sink. Flush with water.

A. Molarity of an Unknown Acid

Obtain a sample of acid of unknown molarity in a clean, dry 125 mL Erlenmeyer flask as directed by your instructor.

With a volumetric pipet, transfer a 10.00 mL sample of the acid to a clean, but not necessarily dry, Erlenmeyer flask. See "Use of the Pipet," on the following page, for instructions on cleaning and using the pipet. Pipet a duplicate 10.00 mL sample into a second flask. (If pipets are not available, a buret which has been carefully cleaned and rinsed may be used to measure the acid samples.)

You will need about 150 mL of base of known molarity (standard solution). Your instructor will give you the exact molarity of the base solution that you used in Experiment 21 or you may be given another sodium hydroxide solution of known molarity. Record the exact molarity of this solution. Keep the flask containing the base stoppered when not in use.

Clean and set up a buret. See "Use of the Buret," in Experiment 21, for instructions on cleaning and using the buret.

Rinse the buret with two 5 to 10 mL portions of the base, running the second rinsing through the buret tip. Discard the rinsings in the sink. Fill the buret with the base, making sure that the tip is completely filled and contains no air bubbles. Adjust the level of the liquid in the buret so that the bottom of the meniscus is at exactly 0.00 mL. Record the initial buret reading (0.00 mL) in the space provided on the report sheet.

Add three drops of phenolphthalein solution and about 25 mL of distilled water to the flask containing the 10.00 mL of acid. Place this flask on a piece of white paper under the buret and lower the buret tip into the flask (see Figure 21.2).

Titrate the acid by adding base until the end-point is reached. During the titration swirl the solution in the flask with the right hand (if you are right handed) while manipulating the stopcock with the left. As the base is added you will observe a pink color caused by localized high base concentration. Near the end-point this color flashes throughout the solution, remaining for increasingly longer periods of time. When this occurs, add the base drop by drop until the end-point is reached, as indicated by the first drop of base which causes the entire solution to retain a faint pink color for at least 30 seconds. Record the final buret reading.

Refill the buret with base and adjust the volume to the zero mark. Titrate the duplicate sample of acid. If the volumes of base used differ by more than 0.20 mL, titrate a third sample. In the calculations, assume that the unknown acid reacts like HCl (one mole of acid reacts with one mole of base).

B.   Acetic Acid Content of Vinegar

Obtain about 40 mL of vinegar in a clean, dry 50 mL beaker. Record the sample number, if any, of this vinegar.

Titrate duplicate 10.00 mL samples of vinegar using exactly the same procedure outlined in Part A. Remember to rinse the pipet with vinegar before pipeting the vinegar samples.

When you are finished with the titrations, empty the buret and rinse it and the pipet at least twice with tap water and once with distilled water.

Use of the Pipet

A volumetric (transfer) pipet (Figure 22.1) is calibrated to deliver a specified volume of liquid to a precision of about ± 0.02 mL in a 10 mL pipet. To achieve this precision, the pipet must be clean and used in a specified manner.

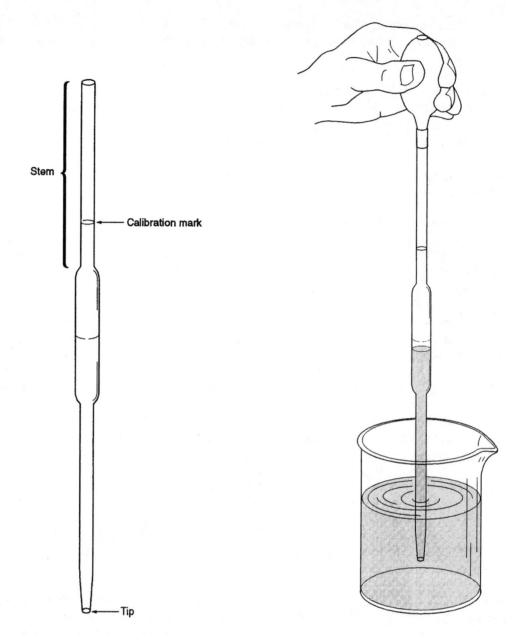

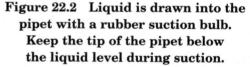

Figure 22.1   A volumetric (transfer) pipet

Figure 22.2   Liquid is drawn into the pipet with a rubber suction bulb. Keep the tip of the pipet below the liquid level during suction.

Liquids are drawn into a pipet by means of a rubber suction bulb (Figure 22.2) or by a rubber tube connected to a water aspirator pump. Suction by mouth has also been used to draw liquids into a pipet, but this a dangerous practice and is not recommended.

1. **Cleaning the Pipet.**   Use a rubber suction bulb to draw up enough detergent solution to fill about two-thirds of the body or bulb of the pipet. Retain this solution in the pipet by pressing the forefinger tightly against the top of the pipet stem (Figure 22.3). Turn the pipet to a nearly horizontal position and gently shake and rotate it until the entire inside surface is wetted. Allow the pipet to drain and rinse it at least three times with tap water and once with distilled water.

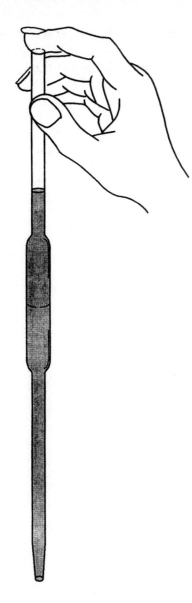

Figure 22.3   Liquid is retained in the pipet by applying pressure with the forefinger to the top of the stem.

Figure 22.4   The pipet is calibrated to deliver the specified volume, leaving a small amount of liquid in the tip.

2. Using the Pipet.   Unless the pipet is known to be clean and absolutely dry on the inside, it must be rinsed twice with small portions of the liquid that is to be pipetted. This is done as in the washing procedure described above. These rinses are discarded in order to avoid contamination of the liquid being pipetted. A pipet does not need to be rinsed between successive pipettings of the same solution.

To transfer a measured volume of a liquid, collapse a suction bulb by squeezing and place it tightly against the top of a pipet. (Do not try to push the bulb on to the pipet.) Draw the liquid into the pipet until it is filled to about 5 cm above the calibration mark by allowing the bulb to slowly expand. Be careful—do not allow the liquid to get into the bulb. Remove the bulb and quickly place your forefinger over the top of the pipet stem. The liquid will be retained in the pipet if the finger is pressed tightly against the top of the stem. Keeping the pipet in a vertical position, decrease the finger pressure very slightly, and allow the liquid level to drop slowly toward the calibration mark. When the liquid level has almost reached the calibration mark, again increase the finger pressure and stop the liquid when the bottom of the meniscus is exactly on the calibration mark. Touch the tip to the wall of the flask to remove the adhering drop of liquid.

Move the pipet to the flask which is to receive the sample and allow the liquid to drain while holding the pipet in a vertical position. About 10 seconds after the liquid has stopped running from the pipet, touch the tip to the inner wall of the sample flask to remove the drop of liquid adhering to the tip. A small amount of liquid will remain in the tip (Figure 22.4). Do not blow or shake this liquid into the sample; the pipet is calibrated to deliver the volume specified without this small residual.

If you have never used a volumetric pipet, it is advisable to practice by pipetting some samples of distilled water until you have mastered the technique.

NAME ______________________

SECTION ______ DATE ______________

REPORT FOR EXPERIMENT 22

INSTRUCTOR ______________________

Neutralization—Titration II

A. Molarity of an Unknown Acid

Data Table

	Sample 1		Sample 2		Sample 3 (if needed)	
	Acid*	Base	Acid*	Base	Acid*	Base
Final buret reading						
Initial buret reading						
Volume used						

*If a pipet is used to measure the volume of acid, record only in the space for volume used.

Molarity of base (NaOH) ______________

CALCULATIONS: In the spaces below, show calculation setups for Sample 1 only. Show answers for both samples in the boxes.

	Sample 1	Sample 2	Sample 3 (if needed)
1. Moles of base (NaOH) (if needed)			
2. Moles of acid used to neutralize (react with) the above number of moles of base			
3. Molarity of acid			

4. Average molarity of acid ______________

5. Unknown acid number ______________

B. Acetic Acid Content of Vinegar

Data Table

	Sample 1		Sample 2		Sample 3 (if needed)	
	Vinegar*	Base	Vinegar*	Base	Vinegar*	Base
Final buret reading						
Initial buret reading						
Volume used						

*If a pipet is used to measure the volume of vinegar, record only in the space for volume used.

Molarity of base (NaOH) _____   Vinegar number _____

CALCULATIONS: In the spaces below, show calculation setups for Sample 1 only. Show answers for both samples in the boxes.

	Sample 1	Sample 2	Sample 3 (if needed)
1. Moles of base (NaOH)			
2. Moles of acid ($HC_2H_3O_2$) used to neutralize (react with) the above number of moles of base			
3. Molarity of acetic acid in the vinegar			

4. Average molarity of acetic acid in the vinegar   _____

5. Grams of acetic acid per liter (from average molarity)   _____

6. Mass percent acetic acid in vinegar sample (density of vinegar = 1.005 g/mL)   _____

EXPERIMENT 23

Chloride Content of Salts

MATERIALS AND EQUIPMENT

Solutions: 0.10 M aluminum chloride ($AlCl_3$), 0.10 M barium chloride ($BaCl_2$), 0.10 M potassium chloride (KCl), 0.10 M silver nitrate ($AgNO_3$), and 0.10 M sodium chloride (NaCl). Dichlorofluorescein indicator and "unknown" solutions.

DISCUSSION

This experiment verifies the ratios of chloride ions in several metal chlorides. It also furnishes us with data for studying the relationship between the molar concentrations of a salt and that of its ions in solution.

We determine the amount of chloride ion in solution by reacting it with silver nitrate solution. the net ionic equation for this reaction is

$$Cl^-(aq) + Ag^+(aq) \rightarrow AgCl(s)$$

If we titrate different chloride salt solutions having the same molarity (moles/liter), the ratio of the amounts of $AgNO_3$ solution required should be the same as the ratio of the chloride contents of the respective salts. Consider the chloride available in 1 liter of each 1 molar NaCl and $BaCl_2$ solutions.

$$1 \text{ mole NaCl} \longrightarrow 1 \text{ mole Na}^+ + 1 \text{ mole Cl}^-$$
$$1 \text{ mole BaCl}_2 \longrightarrow 1 \text{ mole Ba}^{2+} + 2 \text{ moles Cl}^-$$

Thus it is evident that if equal volumes of the same molar concentrations of NaCl and $BaCl_2$ solutions are reacted, the $BaCl_2$ should require twice as much Ag^+ (silver nitrate solution).

In the experiment, silver nitrate solution is added to the chloride solutions containing dichlorofluorescein indicator. A sudden change from white to pink indicates that all the chloride in solution is reacted and marks the end point of the titration.

An unknown solution can be analyzed for its chloride content by the above method. If the data include the volume of the unknown solution and the molarity and volume of the $AgNO_3$ solution, the chloride molarity of the unknown may be calculated.

The following relationships are useful for calculations in this experiment:

$$\text{Moles} = \frac{\text{g of solute}}{\text{molar mass of solute}}$$

$$\text{Molarity} = \frac{\text{moles}}{\text{liters}} = \frac{\text{g of solute}}{\text{molar mass of solute} \times \text{liters of solution}}$$

$$\text{Moles} = \text{liters} \times \text{molarity} = \frac{\text{liters} \times \text{moles}}{\text{liter}}$$

In a chemical reaction where the reactants react in equal molar amounts, such as in $Cl^- + Ag^+ \longrightarrow AgCl$ (these ions react in a 1:1 mole ratio), we may say that

moles Cl^- reacted = moles Ag^+ reacted

Since moles = liters × molarity

liters Cl^- × molarity Cl^- = liters Ag^+ × molarity Ag^+

For the purpose of calculations, when the volumes of the two solutions are in the same units, whether they be liters, mL, or drops, we can write a more general statement equating the two reactants.

Volume Cl^- × molarity Cl^- = volume Ag^+ × molarity Ag^+

Using this relationship, we can calculate the molarity of the chloride ion in solution when data for the other three values are known.

PROCEDURE

Wear protective glasses.

General Instructions

1. Use the same pipet/medicine dropper to dispense the chloride solutions and the silver nitrate solution. Rinse the pipet with distilled water when changing from one solution to another.

2. Due to the different sizes of drops (even with the same pipet/medicine dropper), the volume of silver nitrate may vary about one drop in every ten drops in duplicate titrations. **It is very important to hold the medicine dropper in a vertical position in order to deliver drops of as uniform a volume as possible.**

3. Since less than 2 mL of most reagents are required, obtain only small volumes of reagents in clean, dry beakers.

Submit a small, clean, dry beaker (or test tube) to your instructor for an unknown chloride solution. Record the number of the unknown in the data table.

Place 10 clean test tubes in the rack. From a medicine dropper, add exactly 10 drops of 0.10 M NaCl to tubes one and two; 10 drops of 0.10 M KCl to tubes three and four; 10 drops of 0.10 M $BaCl_2$ to tubes five and six; 10 drops of 0.10 M $AlCl_3$ to tubes seven and eight; and 10 drops of the unknown solution to tubes nine and ten. Add 1 drop of dichlorofluorescein indicator to each tube.

Obtain about 10 mL of 0.10 M $AgNO_3$ solution from the reagent bottle. titrate the chloride samples by adding silver nitrate solution dropwise to each solution. Shake the solution during the titration. The end-point is reached when the color of the precipitate changes from white to pink and remains pink throughout the entire solution for at least ten seconds while shaking. **Count and record the number of drops of AgNO$_3$ solution required to reach the end-point in each tube.** If results for duplicate samples vary by more than one drop for every 10 drops of silver nitrate added, run another sample.

Since all solutions contain silver nitrate, dispose of all solutions in the "heavy metals waste" container.

Complete the data table and answer the questions on the report form.

REPORT FOR EXPERIMENT 23

Chloride Content of Salts

Data Table

Solution	Drops of Silver Nitrate			
	Trial 1	Trial 2	Trial 3 (if needed)	Average
NaCl				
KCl				
$BaCl_2$				
$AlCl_3$				
Unknown No. _____				

QUESTIONS AND PROBLEMS

1. Tabulate the experimental ratios of the chloride ion contents of the salts tested. Obtain these ratios by dividing the average number of drops of $AgNO_3$ required for each solution by the average number of drops required for the NaCl solution. Express the ratios to the nearest tenth.

Salt	NaCl	KCl	$BaCl_2$	$AlCl_3$
Ratio	1.0			

2. Assuming that sodium chloride has one chloride ion per simplest formula unit.

    (a) Do the data of this experiment prove that potassium chloride has one chloride ion per simplest formula unit? Explain.

(b) Do the data of this experiment prove that barium chloride has two chloride ions per simplest formula unit? Explain.

(c) Do the data of this experiment prove that aluminum chloride has three chloride ions per simplest formula unit? Explain.

(d) Do the data of this experiment prove that the simplest formula of aluminum chloride is $AlCl_3$? Explain.

3. Write and balance: (a) the un-ionized equation; (b) the total ionic equation; and (c) the net ionic equation for each of the following reactions:

$$NaCl(aq) + \quad AgNO_3(aq) \longrightarrow$$

$$KCl(aq) + \quad AgNO_3(aq) \longrightarrow$$

$$BaCl_2(aq) + \quad AgNO_3(aq) \longrightarrow$$

$$AlCl_3(aq) + \quad AgNO_3(aq) \longrightarrow$$

4. It was seen in the experiment that about 10 drops of $AgNO_3$ solution were required to react with 10 drops of NaCl solution, and about 20 drops of $AgNO_3$ solution were required to react with 10 drops of $BaCl_2$ solution.

   (a) What two factors are responsible for the fact that equal volumes of NaCl and $AgNO_3$ solutions react with each other?

   (b) What change(s) would you suggest to make equal volumes of $BaCl_2$ and $AgNO_3$ solutions react with each other?

5. What is the molarity of Cl^- ion in each of the solutions tested?

0.10 M NaCl	_____
0.10 M KCl	_____
0.10 M $BaCl_2$	_____
0.10 M $AlCl_3$	_____

6. Calculate the chloride molarity of the unknown solution. Show setups.

Unknown no.	_____
Molarity	_____

7. How many drops of 0.10 M $AgNO_3$ would be required to react with each of the following solutions? Assume each individual solution is 0.10 M.

   (a) 5 drops NaCl + 10 drops KCl   _____

   (b) 10 drops NaCl + 5 drops $BaCl_2$ _____

   (c) 5 drops $BaCl_2$ + 5 drops $AlCl_3$ _____

   (d) 5 drops $SnCl_4$ _____

   (e) 15 drops NH_4Cl _____

8. (a) How many moles of sodium chloride are in 5 drops of 0.10 M NaCl solution? Assume 1 drop is 0.050 mL.

_____

   (b) Calculate the number of milligrams of potassium chloride in 1 drop of 0.10 M KCl solution.

_____

9. How many grams of sodium chloride are required to prepare 350 mL of 0.15 M NaCl solution? Show calculations.

_____

EXPERIMENT 24

Chemical Equilibrium—Reversible Reactions

MATERIALS AND EQUIPMENT

Solid: ammonium chloride (NH_4Cl). **Solutions:** saturated ammonium chloride, concentrated (15 M) ammonium hydroxide (NH_4OH), 0.1 M cobalt(II) chloride ($CoCl_2$), 0.1 M iron(III) chloride ($FeCl_3$), concentrated (12 M) hydrochloric acid (HCl), dilute (6 M) nitric acid (HNO_3), phenolphthalein, 0.1 M potassium chromate (K_2CrO_4), 0.1 M potassium thiocyanate (KSCN), 0.1 M silver nitrate ($AgNO_3$), saturated sodium chloride (NaCl), 10 percent sodium hydroxide (NaOH), and dilute (3 M) sulfuric acid (H_2SO_4).

DISCUSSION

In many chemical reactions the reactants are not totally converted to the products because of a reverse reaction; that is, because the products react to form the original reactants. Such reactions are said to be reversible and are indicated by a double arrow ($\rightleftharpoons$) in the equation. The reaction proceeding to the right is called the **forward reaction**; that to the left, the **reverse reaction**. Both reactions occur simultaneously.

Every chemical reaction proceeds at a certain rate or speed. The rate of a reaction is variable and depends on the concentrations of the reactants and the conditions under which the reaction is conducted. When the rate of the forward reaction is equal to the rate of the reverse reaction, a condition of **chemical equilibrium** exists. At equilibrium the products react at the same rate as they are produced. Thus the concentrations of substances in equilibrium do not change, but both reactions, forward and reverse, are still occurring.

The principle of Le Chatelier relates to systems in equilibrium and states that when the conditions of a system in equilibrium are changed the system reacts to counteract the change and reestablish equilibrium. In this experiment we will observe the effect of changing the concentration of one or more substances in a chemical equilibrium. Consider the hypothetical equilibrium system

$$A + B \rightleftharpoons C + D$$

When the concentration of any one of the species in this equilibrium is changed, the equilibrium is disturbed. Changes in the concentrations of all the other substances will occur to establish a new position of equilibrium. For example, when the concentration of B is increased, the rate of the forward reaction increases, the concentration of A decreases, and the concentration of C and D increase. After a period of time the two rates will become equal and the system will again be in equilibrium. The following statements indicate how the equilibrium will shift when the concentrations of A, B, C, and D are changed.

An increase in the concentration of A or B causes the equilibrium to shift to the right.

An increase in the concentration of C or D causes the equilibrium to shift to the left.

A decrease in the concentration of A or B causes the equilibrium to shift to the left.

A decrease in the concentration of C or D causes the equilibrium to shift to the right.

Evidence of a shift in equilibrium by a change in concentration can easily be observed if one of the substances involved in the equilibrium is colored. **The appearance of a precipitate or the change in color of an indicator can sometimes be used to detect a shift in equilibrium.**

Net ionic equations for the equilibrium systems to be studied are given below. These equations will be useful for answering the questions in the report form.

Saturated Sodium Chloride Solution

$$NaCl(s) \underset{\longleftarrow}{\overset{H_2O}{\longrightarrow}} Na^+(aq) + Cl^-(aq)$$

Saturated Ammonium Chloride Solution

$$NH_4Cl(s) \underset{\longleftarrow}{\overset{H_2O}{\longrightarrow}} NH_4^+(aq) + Cl^-(aq)$$

Iron(III) Chloride plus Potassium Thiocyanate

$$Fe^{3+}(aq) + SCN^-(aq) \rightleftharpoons Fe(SCN)^{2+}(aq)$$
$$\text{Pale} \quad \text{Colorless} \quad \quad \text{Red}$$
$$\text{yellow}$$

Potassium Chromate with Nitric and Sulfuric Acids

The equilibrium involves the following ions in solution:

$$2\ CrO_4^{2-}(aq) + 2\ H^+(aq) \rightleftharpoons Cr_2O_7^{2-}(aq) + H_2O(l)$$
$$\text{Yellow} \quad\quad\quad\quad\quad\quad \text{Orange}$$

Cobalt(II) Chloride Solution

The equilibrium involves the following ions in solutions:

$$Co(H_2O)_6^{2+}(aq) + 4\ Cl^-(aq) \rightleftharpoons CoCl_4^{2-}(aq) + 6\ H_2O(l)$$
$$\text{Pink} \quad\quad\quad\quad\quad\quad\quad \text{Blue}$$

Ammonia Solution

$$NH_3(aq) + H_2O(l) \rightleftharpoons NH_4^+(aq) + OH^-(aq)$$

PROCEDURE

Wear protective glasses.

> **NOTE:** Record observed evidence of equilibrium shifts as each experiment is done.

A. Saturated Sodium Chloride Solution

Add a few drops of conc. hydrochloric acid to 2 to 3 mL of saturated sodium chloride solution in a test tube, and note the results.

B. Saturated Ammonium Chloride Solution

Repeat Part A, using saturated ammonium chloride solution instead of sodium chloride solution.

 Dispose of the solutions in A and B in the sink and flush with water.

C. Iron(III) Chloride plus Potassium Thiocyanate

Prepare a stock solution to be tested by adding 2 mL each of 0.1 M iron(III) chloride and 0.1 M potassium thiocyanate solutions to 100 mL of distilled water and mix. Pour about 5 mL of this stock solution into each of four test tubes.

1. Use the first tube as a control for comparison.

2. Add about 1 mL of 0.1 M iron(III) chloride solution to the second tube and observe the color change.

3. Add about 1 mL of 0.1 M potassium thiocyanate solution to the third tube and observe the color change.

4. Add 0.1 M silver nitrate solution dropwise (less than 1 mL) to the fourth tube until almost all the color is discharged. The white precipitate formed consists of both AgCl and AgSCN. Pour about half the contents (including the precipitate) into another tube. Add 0.1 M potassium thiocyanate solution dropwise (1 to 2 mL) to one tube and 0.1 M iron(III) chloride solution (1 to 2 mL) to the other. Observe the results.

 Dispose of the contents in tubes C.1–3 and the unused stock solutions in the sink and flush with water. Dispose of the contents of both C.4 tubes in the "heavy metals" waste container.

D. Potassium Chromate with Nitric and Sulfuric Acids

Pour about 2 to 3 mL of 0.1 M potassium chromate solution into each of two test tubes. Add 2 drops of dil. (6 M) nitric acid to one tube, add 2 drops of dil. (3 M) sulfuric acid to the other, and note the color change. Now add 10 percent sodium hydroxide solution dropwise to each tube until the original color of potassium chromate is restored.

 Dispose of the contents in these test tubes in the "heavy metals" waste container.

E. Cobalt(II) Chloride Solution

Place about 2 mL (no more) of 0.1 M cobalt(II) chloride solution into each of three test tubes.

1. To one tube add about 3 mL of conc. hydrochloric acid dropwise and note the result. Now add water dropwise to the solution until the original color (reverse reaction) is evident.

2. To the second tube add about 1.5 g of solid ammonium chloride and shake to make a saturated salt solution. Compare the color with the solution in the third tube (control). Place the second and third tubes (unstoppered) in a beaker of boiling water, shake occasionally, and note the results. Cool both tubes under tap water until the original color (reverse reaction) is evident.

 Dispose of these solutions in the "heavy metals" waste container.

F. Ammonia Solution

Prepare an ammonia stock solution by adding 4 drops of conc. ammonium hydroxide and 3 drops of phenolphthalein to 100 mL of tap water and mix. Pour about 5 mL of this stock solution into each of two test tubes.

1. Dissolve a very small amount of solid ammonium chloride in the stock solution in the first tube and observe the result.

2. Add a few drops of dil. (6 M) hydrochloric acid to the stock solution in the second tube. Mix and observe the result.

 Dispose of these solutions in the sink and flush with water.

REPORT FOR EXPERIMENT 24

Chemical Equilibrium—Reversible Reactions

Refer to equilibrium equations in the discussion when answering these questions.

A. Saturated Sodium Chloride

1. What is the evidence for a shift in equilibrium?

2. Which ion caused the equilibrium to shift?   _____

3. In which direction did the equilibrium shift?   _____

4. If solid sodium hydroxide were added to neutralize the hydrochloric acid, would this reverse the reaction and cause the precipitated sodium chloride to redissolve? Explain.

B. Saturated Ammonium Chloride

1. What is the evidence for a shift in equilibrium?

2. In which direction did the equilibrium shift?   _____

3. Which ion caused the equilibrium to shift?   _____

C. Iron(III) Chloride plus Potassium Thiocyanate

1. What is the evidence for a shift in equilibrium when iron(III) chloride is added to the stock solution?

2. What is the evidence for a shift in equilibrium when potassium thiocyanate is added to the stock solution?

3. (a) What is the evidence for a shift in equilibrium when silver nitrate is added to the stock solution? (The formation of a precipitate is not the evidence since the precipitate is not one of the substances in the equilibrium.)

   (b) The change in concentration of which ion in the equilibrium caused this equilibrium shift?

_____

   (c) Write a net ionic equation to illustrate how this concentration change occurred.

   (d) When the mixture in C.4 was divided and further tested, what evidence showed that the mixture still contained Fe^{3+} ions in solution?

D. Potassium Chromate with Nitric and Sulfuric Acids

1. What was the evidence for a shift in equilibrium when the nitric acid and the sulfuric acid were added to the potassium chromate solution?

2. Explain how sodium hydroxide caused the equilibrium to shift back again.

E.   Cobalt(II) Chloride Solution

1. What was the evidence for a shift in equilibrium when conc. hydrochloric acid was added to the cobalt chloride solution?

2. (a) Write the equilibrium equation for this system.

    (b) State whether the concentration of each of the following substances was **increased, decreased, or unaffected** when the conc. hydrochloric acid was added to cobalt chloride solution.

      $Co(H_2O)_6^{2+}$ _____ , Cl^- _____ , $CoCl_4^{2-}$ _____

3. (a) What did you observe when ammonium chloride was added to cobalt chloride solution?

    (b) What did you observe when this mixture was heated?

    (c) Explain why heating the mixture caused the equilibrium to shift.

    (d) What did you observe when the mixture was cooled?

    (e) Explain why cooling the mixture caused the equilibrium to shift.

F.   Ammonia Solution

1.   What is the evidence for a shift in equilibrium when ammonium chloride was added to the stock solution?

2.   Explain, in terms of the equilibrium, the results observed when hydrochloric acid was added to the stock solution.

3.   State whether the concentration of each of the following was increased, decreased, or was unaffected when dilute hydrochloric acid was added to the ammonia stock solution:

   NH_3 _____ , NH_4^+ _____ , OH^- _____ , Pink color _____

4.   (a) In which direction would the equilibrium shift if sodium hydroxide were added to the ammonia stock solution?

   (b) Would the sodium hydroxide tend to decrease the color intensity? Explain.

5.   Would boiling the ammonia solution have any effect on the equilibrium? Explain.

EXPERIMENT 25

Heat of Reaction

MATERIALS AND EQUIPMENT

Solid: ammonium chloride (NH_4Cl). **Solutions:** concentrated (12 M) hydrochloric acid (HCl), concentrated (16 M) nitric acid (HNO_3), 10 percent sodium hydroxide (NaOH), and concentrated (18 M) sulfuric acid (H_2SO_4). A styrofoam cup; a thermometer.

DISCUSSION

All chemical reactions involve a heat effect. If the reaction liberates heat, it is called an **exothermic reaction**. If the reaction consumes heat, it is called an **endothermic reaction**. These heat effects are the result of the formation or breaking of chemical bonds. Energy is required to break bonds, and energy is liberated in the formation of chemical bonds. The **heat of reaction** is the net heat effect for the reaction and is often the combination of several effects.

This experiment will investigate the heat of reaction for three types of reactions: the hydration of a liquid, the dissolving of a solid, and neutralization (the reaction of an acid with a base). The quantity of heat will not be measured in any of the experiments. Instead temperature changes will be observed as indicators of heat effects.

Hydration of Concentrated Sulfuric Acid

A strong heat effect is observed when conc. sulfuric acid is added to water. This reaction is so exothermic that it can lead to dangerous spattering. The general rule "add the acid to the water" will help avoid the danger.

The reaction involved is the hydration of sulfuric acid molecules to form $H_2SO_4 \cdot H_2O$ and $H_2SO_4 \cdot 2H_2O$. The formation of the bonds between the acid and the water is the primary cause of the heat liberated. The usual commercial concentrated sulfuric acid contains less than 5 percent water, so some of the acid molecules are already hydrated.

Dissolving of Ammonium Chloride

When a salt dissolves there are two major competing heat effects. The energy required to break the ionic bonds of the crystal lattice is called the **lattice energy**. As the ions are freed from the crystal lattice they become hydrated, liberating energy. This energy is known as the **hydration energy**. If the lattice energy is greater than the hydration energy, the reaction will be endothermic. If the hydration energy is greater than the lattice energy, the reaction will be exothermic.

Neutralization Reactions

When hydrochloric acid is added to sodium hydroxide solution, heat is evolved. Since the reaction of an acid with a base is called neutralization, this heat is called the **heat of neutralization**. The molecular equation for this reaction is

$$HCl(aq) + NaOH(aq) \longrightarrow H_2O(l) + NaCl(aq) + \text{Heat}$$

Since all the substances are soluble, and all of them except water are strong electrolytes, the total ionic equation is

$$H^+ + Cl^- + Na^+ + OH^- \longrightarrow H_2O + Na^+ + Cl^- + \text{Heat}$$

Therefore, the net ionic equation is

$$H^+(aq) + OH^-(aq) \longrightarrow H_2O(l) + \text{Heat}$$

According to the net ionic equation, a neutralization reaction between a strong acid and a strong base is the reaction of the hydrogen ions (H^+) and the hydroxide ions (OH^-) to form water. Since this reaction is common to all strong acids and bases, and since a definite amount of heat is liberated for each mole of reacting hydrogen ion and hydroxide ion, it should be possible to use different strong acids with a fixed amount of base and obtain the same quantity of heat for each acid neutralized. This experiment will investigate this hypothesis.

Each neutralization will involve the same total volume of liquid. Thus if the same quantity of heat actually is produced, the temperature increase should be the same for each neutralization. This assumes a constant heat capacity for the various solutions used, which is not precisely true, but is satisfactory for the purpose of this experiment. Duplicate runs are made for each acid used. Because of experimental errors, variations occur in the temperature increases obtained in duplicate runs with the same acid. Thus the variations in temperature changes with different acids should be of the same order of magnitude, if our hypothesis of identical reactions is correct. It is necessary to use dilute acids and bases to minimize heats of hydration.

In all the experimental trials the moles of hydroxide ion will be the limiting reactant. Thus there will always be more moles of hydrogen ion available than of hydroxide ion. The excess moles of hydrogen ion will remain unreacted.

PROCEDURE

⚠ **Wear protective glasses.**

Record all temperatures to the nearest 0.1°C.

 Dispose of all solutions in the sink and flush with water.

A. Hydration of Concentrated Sulfuric Acid

⚠ Concentrated sulfuric acid is an extremely corrosive oxidizing acid and a strong dehydrating agent. Wash immediately if you spill any on your skin or clothing. Rinse off any acid that runs down the side of the bottle with a wash bottle and rinse the graduated cylinder used.

Pour 44 mL of tap water into a styrofoam cup. Read and record the temperature of the water. Measure 4.0 mL of conc. sulfuric acid and pour it into the water in the styrofoam cup. Stir with the thermometer and record the maximum temperature reached. Rinse off any acid that runs down the side of the bottle and rinse out the graduated cylinder. Pour this dilute sulfuric acid solution from the styrofoam cup into a 125 mL Erlenmeyer flask and save for Part C.

B. Dissolving of Ammonium Chloride

Rinse the styrofoam cup and pour 20 mL of tap water into it. Read and record the temperature of the water. Weigh approximately 3 g of ammonium chloride crystals and pour then into the styrofoam cup. Stir with the thermometer and record the minimum temperature reached.

C. Neutralization Reactions

You will need four stock solutions. One, the NaOH stock solution, requires a 250 mL Erlenmeyer flask; the other three can be in either 250 or 125 mL flasks. Prepare these stock solutions by adding the acid or base to water as required and mixing. Label each flask for identification.

> ⚠ These concentrated HCl and HNO_3 acids and the NaOH are extremely corrosive. Wash immediately if you spill any on your skin or clothing. Handle with care.

Flask 1 (250 mL)	NaOH stock solution:	65 mL H_2O + 65 mL 10 percent NaOH.
Flask 2	HCl stock solution:	75 mL H_2O + 25 mL concentrated HCl.
Flask 3	HNO_3 stock solution:	21 mL H_2O + 5 mL concentrated HNO_3.
Flask 4	H_2SO_4 stock solution:	H_2SO_4 solution prepared in Part A.

After the four stock solutions are prepared, carry out the five neutralizations using the specified amounts of stock solutions and water. Make a duplicate run for each neutralization. Use the following six-step procedure for each neutralization:

Step 1: Pour the measured volumes of acid and water into the styrofoam cup.

Step 2: Measure out the sodium hydroxide solution.

Step 3: Read and record the temperature of the water-acid mixture.

Step 4: Add the base, stir with the thermometer, and record the maximum temperature reached.

Step 5: Calculate the temperature change.

Step 6: Empty the styrofoam cup; rinse the cup and the graduated cylinders with water.

The amounts of the prepared stock solutions to be used for each neutralization are as follows:

Neutralization 1.   10 mL HCl, 20 mL H_2O, 10 mL NaOH

Neutralization 2.   10 mL HNO_3, 20 mL H_2O, 10 mL NaOH

Neutralization 3.   10 mL H_2SO_4, 20 mL H_2O, 10 mL NaOH

Neutralization 4.   20 mL HCl, 10 mL H_2O, 10 mL NaOH

Neutralization 5.   10 mL HCl, 10 mL H_2O, 20 mL NaOH

REPORT FOR EXPERIMENT 25

Heat of Reaction

A. Hydration of Concentrated Sulfuric Acid

1. Temperature of water.   _____

    Temperature of water after adding acid.   _____

    Temperature change.   _____

2. Is this an endothermic or exothermic reaction?   _____

3. In this reaction, what is the major change, in terms of bonds made or broken, that causes
    the heat effect?

4. If you had used dilute sulfuric acid rather than concentrated sulfuric acid, would you ex-
    pect the temperature change to be greater or less?

                                                                      _____
    Why?

B. Dissolving of Ammonium Chloride

1. Temperature of water.   _____

    Temperature of water after adding salt.   _____

    Temperature change.   _____

2. Is this an endothermic or exothermic reaction?   _____

3. In this reaction, what is the major change, in terms of bonds made or broken, that causes
    the heat effect?

4. What other heat effect is present in this reaction, acting in the opposite direction?

5. If you had used 40 mL of water and 6 g of ammonium chloride, rather than the 20 mL and 3 g in the experiment, would you expect to get a larger, smaller, or identical temperature change?

    _____

    Why?

C. Neutralization Reactions

Data Table

		Temp. Before Adding NaOH	Temp. After Adding NaOH	Temp. Change
1. 10 mL HCl, 20 mL H_2O, 10 mL NaOH	Trial 1	_____	_____	_____
	Trial 2	_____	_____	_____
	Average	_____	_____	_____
2. 10 mL HNO_3, 20 mL H_2O, 10 mL NaOH	Trial 1	_____	_____	_____
	Trial 2	_____	_____	_____
	Average	_____	_____	_____
3. 10 mL H_2SO_4, 20 mL H_2O, 10 mL NaOH	Trial 1	_____	_____	_____
	Trial 2	_____	_____	_____
	Average	_____	_____	_____
4. 20 mL HCl, 10 mL H_2O, 10 mL NaOH	Trial 1	_____	_____	_____
	Trial 2	_____	_____	_____
	Average	_____	_____	_____
5. 10 mL HCl, 10 mL H_2O, 20 mL NaOH	Trial 1	_____	_____	_____
	Trial 2	_____	_____	_____
	Average	_____	_____	_____

QUESTIONS AND PROBLEMS

The following pertain to Part C:

1. (a) How do the average temperature changes observed in Parts 1, 2, and 3 compare?

    (b) What do you conclude from this?

2. (a) How does the average temperature change in Part 4 compare with that in Parts 1, 2, and 3?

    (b) What do you conclude from this?

3. (a) How does the average temperature change in Part 5 compare with that in Part 1?

    (b) What do you conclude from this?

EXPERIMENT 26

Distillation of Volatile Liquids

MATERIALS AND EQUIPMENT

Liquids: Distilled water, ethanol (denatured). **Solutions:** ethanol/water(50/50%), red wine.

Special equipment: Distillation flask, 125 or 250 mL; thermometer, 100°C; condenser with bent adapter, heat source without open flame (hot plate or heating mantle with rheostat), boiling stones, and pot holders or mitts.

DISCUSSION

A. Background

The most commonly used method for purification of liquids is distillation, a process by which one liquid can be separated from another liquid or from a nonvolatile solid. There are many applications of distillation. For example, in many arid regions, distillation is used to obtain potable water from seawater. In the petroleum industry, distillation is used to separate the components of crude oil into fractions such as gasoline, kerosene, and diesel fuel.

Evaporation is the escape of molecules from the liquid state to the vapor (gaseous) state.

$$\text{liquid} \xrightarrow{\text{evaporation}} \text{vapor}$$

The rate of evaporation is dependent on the temperature and vapor pressure of the liquid. The vapor pressure of a liquid is the pressure exerted by a vapor in equilibrium with its liquid. When a liquid is heated, its vapor pressure increases until it equals the pressure of the atmosphere above it, at which point the liquid begins to boil. The normal boiling point (bp) of a liquid is the temperature at which the liquid boils when the atmospheric pressure is 1 atm (760 torr), the average atmospheric pressure at sea level.

> **Note**: The temperatures in this discussion and procedure refer to the normal boiling points at 1 atm pressure. If the atmospheric pressure at your laboratory is lower than 1 atm, experimental temperatures will be lower than those given in this experiment.

A volatile liquid has relatively weak intermolecular attractions. When such a liquid is placed in an open dish, the liquid evaporates and the vapor diffuses into the atmosphere. During a distillation, the vapor forms in the distilling flask and is condensed back to a liquid and collected in a separate vessel. Thus, in distillation

$$\text{liquid} \xrightarrow[\text{heat}]{\text{evaporation}} \text{vapor} \xrightarrow[\text{cooling}]{\text{condensation}} \text{liquid}$$

Consider the three liquids, ethyl ether (bp 34.5°C), ethanol (bp 78.5°C), and water (bp 100°C). Ethyl ether is more volatile than ethanol, which is more volatile than water. This means that molecules of ethyl ether have less intermolecular attraction than the molecules of the other two compounds. It also means that ethyl ether has a higher vapor pressure than the other two compounds at any temperature up to its boiling point. All three substances boil at atmospheric pressure, but at different temperatures.

When a solution of two volatile liquids is heated in the distilling apparatus shown in Figure 26.1, both compounds are present in the vapor. In an efficient distilling apparatus, the vapor condenses to a liquid and revaporizes many times, continually becoming richer in the lower boiling component. When the vapor reaches the thermometer and passes into the condenser, it is possible to have separated one compound from the other. The apparatus in Figure 26.1 is a simple distillation setup that is not very efficient for separating two liquids unless there is a large difference in their boiling points. However, it serves the purpose for demonstrating the principles of distillation.

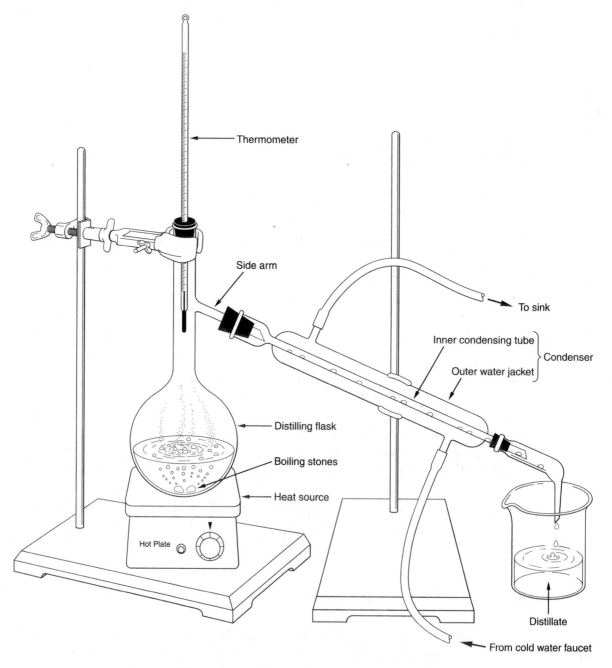

Figure 26.1   A simple distillation setup

As a liquid boils, hot vapor moves up the distilling flask, past the thermometer measuring its temperature, and through the sidearm into the condenser. If the vapor cools off before it reaches the thermometer, it condenses on the sides of the flask and does not reach the thermometer or the condenser. The condenser consists of an inner tube surrounded by a water jacket for cooling the vapor. The vapor loses heat to the flowing water and condenses back to a liquid on the cool surface of the inner tube. The condensed liquid, called the distillate, flows down the inner tube and is collected in a beaker.

Distillation should be conducted slowly and steadily and at such a rate that an equilibrium is maintained between the vapor and the condensate (liquid) where the thermometer is located. A drop of condensate should always appear on the bulb of the thermometer while vapor is flowing into the condenser. The distilling flask should not be heated to dryness.

B. Boiling Points and Heating Curves.

During the first part of this experiment you will observe the boiling points and obtain data to plot heating curves for two liquids, ethanol and water. Each liquid is heated separately in a distilling flask and the temperature is recorded at intervals of 0.5 minute for three minutes and at 1-minute intervals thereafter. A temperature plateau occurs even as heat continues to be added to the system. The time/temperature data, when plotted as a graph, is known as a heating curve. The experimental boiling point is at the plateau where the temperature remains constant over a narrow range of 1°–2°C or less.

C. Distillation of Solutions Containing Two Volatile Liquids.

In the second part of this experiment, the liquids being distilled are solutions of alcohol and water. When there is more than one volatile liquid in a solution, the distillation is more complex. In theory, the lower-boiling, more volatile liquid will distill first, followed by the higher-boiling liquid. Any nonvolatile components will remain in the distilling flask. The theoretical heating curve for two liquids with boiling points of 80°C and 110°C should look like Figure 26.2.

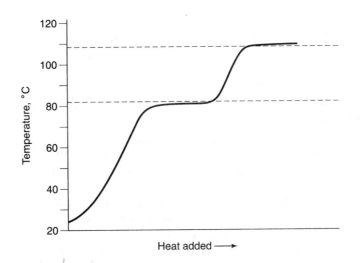

Figure 26.2   Theoretical heating curve for two volatile liquids

In reality, it is difficult to get the two distinct plateaus necessary to separate the two volatile liquids. When you distill a solution of two volatile liquids, the vapor will contain both compounds over much of the temperature range. Due to their relative vapor pressures at the lower temperatures, the vapor will contain a higher concentration of the lower boiling (higher vapor pressure) liquid and the distillate will be mostly the lower boiling component. Later, at the higher end of the boiling range, the vapor will have more of the less volatile compound and the distillate will be mostly the higher boiling component. Thus, in a simple distillation setup such as the one we are using, the temperature plateaus may not be as flat as when distilling pure compounds. In fact, depending on the concentration of the solution and the rate of heating, you may not observe much of a plateau at all.

PROCEDURE

Wear protective glasses.

1. Set the heat source aside and set up the rest of the apparatus as shown in Figure 26.1. For the first distillation you must start with a dry distilling flask. Because the heat source cannot be an open flame, you will use a hot plate or heating mantle. Your instructor will explain the use of the heat source available to you. Set the dial as instructed, keep it away from the flask, and allow it to preheat for five minutes.

 From this point on, remember that the heat source is hot and can burn you when you touch its surface.

2. Every team of students will do three distillations: ethanol, distilled water, and one of the two alcohol/water solutions. Assignments will be made by the instructor.

    a. ethanol   }   every team will do
    b. deionized water }   these two distillations

    c. 50/50 ethanol/water solution }   every team will do
    d. alcoholic beverage, red wine }   one of these distillations

First Distillation: Ethanol

3. Fill a dry graduated cylinder with 50 ml of ethanol. Use a funnel to pour the liquid from the graduated cylinder into the distilling flask. Be very careful to avoid pouring the liquid into the side arm. Add two boiling stones. Boiling stones are used to ensure uniform boiling and to avoid superheating.

4. Put the stopper into the distilling flask and position the thermometer just below the side arm. Make sure the apparatus is put together securely and place a beaker under the receiving end of the condenser to collect the distillate. The heat source is not under the flask yet.

5. Attach the bottom hose on the condenser to a cold water spigot and put the top hose into a sink. **Carefully** turn on the spigot to pass water through the condenser.

6. Before beginning the distillation, work out plans with a partner to observe and record data. Once the heating begins, you need to be very observant. One person should watch the thermometer bulb, the temperature, and the clock, and call out the temperatures, while the other should record temperatures that are being called out. Record all temperatures to the nearest 0.1°C.

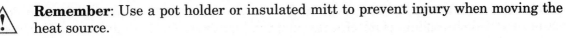 **Remember**: Use a pot holder or insulated mitt to prevent injury when moving the heat source.

7. Position the preheated heat source so it is touching the bottom of the distilling flask. Observe the temperature and when the thermometer reads about 34°C, record this temperature for time 0.0 minutes. As you wait for the temperature to reach this point, notice that bubbles begin to emerge from the boiling stones. These bubbles will get more vigorous as the temperature in the liquid increases. Remember, however, the thermometer is not in the liquid. It is measuring the temperature near the side arm, relatively far away from the heat source and the liquid.

8. Continue heating and recording the temperature at 30-second intervals for three minutes and then at 1-minute intervals. Be very alert because sometimes the temperature can shoot up very suddenly once the hot vapor reaches the top of the flask where the thermometer is located. Be very patient because at other times the temperature can change very slowly.

9. Pay attention to the formation of the first drop of liquid on the thermometer. When this drop appears, the liquid-vapor equilibrium has been established. If this drop of liquid evaporates completely, the equilibrium has been lost. Make a note on the data table and adjust the heat source. The heat will need to be turned down slightly if it is too hot near the thermometer bulb for condensation to occur. (You will see the drop of liquid dripping from the side arm as vapor enters the condenser.) The heat will need to be turned up slightly if the vapor is condensing before it reaches the thermometer bulb.

10. Continue recording the temperature for fifteen minutes. When the temperature remains constant for five minutes before the full fifteen minutes have elapsed you may assume that it will remain constant for the remaining time. If the temperature has not stabilized after fifteen minutes, add more time on the data table and continue recording. Turn off the hot plate and separate the distilling flask from the condenser using pot holders to avoid burning yourself. Run cold water over the round bottom of the flask to cool it and the liquid inside.

 11. Empty the ethanol remaining in the distilling flask and the distillate into the recycling bottle provided.

Second Distillation: Water

12. Rinse the distillation flask and thermometer with distilled water and add 50 mL of distilled water and two boiling stones to the flask. Repeat steps 7–10 again. Ask your instructor if it is necessary to increase the temperature of the heat source.

 Pour the distillate and water remaining in the distilling flask into the sink.

Third Distillation: A Solution of Ethanol and Water

13. Use a cooled distilling flask as before. Add 50 mL of the ethanol/water solution that has been assigned to your group. Before you begin to heat, note that the time intervals on the data table for this distillation are different. Begin recording at 0.0 minutes when the temperature reaches about 34°C. You will record at 1-minute intervals during the first six minutes and at 2-minute intervals thereafter. It may take about 45 minutes to do this distillation. Your instructor may choose to terminate these distillations before the final plateau is reached in order to save time. If the drop of liquid on the thermometer disappears before the boiling point of

water is reached, turn up the heat slightly. Stop recording after 45 minutes. When the temperature reaches the boiling point of water and remains there for 2–4 minutes, you may stop the distillation sooner than 45 minutes.

Turn off and unplug the heat source. Use the pot holders to separate the distilling flask from the condenser and the heat source.

WASTE DISPOSE OF PROPERLY   Empty the water/alcohol mixtures into the sink and flush with water.

14. Exchange data for the third distillation with another team that distilled the alternate solution.

15. Complete two graphs using time as the independent variable and temperature as the dependent variable. Use the graph paper provided or make a computer graph. Review the instructions in Study Aid 3 if necessary.

Graph 1. Plot temperature vs. time data for distillations 1 and 2.

Graph 2. Plot temperature vs. time data for the other two distillations (solutions).

REPORT FOR EXPERIMENT 26

Distillation of Volatile Liquids

Data Table

time, minutes	temp, °C ethanol	temp, °C H_2O		time, minutes	temp, °C 50% ethanol	temp, °C red wine	
0.0				0.0			
0.5				1.0			
1.0				2.0			
1.5				3.0			
2.0				4.0			
2.5				5.0			
3.0				6.0			
4.0				8.0			
5.0				10.0			
6.0				12.0			
7.0				14.0			
8.0				16.0			
9.0				18.0			
10.0				20.0			
11.0				22.0			
12.0				24.0			
13.0				26.0			
14.0				28.0			
15.0				30.0			
				32.0			
				34.0			
				36.0			
				38.0			
				40.0			
				42.0			
				44.0			

Questions and Problems

1. The boiling points of three liquids are provided below.

    Acetone, bp = 56.2°C   Methanol, bp = 65°C   Ethylene glycol, bp = 198°C

    Which liquid is the least volatile? (circle your choice)

    Which liquid has the weakest attractive forces between its molecules? (underline your choice)

2. If the drop of liquid on the thermometer disappears while the vapor is visibly condensing in the side arm, which process is proceeding faster in the flask, evaporation or condensation? (Circle your choice) How would you get the drop of liquid back?

3. Define boiling point.

4. Why is the thermometer bulb placed above the liquid beside the side arm of the distilling flask and not in the liquid?

5. What is the purpose of

    a. the inner tube of the condenser?

    b. the outer jacket of the condenser?

6. When the red wine is distilled, why does the distillate remain colorless throughout the full temperature range?

7. Explain *briefly* how distillation can be used to purify seawater to make salt-free water.

8. One team collected distillate from the wine in four 10 mL fractions in a graduated cylinder. What is the difference between the composition of the first 10 mL collected and the last 10 mL?

   Explain this difference.

Use your graphs to answer the following questions:

9. a. What is the experimental boiling point of ethanol? _____ water? _____

   b. What is the boiling point range for the 50% ethanol solution? _____

10. Why did it take so much longer to reach the boiling point of water in the solutions of ethanol and water than in the pure water?

Graph 1

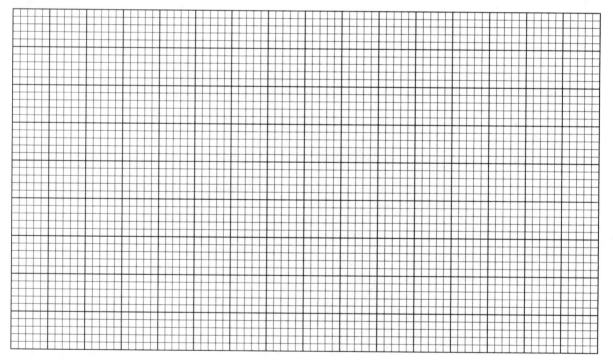

Graph 2

EXPERIMENT 27

Boiling Points and Melting Points

MATERIALS AND EQUIPMENT

Solids: Boiling stones, benzoic acid (C_6H_5COOH), benzophenone ($C_6H_5COC_6H_5$), cholesterol ($C_{27}H_{45}OH$), diphenylacetic acid ($C_{14}H_{12}O_2$) trans-cinnamic acid ($C_9H_8O_2$), urea (CH_4N_2O), 50% urea-50% trans-cinnamic acid, 50% diphenylacetic acid-50% cholesterol, and an unknown.
Liquids: Isopropyl alcohol [$CH_3CH(OH)CH_3$], ethyl alcohol (C_2H_5OH), and methyl alcohol (CH_3OH). Water and oil (mineral or vegetable oil) baths, wire stirrer, 110°C and 250°C thermometers, capillary tubes (sealed at one end), rubber bands cut from 3/16 inch rubber tubing.

DISCUSSION

Organic chemistry is the branch of chemistry that deals with carbon-containing molecules. Carbon bonds not only to other atoms but to itself, forming multiple chains and rings. This feature of carbon allows almost an unlimited number of organic compounds to exist. Organic chemists are faced with the difficult task of identifying and categorizing this vast array of molecules. Each organic compound has many chemical and physical properties associated with its structure. If enough of the properties are identified, we are able to say that we are dealing with one specific compound and not another.

This experiment will introduce two physical properties, the boiling point and the melting point, that organic chemists use to identify organic molecules. The boiling point of a liquid is the temperature at which the vapor pressure of the liquid equals the surrounding atmospheric pressure. Since the boiling point is dependent upon atmospheric pressure, a liquid will boil at different temperatures depending on the altitude above sea level. Consequently, water boils at a lower temperature in the mountains than it does at the seashore.

The second property we will measure is the melting point of a substance. The melting point is the temperature at which the solid becomes a liquid. Unlike the boiling point, normal atmospheric pressures do not significantly affect the melting point.

Most pure substances have sharp melting points. If a substance is made impure by mixing it with another solid chemical, the temperature at which the mixture melts is no longer sharp but is broadened over a range. Then the melting temperature is known as the "melting point range." Mixtures also have melting point ranges that are usually lower than the melting points of the pure components. This broadening and lowering of the melting point can be used to identify unknown substances. Two chemicals can have the same melting point, e.g., urea and trans-cinnamic acid. Thus, if we have an unknown sample that is either urea or trans-cinnamic acid, and mix it with urea and obtain a sharp melting point, we know the unknown must be urea. But if the unknown when mixed with urea gives a broad and lowered melting point range, then the unknown must be trans-cinnamic acid. This procedure is known as the "method of mixed melting points."

PROCEDURE

Wear protective glasses.

 CAUTION: All three alcohols are flammable. The quantities used are small, but be sure to keep the flame as far as possible from the mouth of the test tube (your neighbor's tube also!).

A. Boiling Points

1. Assemble the apparatus illustrated in Figure 27.1.

2. Determine the boiling point of these three liquids.

    (a) methyl alcohol (methanol)

    (b) ethyl alcohol (ethanol)

    (c) isopropyl alcohol (2-propanol)

3. Pour 5 mL of methanol into a clean dry test tube, add two small boiling stones, and secure the tube to a ring stand so that

    (a) the surface of the methanol is slightly below the water level in the beaker and

    (b) the tip of the thermometer is about 1 cm above the surface of the methanol (see Figure 27.1).

4. Heat the water. Check the temperature of the water bath with a second thermometer. Do not allow the bath's temperature to go above 80°C for this methanol sample. As the sample boils, the vapor will cause the temperature reading of the thermometer to increase until it remains constant. This constant value is the boiling point of the sample. Record this temperature.

5. Repeat Steps 3 and 4 for other liquids. Do not allow the temperature of the bath to exceed 90°C for the ethanol sample. Allow the bath to cool to about 75°C before you place the third sample in the bath.

6. Dispose of the used alcohols in the organic waste bottle.

B. Melting Points

You will measure the melting points of several pure compounds and the melting point ranges for two mixtures. You also will determine the identity of an unknown solid by the method of mixed melting points.

1. Assemble the apparatus shown in Figure 27.2. (Other methods may be used in your laboratory as indicated by your instructor.) Note: A 250°C thermometer is used.

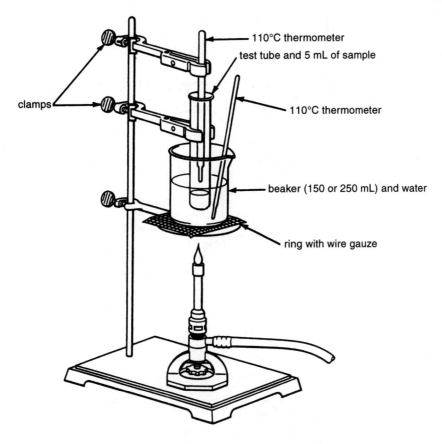

Figure 27.1. Boiling Point Apparatus.

2. Measure the melting points (or ranges) for the following compounds and mixtures. Take the melting points in the order given below:

(a) benzophenone   (e) diphenylacetic acid

(b) benzoic acid   (f) cholesterol

(c) urea   (g) 50% urea-50% trans-cinnamic acid

(d) trans-cinnamic acid   (h) 50% diphenylacetic acid-50% cholesterol

3. You will need a separate capillary tube (sealed at one end) for each sample. To fill the capillary tube, make a small pile of the solid on a watch glass and force the open end of the capillary through the crystals to get some of the sample into the tube. Then tap the sealed end of the tube lightly on the table top to get the sample down into the sealed end. Repeat this procedure until you have about 0.5 cm of the sample in the capillary.

4. Use a small rubber band to attach the capillary tube to a 250°C thermometer. The bottom of the capillary tube should be level with the bottom of the thermometer.

5. Place the thermometer and capillary tube in the oil bath as shown in Fig. 27.2.

6. It is important to stir the oil to produce a uniform temperature. Use a wire that has a loop at one end and a handle at the other (Figure 27.2). Stir the oil by raising and lowering the wire loop as you slowly heat the oil.

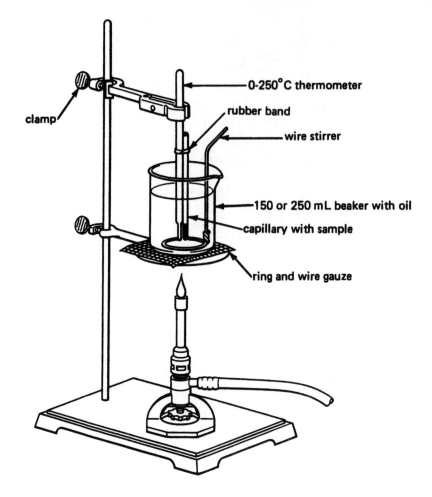

Figure 27.2. Melting Point Apparatus. (Other types are available)

7. Watch **both** the temperature and the sample as you slowly heat the oil bath. Record the temperature when the crystals first begin to melt and also the temperature at which the last crystals melt. The melting process may occur over a range of several degrees. (Be sure that you stir the oil constantly.)

8. Repeat Steps 3–7 for the remaining samples. Be sure the oil bath has cooled below the expected melting point of your next sample.

9. Obtain an unknown sample from your instructor.

10. Determine the melting point range for your unknown. You should now have a good idea as to its identity by comparing its melting point to those that you have already determined.

11. Determine, by the "method of mixed melting points," if your unknown is what you believe it to be. Record your results. (Be sure your sample and the known are **very thoroughly mixed** before determining the mixed melting point range.)

 Dispose of capillary tubes and samples in the container provided.

REPORT FOR EXPERIMENT 27

Boiling Points and Melting Points

For this report you will need to obtain literature values (from Handbooks) for the accepted boiling and melting points.

A. Boiling Points

Sample	Handbook values of boiling points, °C	Experimental values of boiling points, °C
Methyl alcohol		
Ethyl alcohol		
Isopropyl alcohol		

B. Melting Point Ranges

Sample	Handbook values of melting points, °C	Experimental values melting point range, °C
Benzophenone		
Benzoic acid		
Urea		
Trans-cinnamic acid		
Diphenylacetic acid		
Cholesterol		
50% urea and 50% trans-cinnamic acid	none	
50% cholesterol and 50% diphenylacetic acid	none	
Unknown	none	

Results for the "method of mixed melting points":

(a) With what compound did you mix your unknown?   _____

(b) Observed melting point range   _____

(c) What is your unknown number?   _____

(d) What is the name of your unknown?   _____

QUESTIONS AND PROBLEMS

1. Why must you allow the oil bath to cool between melting point determinations?

2. (a) If you performed a melting point experiment on the same compound in San Francisco and on top of Mt. Everest, would your results differ? Explain.

    (b) If you performed a boiling point experiment on the same compound in San Francisco and on top of Mt. Everest, would your results differ? Explain.

3. Cocaine melts at 98°C and glucose melts at 146°C. As a chemist for the government it is your task to quickly identify the contents of three vials. One contains pure cocaine, another pure glucose, and the third a mixture of cocaine and glucose. How would you accomplish your assignment?

4. Suppose you determine the melting point of cortisone (a hormone) to be 230°C. Your two neighbors obtain values of 226°C and 233°C. Why might their values differ from your value?

EXPERIMENT 28

Hydrocarbons

MATERIALS AND EQUIPMENT

Solid: calcium carbide (about 3/8 in. lumps) (CaC_2). **Liquids:** pentene (amylene) (C_5H_{10}), heptane (C_7H_{16}), kerosene, and toluene $(C_6H_5CH_3)$. **Solutions:** 5% bromine (Br_2) in 1,1,1-trichloroethane (CCl_3CH_3) and 0.1 M potassium permanganate $(KMnO_4)$. Wooden splints.

Hydrocarbons

Hydrocarbons are organic compounds made up entirely of carbon and hydrogen atoms. Their principal natural sources are coal, petroleum, and natural gas. Hydrocarbons are grouped into several series by similarity of molecular structure. Some of these are the alkanes, alkenes, alkynes, and aromatic hydrocarbons.

Alkanes

Also known as the paraffins or **saturated hydrocarbons,** the **alkanes** are straight- or branched-chain hydrocarbons having only single bonds between carbon atoms. They are called saturated hydrocarbons because all their carbon-carbon bonds are single bonds.

The first 10 members of the alkane series and their molecular formulas are listed below:

Methane CH_4	Hexane C_6H_{14}
Ethane C_2H_6	Heptane C_7H_{16}
Propane C_3H_8	Octane C_8H_{18}
Butane C_4H_{10}	Nonane C_9H_{20}
Pentane C_5H_{12}	Decane $C_{10}H_{22}$

Like most organic substances, the alkanes are combustible. The products of their complete combustion are carbon dioxide and water. The reactions of the alkanes are of the substitution type; that is, some atom or group of atoms is substituted for one or more of the hydrogen atoms in the hydrocarbon molecule. For example, in the bromination of methane, a bromine atom is substituted for a hydrogen atom. This reaction does not occur appreciably in the dark at room temperature but is catalyzed by ultraviolet light. The equation is

$$CH_4 + Br_2 \xrightarrow[\text{light}]{\text{Ultraviolet}} CH_3Br + HBr$$

Methane   Methyl bromide

Alkenes

Also known as the olefins, the alkenes are a series of straight- or branched-chain hydrocarbons containing a carbon-carbon double bond in their structures. They are considered to be unsaturated hydrocarbons. The first two members of the series are ethene (C_2H_4) and propene (C_3H_6). Their structural and condensed structural formulas are:

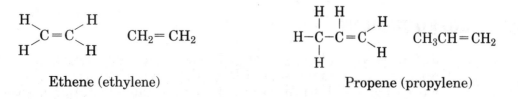

Ethene (ethylene)   Propene (propylene)

The functional group of this series is the carbon-carbon double bond (C=C); it is a point of high reactivity. Alkenes undergo addition-type reactions; that is, other groups are added to the double bond, causing the molecule to become saturated. For example, when hydrogen is added, one H atom from H_2 is added to each carbon atom of the double bond to saturate the molecule, forming an alkane:

$$CH_3CH=CH_2 + H_2 \xrightarrow[\text{Heat and Pressure}]{\text{Ni Catalyst}} CH_3CH_2CH_3$$

Propene   Propane

When a halogen such as bromine is added, one Br atom from Br_2 is added to each carbon atom of the double bond to saturate the molecule:

$$CH_2=CH_2 + Br_2 \longrightarrow \underset{\substack{| \quad | \\ Br \quad Br}}{CH_2CH_2}$$

1,2-Dibromoethane
(Ethylene dibromide)

Evidence that bromine has reacted is the disappearance of the red-brown color of free bromine. Other reactions of olefins also show the increased reactivity of the alkenes over the alkanes.

Unsaturated hydrocarbons can be oxidized by potassium permanganate. The reaction is known as the Baeyer test for unsaturation. Evidence that reaction has occurred is the rapid disappearance (within a few seconds) of the purple color of the permanganate ion. The resulting reaction products will not be colorless. Potassium permanganate is a very strong oxidizing agent and gives similar results when reacted with other oxidizable substances, such as alcohols.

Alkynes

Also called the **acetylenes,** the alkynes are another class of **unsaturated hydrocarbons,** but they contain a carbon-carbon triple bond in their structures. The first two members of this series are acetylene (ethyne) and propyne:

$$H-C\equiv C-H \qquad\qquad CH_3C\equiv CH$$

Ethyne (Acetylene)   Propyne

Acetylene is the most important member of this series and can be prepared from calcium carbide and water. The equation for this reaction is

$$CaC_2(s) + 2H_2O(l) \rightarrow CH \equiv CH(g) + Ca(OH)_2(aq)$$

Mixtures of acetylene and air are explosive. The alkynes undergo addition-type reactions similar to those of the alkenes.

Aromatic Hydrocarbons

The parent substance of this class of hydrocarbons is benzene (C_6H_6). From its formula benzene appears to be a highly unsaturated molecule; the corresponding six-carbon alkane contains 14 hydrogen atoms per molecule (C_6H_{14}). However, the chemical reactions of benzene show that its behavior is like that of the saturated hydrocarbons in many respects. Its reactions are primarily of the substitution type. In the past, benzene was used extensively in student laboratories to illustrate the properties of aromatic hydrocarbons. Studies have shown benzene to be a cancer-causing substance and it is being eliminated from many experiments. We will use toluene in this experiment instead of benzene.

The carbon atoms in a benzene molecule are arranged in a six-membered ring structure, with one hydrogen atom bonded to each carbon atom. The following diagrams represent the benzene molecule; in the second and third structures it is understood that a carbon and a hydrogen atom are present at each corner of the hexagon or benzene ring.

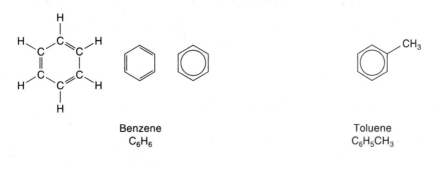

Benzene
C_6H_6

Toluene
$C_6H_5CH_3$

PROCEDURE

Wear protective glasses.

In the following reactions, heptane will be used to represent the saturated hydrocarbons; pentene (amylene), the unsaturated hydrocarbons; and toluene, the aromatic hydrocarbons. (Toluene is *only* reacted with potassium permanganate and *not* bromine.)

Note: As you can see from its structure, toluene is not only aromatic but also contains a methyl substituent group. As seen above, bromine reacts with alkanes in the presence of ultraviolet light to produce a substitution derivative. The overhead lighting found in most laboratories supplies sufficient levels of ultraviolet light to cause a bromine substitution reaction to occur slowly with alkanes. Thus, with toluene, bromine does not react at the aromatic ring, but does react, slowly, at the methyl substituent group. Hence, the lack of reactivity between aromatic benzene and bromine can not be effectively demonstrated by mixing aromatic toluene and bromine (the methyl substitution reaction does slowly occur). Further, the product formed, in the presence of ultraviolet light, between bromine and toluene is a mild lachrymator (a tear-producing compound similar to those found in onions) and may be bothersome to some students or very irritating to more susceptible students.

Hydrocarbons are extremely flammable and should not be handled near open flames. Avoid inhaling the vapors, contact with skin and clothing, and do not ingest.

WASTE DISPOSE OF PROPERLY Dispose of waste and reaction products in the "organic solvent" waste container provided.

A. Combustion

Obtain about 1 mL (no more) of heptane in an evaporating dish and start it burning by carefully bringing a lighted match or splint to it. Repeat with an equally small volume of pentene. Note the characteristics of the flames produced.

B. Reaction with Bromine

CAUTION: Dispense bromine solution under the hood, and be especially careful not to spill bromine on your hands.

Take two clean dry test tubes. Place about 1 mL of heptane in the first tube and 1 mL of pentene in the second. Add 3 drops of 5 percent bromine in trichloroethane solution to each sample; stopper the tubes and note the results. Any tube that still shows bromine color after 1 minute should be exposed to sunlight or to a strong electric light for an additional 2 minutes.

C. Reaction with Potassium Permanganate

The Baeyer test for unsaturation in hydrocarbons involves the reaction of hydrocarbons with potassium permanganate solution. Evidence that reaction has occurred is the rapid disappearance (within a few seconds) of the purple color of the permanganate ion. Potassium permanganate is a very strong oxidizing agent and gives similar results when reacted with other oxidizable substances, such as alcohols.

Add 2 drops of potassium permanganate solution to about 1 mL each of heptane, pentene, and toluene in test tubes. Mix and note the results.

D. Kerosene

Determine which class of hydrocarbons (alkanes, alkenes, or aromatic) kerosene belongs to by reacting it with bromine and with potassium permanganate, as in Tests B and C.

E. Acetylene

In this part of the experiment you will prepare acetylene and test its combustibility.

Fill a 400 mL beaker nearly full of tap water. Fill three test tubes (18 x 150 mm) with water as follows: Tube (1) completely full; Tube (2) 15 mL; and Tube (3) 6 mL.

Obtain a small lump of calcium carbide from the reagent bottle and drop it into the beaker of water. (See Figure 28.1.) Place your thumb over the full test tube of water (Tube 1) and invert it in the beaker. Hold the tube over the bubbling acetylene and, when it is full of gas, stopper it while the tube is still under the water. Displace the water in the other two tubes in the same manner; **stopper them immediately after the water is displaced.**

Test the contents of each tube as follows:

Tube 1. Bring the mouth of the tube to the burner flame as you remove the stopper. After the acetylene ignites, tilt the mouth of the tube up and down.

Tube 2. Bring the mouth of the tube to the burner flame as you remove the stopper.

 Tube 3. Wrap the tube in a towel and bring the mouth of the tube to the burner flame as you remove the stopper. This sample is the most highly explosive of the three samples tested.

F. Solubility Tests

Test the solubility of heptane, pentene, and toluene in water by adding 1 mL (or less) of each hydrocarbon to about 5 mL portions of water. Shake each mixture for a few seconds and note whether they are soluble. For any that are not soluble note the relative density of the hydrocarbon with respect to water.

Test the miscibility of these three hydrocarbons with each other by mixing about 1 mL of each in a **dry** test tube.

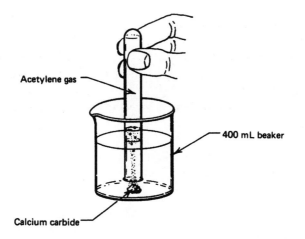

Figure 28.1   Collecting acetylene from calcium carbide-water reaction

REPORT FOR EXPERIMENT 28

Hydrocarbons

A. Combustion

1. Describe the combustion characteristics of heptane and pentene.

2. (a) Write a balanced equation to represent the complete combustion of heptane.

   (b) How many moles of oxygen are needed for the combustion of 1 mole of heptane in this reaction?

   _____

B. and C. Reaction with Bromine and Potassium Permanganate

Data Table: Place an X in the column where a reaction was observed.

	Heptane (Saturated Hydrocarbon)	Pentene (Unsaturated Hydrocarbon)	Toluene (Aromatic Hydrocarbon)
Immediate reaction with Br_2 (*without* exposure to *light*)			
Slow reaction with Br_2 (*or only after exposure to light*)			
Reaction with $KMnO_4$			

1. Which of the two hydrocarbons reacted with bromine (without exposure to light)?

2. Write an equation to illustrate how heptane reacts with bromine when the reaction mixture is exposed to sunlight.

3. Write an equation to illustrate how pentene reacts with bromine. Assume the pentene is $CH_3CH_2CH=CHCH_3$ and use structural formulas.

4. Which of the three hydrocarbons tested gave a positive Baeyer test?

_____

D. Kerosene

1. (a) Did you observe any evidence of reaction with bromine before exposure to light? If so, describe.

   (b) Did you observe any evidence of reaction with bromine after exposure to light? If so, describe.

2. Did you observe any evidence of reaction with potassium permanganate? If so, describe.

3. Based on these tests (bromine and potassium permanganate), to which class of hydrocarbon does kerosene belong?

_____

E. Acetylene

1. Describe the combustion characteristics of acetylene:

   (a) Tube 1.

   (b) Tube 2.

   (c) Tube 3.

2. Write an equation for the complete combustion of acetylene.

F.   Solubility Tests

1. Which of the three hydrocarbons tested are soluble in water?

2. From your observations what do you conclude about the density of hydrocarbons with respect to water?

QUESTIONS AND PROBLEMS

1. Do you expect that acetylene would react with bromine without exposure to light? Explain your answer.

2. Write structural formulas for the three different isomers of pentane, all having the molecular formula C_5H_{12}.

3. Write structural formulas for (a) ethene, (b) propene, and (c) the three different isomeric butenes (C_4H_8). See Study Aid 6 for help if necessary.

EXPERIMENT 29

Alcohols, Esters, Aldehydes, and Ketones

MATERIALS AND EQUIPMENT

Solids: Copper wire (No. 18, with spiral); salicylic acid [$C_6H_4(COOH)(OH)$]. **Liquids:** acetic acid, glacial (CH_3COOH); acetone [$(CH_3)_2C=O$]; ethyl alcohol (95% C_2H_5OH); isoamyl alcohol ($C_5H_{11}OH$); isopropyl alcohol (iso-C_3H_7OH); methyl alcohol (CH_3OH). **Solutions:** dilute (6 M) ammonium hydroxide (NH_4OH), 10 percent glucose ($C_6H_{12}O_6$), 10 percent formaldehyde ($H_2C=O$), 0.1 M potassium permanganate ($KMnO_4$), 0.1 M silver nitrate ($AgNO_3$), 10 percent sodium hydroxide (NaOH), and dilute (3 M) and concentrated sulfuric acid (H_2SO_4).

DISCUSSION

In this experiment we will examine some of the properties and characteristic reactions of four classes of organic compounds: alcohols, esters, aldehydes, and ketones.

Alcohols

The formulas of alcohols may be derived from alkane hydrocarbon formulas by replacing a hydrogen atom with a hydroxyl group (OH). In the resulting alcohols the OH group is bonded to the carbon atom by a covalent bond and is not an ionizable hydroxide group. Examples follow:

Alkane	Alcohol	Name of Alcohol*
CH_4	CH_3OH	Methyl alcohol (Methanol)
CH_3CH_3	CH_3CH_2OH	Ethyl alcohol (Ethanol)
$CH_3CH_2CH_3$	$CH_3CH_2CH_2OH$	n-Propyl alcohol (1-Propanol)
$CH_3CH_2CH_3$	CH_3CHCH_3 $\mid$ OH	Isopropyl alcohol (2-Propanol)

*IUPAC name in parentheses

Thus there is an entire homologous series of alcohols. The functional group of the alcohols is the hydroxyl group, OH.

Esters

This class of organic compounds may be formed by reacting alcohols with organic acids. Esters generally have a pleasant odor; many of them occur naturally, being found mainly in fruits and fatty material.

Methyl acetate will be used as an example illustrating the formation of an ester. When acetic acid and methyl alcohol are reacted together, using sulfuric acid as a catalyst, a molecule of water is split out between a molecule of the acetic acid and a molecule of the alcohol, forming the ester. The equation is

$$CH_3\overset{O}{\overset{\|}{C}}-[OH] + CH_3-O[H] \xrightarrow[\Delta]{H_2SO_4} CH_3\overset{O}{\overset{\|}{C}}-O-CH_3 + H_2O$$

Acetic acid   Methyl alcohol   Methyl acetate

The functional group characterizing organic acids is

$$-\overset{O}{\overset{\|}{C}}-OH \quad \text{or} \quad -COOH$$

It is called the **carboxyl group.**

Esters are named in the following manner. The first part of the name is taken from the name of the alcohol, the second part is derived by adding the suffix **ate** to the identifying stem of the acid. Thus **acetic** becomes **acetate,** and the name of the ester derived from methyl alcohol and acetic acid is methyl acetate.

Aldehydes and Ketones

The functional groups of the aldehydes and ketones are

$$-\overset{H}{\overset{|}{C}}=O \qquad\qquad R-\overset{}{\underset{O}{\overset{}{C}}}-R$$

Aldehyde   Ketone

Aldehydes and ketones may be obtained by oxidizing alcohols. One major difference between aldehydes and ketones is that aldehydes are very easily oxidized to acids, but ketones are not easily further oxidized. Thus aldehydes are good reducing agents. Chemical reactions for distinguishing aldehydes and ketones are based on this difference.

Alcohol	Aldehyde		Ketone
CH_3OH	$H-\overset{H}{\overset{\|}{C}}=O$	formaldehyde (Methanal)	——
CH_3CH_2OH	$CH_3\overset{H}{\overset{\|}{C}}=O$	acetaldehyde (Ethanal)	——
$CH_3CH_2CH_2OH$	$CH_3CH_2\underset{H}{\overset{}{C}}=O$	propionaldehyde (Propanal)	——
$CH_3\underset{OH}{\overset{}{C}}HCH_3$	——	——	$CH_3\underset{O}{\overset{}{C}}CH_3$ Acetone (propanone)

PROCEDURE

Wear protective glasses.

> Record your observations immediately on the report form as you work through the procedure.
>
> ⚠ Reagents used in this experiment are highly flammable and several are poisonous. Work cautiously away from heat and open flames, and avoid inhalation of vapors, contact with skin and clothing, and do not ingest.
>
> Dispose of all reagents in the "organic solvent" waste container provided. The symbol is used throughout this procedure to remind you of this disposal requirement.

A. Combustion of Alcohols

Obtain about 1 mL (no more) of methyl alcohol in an evaporating dish and ignite the alcohol with a match or burning splint. Repeat with equally small volumes of ethyl alcohol and isopropyl alcohol.

B. Oxidation of Alcohols

1. **Oxidation with Potassium Permanganate.** Mix 3 mL of methyl alcohol with 12 mL of water and divide the solution into three equal portions, placing them in three test tubes. To a fourth tube add 5 mL water. Add 1 drop of 10 percent sodium hydroxide to the first tube and 1 drop of dilute sulfuric acid to the second. Now add 1 drop of potassium permanganate solution to each of the four tubes. Mix and note how long it takes for the reaction to occur in each of the first three tubes, using the fourth tube as a reference tube. Disappearance of the purple permanganate color is evidence of reaction. (Patience, some reactions take a long time.)

Repeat this oxidation procedure, using isopropyl alcohol instead of methyl alcohol.

2. **Oxidation with Copper(II) Oxide.** Put about 2 mL of methyl alcohol in a test tube. Obtain from the reagent shelf about a 20 cm piece of copper wire with a four- or five-turn spiral at one end. Warm the alcohol slightly to promote alcohol vapors in the tube. Heat the copper spiral in the hottest part of the burner flame to get a good copper(II) oxide coating. Do not overheat the copper or it will melt. While the copper spiral is very hot, lower it part way into the tube (not to the liquid) and note the results. Heat the wire again and lower it into the tube, finally dropping it into the liquid alcohol. Remove the wire and gently waft the vapors from the tube to your nose to detect the odor of formaldehyde resulting from the oxidation of the methyl alcohol. **Return the copper wire to the reagent shelf.**

C. Formation of Esters

Take three test tubes and mix the following reagents in them.

Tube 1: 3 mL ethyl alcohol, 0.5 mL glacial acetic acid, and 10 drops of concentrated sulfuric acid.

Tube 2: 3 mL isoamyl alcohol, 0.5 mL glacial acetic acid, and 10 drops concentrated sulfuric acid.

Tube 3: Salicylic acid crystals (about 1 cm deep in the tube), 2 mL methyl alcohol, and 10 drops concentrated sulfuric acid.

Heat the tubes by placing them in boiling water for 3 minutes.

Products formed:

Tube 1: Ethyl acetate.

Tube 2: Isoamyl acetate.

Tube 3: Methyl salicylate.

After heating, pour a small amount of each product onto a piece of filter paper and **carefully** smell it and describe the odor. [WASTE DISPOSE OF PROPERLY] Dispose of the filter paper in the solid trash.

D. Tollens Test for Aldehydes

This test is based on the ability of the aldehyde group to reduce silver ion in solution, forming either a black deposit of free silver or a silver mirror. The aldehyde group is oxidized to an acid in the reaction. Tollens reagent is made by reacting silver nitrate solution with 10% NaOH and dilute NH_4OH. Rinse all glass equipment with distilled water before use.

1. Preparation of Tollens Reagent: **Thoroughly** clean three test tubes. To 8 mL of 0.1 M silver nitrate solution in one of these tubes add one drop of 10% NaOH to generate a brown precipitate of silver oxide. Now add dilute ammonium hydroxide 1 drop at a time until the brown precipitate of silver oxide that was formed just dissolves (mix after each drop is added). Now add 7 mL of distilled water, mix, and divide the solution (Tollens reagent) equally among the three test tubes.

2. To the tubes containing the freshly prepared Tollens reagent, add the following and mix.

Tube 1: 2 drops 10% formaldehyde

Tube 2: 2 drops acetone

Tube 3: 5 drops 10% glucose

Allow the tubes to stand undisturbed and note the results. The solution containing the glucose may take 10 to 15 minutes to react. [WASTE DISPOSE OF PROPERLY] Heavy metals waste container.

REPORT FOR EXPERIMENT 29

Alcohols, Esters, Aldehydes, and Ketones

A. Combustion of Alcohols

1. Compare the combustion characteristics of methyl, ethyl, and isopropyl alcohols, in terms of color and luminosity of their flames.

2. What type of flame would you predict for the combustion of amyl alcohol ($C_5H_{11}OH$)?

3. Write and balance the equation for the complete combustion of ethyl alcohol.

B. Oxidation of Alcohols

1. **Oxidation with Potassium Permanganate**

   (a) Time required for oxidation of methyl alcohol by potassium permanganate:

   Tube 1: Alkaline solution.   _____

   Tube 2: Acid solution.   _____

   Tube 3: Neutral solution.   _____

   (b) Time required for oxidation of isopropyl alcohol by potassium permanganate:

   Tube 1: Alkaline solution.   _____

   Tube 2: Acid solution.   _____

   Tube 3: Neutral solution.   _____

   (c) Balance the equation for the oxidation of methyl alcohol:

   $$CH_3OH + \quad KMnO_4 \longrightarrow \quad H_2C{=}O + \quad KOH + \quad H_2O + \quad MnO_2$$

2. **Oxidation with Copper(II) Oxide**

   (a) Write the equation for the oxidation reaction that occurred on the copper spiral when it was heated.

   (b) What evidence of oxidation or reduction did you observe when the heated Cu spiral was lowered into methyl alcohol vapors?

   (c) Write and balance the oxidation-reduction equation between methyl alcohol and copper(II) oxide.

C. **Formation of Esters**

   1. Describe the odor of:

      (a) Ethyl acetate.

      (b) Isoamyl acetate.

      (c) Methyl salicylate.

   2. Write an equation to illustrate the formation of ethyl acetate from ethyl alcohol and acetic acid.

3. (a) The formula for isoamyl alcohol is $CH_3CH(CH_3)CH_2CH_2OH$. Write the formula for isoamyl acetate.

(b) The formula for salicylic acid is

Write the formula for methyl salicylate.

D. Tollens Test for Aldehydes

1. How is a positive Tollens test recognized?

2. Which of the substances tested gave a positive Tollens test?

3. Circle the formula(s) of the compounds listed that will give a positive Tollens test:

$$CH_3OH, \quad C_2H_5OH, \quad CH_3\overset{\overset{\displaystyle H}{|}}{C}=O, \quad CH_3\overset{\overset{\displaystyle }{}}{\underset{\underset{\displaystyle O}{\|}}{C}}CH_3, \quad Na_2CO_3$$

4. Write the formula for the oxidation product formed from formaldehyde in the Tollens test.

QUESTIONS AND PROBLEMS

1. There are four butyl alcohols of formula C_4H_9OH. Write their structural formulas.

2. Write the name of the ester that can be derived from the following pairs of acids and alcohols:

Alcohol	Acid	Ester
Methyl alcohol	Acetic acid	
Ethyl alcohol	Formic acid	
Isopropyl alcohol	Butyric acid	

3. Write structural formulas for all the aldehyde and ketone isomers having the molecular formula $C_5H_{10}O$.

EXPERIMENT 30

Esterification–Distillation:
Synthesis of n-Butyl Acetate
(A Two Laboratory Period Experiment)

MATERIALS AND EQUIPMENT

Solids: Magnesium sulfate, anhydrous ($MgSO_4$), boiling stones, cotton, ice. **Liquids:** n-butyl alcohol ($CH_3CH_2CH_2CH_2OH$), glacial acetic acid ($HC_2H_3O_2$), concentrated sulfuric acid (H_2SO_4). **Solution:** saturated sodium bicarbonate ($NaHCO_3$). **Reflux and distillation equipment (to be used and returned at the end of the laboratory period):** 100 mL or 250 mL round bottom distilling flask, distilling column, distillation take-off head, condenser, 200° or 250°C thermometer, 250 mL separatory funnel.

DISCUSSION

Carboxylic acids react with alcohols to form esters through a condensation reaction known as esterification. The general equation for the reaction is

$$R - \overset{\overset{\displaystyle O}{\|}}{C} - OH \; + \; R'OH \; \rightleftharpoons \; R - \overset{\overset{\displaystyle O}{\|}}{C} - O - R' \; + \; H_2O$$

$$\text{Acid} \qquad \text{Alcohol} \qquad \text{Ester}$$

The reaction proceeds slowly, usually requiring many hours to reach equilibrium. However, when the reaction is catalyzed with a small amount of mineral acid, such as sulfuric acid, equilibrium is established in a few hours.

According to Le Chatelier's principle, the yield of ester may be increased by increasing the concentration of either reactant. In this experiment, an excess of acid is used.

Since reactions go faster at higher temperatures, the esterification is conducted at the boiling point of the reaction mixture. This is accomplished by refluxing, that is, by boiling the mixture, condensing the vapor in a water-cooled condenser, and returning the liquid to the reaction flask. Figure 30.1 illustrates the apparatus setup for refluxing.

In this experiment you will prepare n-butyl acetate by refluxing n-butyl alcohol with an excess of acetic and a small amount of sulfuric acid as catalyst.

$$CH_3COOH + CH_3CH_2CH_2CH_2OH \; \xrightarrow{H^+} \; CH_3\overset{\overset{\displaystyle O}{\|}}{C} - OCH_2CH_2CH_2CH_3 + H_2O$$

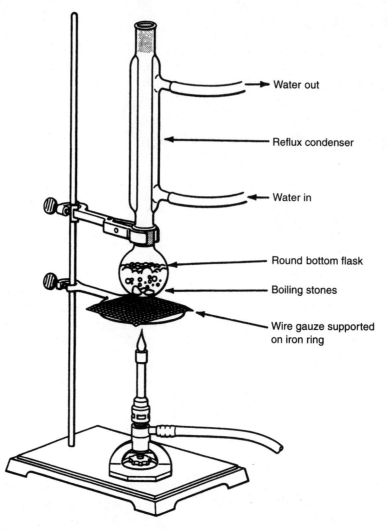

Figure 30.1 Apparatus for refluxing a reaction mixture

Since the mixture is refluxed for only 1 hour, equilibrium will not be reached. The excess acetic acid and unreacted alcohol are separated from the product by extracting them with water using a separatory funnel. In this extraction, we take advantage of the relative insolubility (see Table 30.1) of n-butyl acetate in water. To remove the last traces of acid, the product is washed with sodium bicarbonate solution. This is followed by a water wash to remove sodium bicarbonate, since esters are readily hydrolyzed in alkaline solutions. The ester is finally dried with anhydrous magnesium sulfate and distilled. The drying agent removes water that is dissolved in the ester forming the hydrate $MgSO_4 \cdot 7H_2O$.

Table 30.1 Physical properties of the organic reactants and products in this experiment.

	Molar Mass g/mole	Density g/mL	Boiling Point °C	Solubility g/100 g H_2O
Acetic acid	60.0	1.05	118	Infinite
n-Butyl alcohol	74.0	0.81	117	$9^{15°}$
n-Butyl acetate	116.0	0.88	126	$0.7^{20°}$

Distillation

Distillation is a widely used technique for separation and for purification of compounds. The separation depends on the differences in vapor pressure of the components in a solution. As the temperature of a liquid is raised, the vapor pressure increases until it becomes equal to the pressure of the atmosphere above the liquid. At that temperature the liquid boils. The temperature where the vapor pressure of the liquid equals the atmospheric pressure is called the boiling point of the liquid. The boiling point of a pure liquid is constant and, at 1 atmosphere pressure, is known as the normal boiling point.

Vapor pressure and boiling point are inversely related. A liquid that has a high vapor pressure at room temperature will have a lower boiling point than a liquid with a low vapor pressure at room temperature. Both, of course, will have a vapor pressure of 760 torr (mm Hg) at their respective normal boiling points. Compare the examples given below.

	Boiling point, °C	Vapor Pressure torr (20°C)
Ethyl ether	34.6	442
Water	100	17.5
Acetic acid	118	11.7

When a solution containing two miscible liquid compounds is boiling, the vapor above the solution will contain molecules of both compounds. However, this vapor is richer in the lower boiling compound, that is, the compound with the higher vapor pressure. When this vapor is condensed to a liquid and revaporized a number of consecutive times in a fractional distillation column, it continually becomes richer in the lower boiling compound and is eventually separated (distilled) as a purified or pure compound. With good control of heat input and the rate of distillation, the temperature at the top of the distilling column will remain fairly constant at the boiling point of the first compound being distilled. After the first compound has been removed from the solution, the temperature will rise to the boiling point of the second compound and it can then be collected by continued distillation.

Use of the Separatory Funnel

The separatory funnel (Figure 30.2) is designed for separating immiscible liquids. Thus, it is narrow at the base, near the stopcock, so that the interface between the immiscible liquid layers is confined in a small space for accurate phase separation. The separatory funnel is used for liquid-liquid extractions, that is, extracting a substance from one liquid phase into another liquid phase. This is accomplished by placing both liquids in the funnel, mixing thoroughly by shaking, and allowing the two liquid layers to separate. The lower layer is then drawn off through the stopcock. Multiple extractions can be done without removing the upper layer from the funnel by adding more extracting solvent and repeating the process.

The location of the two liquid layers depends on the relative densities of the solutions. The solution that has the higher density will be the lower layer. When the densities of the two immiscible liquids are close to each other, separation after shaking is often slow, and even difficult because of the formation of emulsions.

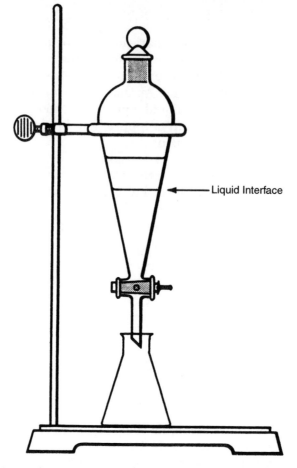

Figure 30.2　Separatory funnel containing two liquid phases

The extraction process is done with the separatory funnel in the inverted position. The extracting solvent and the solution to be extracted are placed into the funnel and the stopper is inserted. (Be certain that the stopcock is closed when adding these liquids.) With the forefinger firmly on top of the stopper, the separatory funnel is inverted and the stopcock opened to relieve any pressure that may have built up inside the funnel. The stopcock is then closed and the contents shaken vigorously for about 40 seconds, stopping occasionally to relieve any pressure built up by carefully opening the stopcock. (**Be careful not to point the separatory funnel at yourself or at anyone else when you open the stopcock.**) Close the stopcock and place the separatory funnel upright in the iron ring for the separation of the two liquid phases.

PROCEDURE

Wear protective glasses.

A.　Preparation of n-butyl acetate

Assemble the apparatus in Figure 30.1 for refluxing according to the following direction. If your equipment has ground glass joints, lightly grease the male joint before putting the apparatus together.

1. Clamp a 100 mL (or 200 mL) round bottom flask in place above the wire gauze.

2. Add 27.5 mL (22.3 g) of n-butyl alcohol and, with the same graduated cylinder measure, add 40.0 mL (42.0 g) of glacial acetic acid. Swirl the flask to mix the reagents.

3. Slowly add 1 mL of concentrated H_2SO_4 and swirl to mix the reagents thoroughly.

4. Add 3 boiling stones.

5. Place the reflux condenser on the flask and start a slow stream of water flowing through the condenser.

6. Heat the mixture with a small flame. Adjust the flame to control the rate of boiling so that the vapor is condensing no further than one-fourth the way up in the condenser. Continue boiling and refluxing for 1 hour.

7. During the 1-hour reflux time, get the equipment ready to separate the product.

8. After the mixture has refluxed for 1 hour, turn off the burner and allow the mixture to cool for about 5 minutes. Then cool the mixture further in an ice-water bath.

B. Separation of n-butyl acetate

1. Pour the cold reaction mixture through a funnel into a 250 mL separatory funnel.

2. Add 100 mL of water to the separatory funnel and close it with the stopper.

3. Place your forefinger firmly on top of the stopper, invert the separatory funnel and open the stopcock to relieve any pressure that may have built up inside the funnel. Close the stopcock.

4. With the funnel still inverted and held in both hands, shake the contents vigorously for about 40 seconds, stopping occasionally to relieve any pressure build-up by carefully opening the stopcock.

5. Close the stopcock and place the separatory funnel upright in the iron ring for separation of the aqueous layer from the ester layer.

6. After the two liquid layers have separated, remove the stopper and withdraw the lower layer through the stopcock, stopping just short of removing all of the aqueous layer. The ester should still be in the separatory funnel.

7. Add another 100 mL of water to the separatory funnel and repeat Steps B. 3 through 6.

8. Add 25 mL of saturated $NaHCO_3$ solution and repeat Steps B. 3 through 6. If the solution is still acidic at this time, carbon dioxide will be formed and pressure will build up inside the funnel. Test the aqueous solution being removed from the funnel with red and blue litmus paper. If this solution is acidic, wash with another 15 mL of the $NaHCO_3$ solution.

9. Add 25 mL of water to the funnel and repeat Steps B. 3 through 6. This time carefully remove all the aqueous layer through the stopcock.

10. Pour the crude n-butyl acetate out the top of the separatory funnel into a clean, dry, 125 mL Erlenmeyer flask. Add 2.5 g of anhydrous MgSO$_4$ powder, close the flask with a cork stopper, and swirl the contents several times to bring the liquid in contact with the drying agent (MgSO$_4$).

11. Label and store the flask in your laboratory locker for distillation at the next laboratory period.

 **Dispose of the aqueous wash solutions down the sink drain.**

C. Distillation of n-butyl acetate (second laboratory period)

NOTE: All equipment used for the distillation must be **dry**.

1. Assemble the distillation apparatus shown in Figure 30.3 using a 100 mL distilling flask. If ground glass equipment is used, lightly grease the male joints. Be sure that all joints fit tightly and that the distilling flask and condenser are clamped to the ring stand. A 150°C (or greater) thermometer is needed. The height of the thermometer is adjusted so that the mercury bulb is just below the side arm of the distillation head. After attaching the rubber tubing to the condenser, turn on the water so that a slow stream is flowing through the condenser.

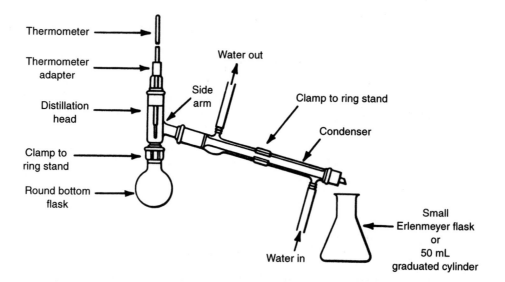

Figure 30.3   Simple distillation apparatus

2. Pour the dried product into the round bottom flask through a funnel that is fitted with a very small plug of cotton just enough to keep the drying agent from entering the flask.

 **Dispose of the drying agent in the waste jar provided.**

3. Add 3 boiling stones.

4. Heat the flask with a small flame. Adjust the heat input so that the rate of distillation is about 1 drop of distillate per second. (An ideal distillation is at equilibrium when vapor is condensing and liquid is dripping off the thermometer back into the distilling flask.)

5. Collect the first distillate up to 118°C in a small Erlenmeyer flask. The first distillate that comes over may be cloudy because of traces of water in the system.

6. At 118°C replace the Erlenmeyer flask with a weighed, dry, 50 mL graduated cylinder. Collect all the distillate (n-butyl acetate) that comes over up to 127°C. Record the boiling range of the ester collected.

⚠ CAUTION: DO NOT heat the distilling flask to dryness.

7. Turn off the burner when the temperature exceeds 127°C or when the volume in the distilling flask decreases to 2 or 3 mL.

8. Weigh the graduated cylinder. Record the volume and the mass of the n-butyl acetate in the data table.

9. **Hand in (properly labeled) or dispose of the n-butyl acetate as directed by your instructor.** WASTE DISPOSE OF PROPERLY

NAME ____________________

SECTION ______ DATE ____________

REPORT FOR EXPERIMENT 30

INSTRUCTOR ____________________

Esterification–Distillation: Synthesis of n-Butyl Acetate

Data Table

	Volume mL	Mass g	Molar Mass g/mole	Moles	Boiling Range
n-Butyl alcohol					
Acetic acid					
n-Butyl acetate					

CALCULATIONS

Show calculation setups and answers.

1. Calculate the limiting reactant in this reaction.

____

2. Calculate the theoretical yield of n-butyl acetate.

____

3. Calculate the percentage yield of n-butyl acetate.

____

QUESTIONS AND PROBLEMS

1. Why was an excess of acetic acid used in this synthesis?

2. At approximately what temperature was the solution boiling during the reflux period?

    _____

3. Explain why it is possible to separate n-butyl acetate from excess and unreacted acetic acid and n-butyl alcohol.

4. Why is the n-butyl acetate the top layer of the two liquid phases in the separatory funnel?

5. The volume of water and $NaHCO_3$ solution used to wash the n-butyl acetate was 250 mL. Calculate the maximum loss of n-butyl acetate due to its solubility in this volume of water.

    _____

6. Write an equation to show how anhydrous $MgSO_4$ behaves as a drying agent. Name the product.

    _____

7. Why can a simple distillation setup be used to distill the n-butyl acetate as prepared in this experiment?

8. Methyl salicylate (oil of wintergreen) is an ester and can be prepared from salicylic acid and methyl alcohol by the same method used in this experiment. Write an equation for this reaction.

EXPERIMENT 31

Synthesis of Aspirin

MATERIALS AND EQUIPMENT

Solids: Salicylic acid ($HO - C_6H_4 - COOH$), ice. **Liquids:** Acetic anhydride [$(CH_3CO)_2O$], ethyl alcohol (C_2H_5OH), mineral oil, 85% phosphoric acid (H_3PO_4). Buchner funnel and suction flask, rubber suction tubing, melting point tubes, filter paper to fit the Buchner funnel.

DISCUSSION

Aspirin (acetyl salicylic acid, or A.S.A.) is the drug that is most widely used for self-medication. The familiar aspirin tablet contains 5 grains (about 325 milligrams) of acetyl salicylic acid and a small amount of an inert binding material such as starch. More than four million pounds of aspirin is manufactured each year in the United States. As a drug, aspirin has analgesic, anti-pyretic, and anti-inflammatory properties; that is, it can relieve pain, lower fever, and reduce inflammation.

Aspirin belongs to a group of drugs called salicylates because of their structural relationship to salicylic acid (SA),

Salicylic acid

Salicylic acid is both an aromatic carboxylic acid and a phenol. Aspirin is represented by this structural formula:

Acetyl salicylic acid (Aspirin)

This formula shows that aspirin is an ester formed between acetic acid and the phenol –OH group of salicylic acid. Aspirin is a weak acid because of its carboxyl group. Acetyl salicylic acid is practically insoluble in cold water, but its sodium salt is soluble in water.

Although it is clearly a derivative of salicylic and acetic acids, aspirin usually is not made from these substances. It can be prepared by reacting either acetyl chloride or acetic anhydride with salicylic acid. Generally, acetic anhydride, the acid anhydride of acetic acid, is used in the synthesis. Phosphoric acid catalyzes the reaction.

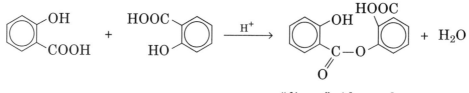

	Acetic anhydride	Acetyl salicylic acid	Acetic acid
Salicylic acid	(Ethanoic anhydride)	(Salicylyl ethanoate)	(Ethanoic acid)
Molar mass = 138.1	Molar mass = 102.1	Molar mass = 180.2	Molar mass = 60.06

Aspirin can be removed from the reaction mixture by adding cold water and filtering out the precipitated aspirin. The crude aspirin crystals are contaminated with small amounts of impurities, chiefly acetic and phosphoric acids. These impurities can be removed by recrystallization of the aspirin. To accomplish this, the crude aspirin is dissolved in hot alcohol, water is added, and the mixture is cooled. The crystals that form are filtered from the solution and dried to yield a product of high purity.

Why are we using acetic anhydride in this reaction and not the more abundant and less costly acetic acid? As noted above, in this experiment the phenolic group of salicylic acid is reacted with the acetyl $\left(CH_3C \diagdown^O \right)$ group of acetic acid to produce aspirin. However, notice that salicylic acid has both a phenolic group and a carboxylic acid group. The phenolic group of one salicylic acid molecule could react with the carboxylic acid group of another salicylic acid molecule to produce a "dimer" side-product and not the desired aspirin product.

"dimer" side-product

To minimize the production of the "dimer" side-product, acetic anhydride is utilized in place of acetic acid. Acetic anhydride is more reactive than acetic acid and thus minimizes the amount of "dimer" side-product by quickly consuming the salicylic acid before much can react with itself.

Yields in Organic Synthesis

In an organic synthesis the actual amount of finished product is almost always less—often a great deal less—than the amount theoretically obtained from the reactants used. Incomplete reactions or side reactions, that is, reactions that do not produce the desired product, can reduce the amount of product obtained. In addition, losses invariably occur in the recovery and purification steps (crystallization and recrystallization in this case). Thus, the percent of the theoretical amount of product actually obtained, or percentage yield, is a number that can be used to judge the success of a synthesis. It is calculated by this formula:

$$\text{Percentage yield} = \frac{\text{Actual mass of product obtained}}{\text{Theoretical mass of product obtainable}} \times 100$$

PROCEDURE

Wear protective glasses.

Accurately weigh and transfer to a 125 mL **dry** Erlenmeyer flask 3.00 g of salicylic acid. Next add 6 mL of acetic anhydride and 5 to 8 drops of 85% phosphoric acid to the flask. Swirl the flask gently to mix the reagents and place it in a beaker of warm (70–80°C) water for 15 minutes.

Then, while the reaction mixture in the flask is still warm, **carefully** add, drop by drop, 20 drops of cold water from a medicine dropper to destroy the excess acetic anhydride.

 CAUTION: The vigorous reaction of the excess acetic anhydride with water may cause splattering.

Add 20 mL of water to the flask; then put the flask in an ice bath to cool the reaction mixture and speed the crystallization of aspirin. When the crystallization seems to be complete, collect the crystals by suction filtration using a small Buchner funnel (Fig. 31.1 on page 268) or an alternate filtering device. Wash the crystals on the filter once with a few millimeters of cold, preferably iced, water.

Recrystallization: Transfer the crystals to a 100 mL beaker, add 10 mL of ethyl alcohol and stir to dissolve. If necessary warm the beaker in a 250 mL beaker containing warm water to complete the solution formation. When all of the crystals have dissolved, pour 25 mL of warm (60–70°C) water into the alcohol solution. Cover the beaker with a watch glass and set aside to cool until crystals begin to form. Then place the beaker in an ice bath for about 10 minutes to complete crystallization.

After crystallization is complete, filter the crystals using a suction filter. Spread the crystals out on the filter with a spatula and press them with dry filter paper to aid in drying. Transfer the crystals to a watch glass, cover with filter paper and finish drying by storing in your locker for at least one day. Transfer the dried aspirin to a weighed vial or test tube. Then reweigh and determine the mass of aspirin.

Determine the melting point of your dried aspirin crystals using the method given in Experiment 27. Look up the melting point given for aspirin in the chemical literature. The *Handbook of Chemistry and Physics* or *Lange's Handbook of Chemistry* are suitable sources.

 Dispose of all solutions down the drain.

Hand in (properly labeled) or dispose of the product as directed by your instructor.
Your label should contain your name, name of product, mass of product, and percent yield.

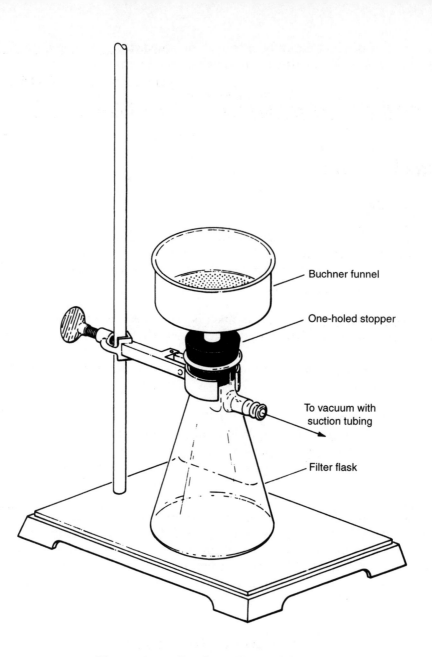

Buchner funnel

One-holed stopper

To vacuum with
suction tubing

Filter flask

Figure 31.1   Buchner Funnel Setup

REPORT FOR EXPERIMENT 31

Synthesis of Aspirin

Percentage yield of aspirin

1. Mass of salicylic acid used _____

2. Mass of aspirin crystals obtained _____

3. Mass of aspirin theoretically obtainable from salicylic acid used. Show calculation setup.

_____

4. Percentage yield of aspirin based on theoretical mass from 3. Show calculation setup.

_____

Melting point:

1. Melting point range of your aspirin _____

2. Melting point of aspirin from the literature _____

QUESTIONS AND PROBLEMS

1. Write the chemical equation representing the synthesis of aspirin from acetyl chloride.

2. Could aspirin be prepared by the method used in this experiment from these compounds? Explain your answer.

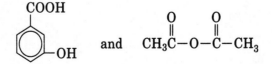

3. (a) How many moles of acetic anhydride were represented by the 6.00 mL (density = 1.08 g/mL) used in the synthesis?   (b) How many excess moles of acetic anhydride were used—over and above the amount needed to react with 3.00 g of salicylic acid? Show calculations.

(a)_____

(b)_____

4. Aspirin is insoluble in water. Explain why it is soluble in sodium hydroxide solution.

5. If the annual production of acetyl salicylic acid in the U.S.A. were converted to 5-grain aspirin tablets, approximately how many tablets would there be for each person? Assume that the population is about 280,000,000. Show calculation.

_____

6. What is (a) an analgesic and (b) an antipyretic?

EXPERIMENT 32

Amines and Amides

MATERIAL AND EQUIPMENT

Solids: Urea (CH_4N_2O), ethanamide (acetamide, C_2H_5NO), 1-naphthol (α-naphthol, $C_{10}H_8O$), 4-nitroaniline (p-nitroaniline, $C_6H_6N_2O_2$), sodium nitrite ($NaNO_2$). **Liquids:** concentrated ammonium hydroxide (NH_4OH), aniline (C_6H_7N), pyridine (C_5H_5N), diethylamine ($C_4H_{11}N$), n-hexylamine ($C_6H_{13}NH_2$), 6 M HCl, 3 M H_2SO_4, 1 M NaOH. Ice bath.

DISCUSSION

Many organic and biological molecules contain nitrogen. Amines and amides are two such nitrogen-containing molecules. In general, amines can be thought of as ammonia (NH_3) in which the hydrogens have been removed and replaced by organic "R"-groups (aromatic or aliphatic) to produce primary (1°), secondary (2°), and tertiary amines (3°).

$$RNH_2 \qquad R_2NH \qquad R_3N$$
$$1° \qquad\qquad 2° \qquad\quad 3°$$

Since the nitrogen atom in an amine has an unshared pair of electrons, a fourth R-group can be added to produce a positively charged quarternary (4°) ammonium salt.

$$R_3N + RCl \longrightarrow R_4N^+Cl^- \quad \text{(quaternary ammonium salt)}$$

Examples of amines include:

$$CH_3CH_2NH_2 \qquad\qquad \underset{\underset{H}{|}}{CH_3CH_2-N-CH_3} \qquad\qquad \underset{H_3C}{\overset{H_3C}{{>}}} N-CH_2CH_2CH_3$$

aminoethane (1°)   ethylmethylamine (2°)   dimethylpropylamine (3°)
(ethylamine)

aminocyclohexane   aniline   4-nitroaniline   3-methylaniline
(cyclohexylamine)   (p-nitroaniline)   (m-methylaniline)

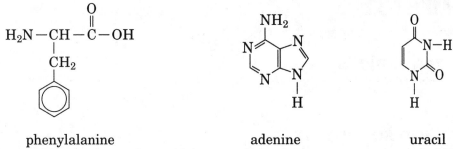

| phenylalanine | adenine | uracil |
| (one of the twenty common amino acids) | (a purine) | (a pyrimidine) |

Amines are basic substances and can be protonated, in acid, to form positively charged substituted ammonium ions.

$$RNH_2 + H^+ \longrightarrow RNH_3^+$$

$$R_2NH + H^+ \longrightarrow R_2NH_2^+$$

$$R_3N + H^+ \longrightarrow R_3NH^+$$

Even water is acidic enough to protonate ammonia and many amines. For example

$$NH_3 + H_2O \rightleftharpoons NH_4^+ + OH^-$$

$$RNH_2 + H_2O \rightleftharpoons RNH_3^+ + OH^-$$

Due to their hydrogen bonding ability, aliphatic amines containing up to five carbons are quite soluble in water. Aromatic amines are considerably less soluble than aliphatic amines. Both the size and the number of R-groups attached to the nitrogen atom affect the solubility of amines in water. Generally, solubility decreases with larger R-groups and with increased numbers of R-groups attached to the nitrogen.

Amines are utilized as starting materials in many medical and commercial chemical syntheses. One such starting material is aniline. Also, substituted aniline derivatives such as 2, 3, and 4-nitroanilines can be used. Several important dyes can be produced by reacting aniline with sodium nitrite to yield a diazonium salt, which is then reacted with a suitable reactant such as phenol and similar compounds. These syntheses generally employ a diazonium salt as a critical intermediate.

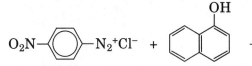 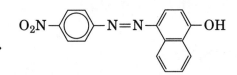

diazonium salt	1-naphthol	final "coupled" dye
(4-nitrobenzene		
diazonium chloride)		

Amides are neutral nitrogen-containing compounds in which the nitrogen is bonded directly to a carbonyl carbon. Mono- or disubstituted amides have one or two R-groups instead of H on the –NH$_2$ group.

$$\overset{O}{\overset{\|}{R C}}-NH_2 \qquad \overset{O}{\overset{\|}{R C}}-NHR \qquad \overset{O}{\overset{\|}{R C}}-NR_2$$

Examples of amides include:

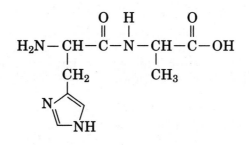

ethanamide   N-methylethanamide   N-ethyl-N-methylbenzamide
(acetamide)

histidylalanine (a dipeptide)

Simple amides can synthesized by reacting a carboxylic acid with ammonia to produce an ammonium salt and then heating the ammonium salt to generate the corresponding amide.

$$RCOOH + NH_3 \longrightarrow RCOO^- NH_4^+$$

$$RCOO^-NH_4^+ \overset{\Delta}{\longrightarrow} RCONH_2 + H_2O$$

Amides are hydrolyzed under acid conditions to yield a carboxylic acid and an ammonium salt. Under basic conditions the products are the salt of a carboxylic acid and ammonia or an amine.

$$RCONH_2 + H_2O + HCl \longrightarrow RCOOH + NH_4Cl$$

$$RCONH_2 + H_2O + NaOH \longrightarrow RCOO^- Na^+ + NH_3$$

PROCEDURE

Wear protective glasses.

A. Amine Solubility and Acidity Tests

Clean five (5) test tubes. Place about 1 mL of the following amines in each test tube (one amine per test tube): (1) concentrated ammonium hydroxide; (2) aniline; (3) pyridine; (4) diethyl-amine; (5) n-hexylamine. Add approximately 4 mL of water to each test tube and **thoroughly** mix each test tube. Record your results concerning the solubility of each amine.

Using pH indicator paper, test the pH of the water layer (be sure you understand which layer, if any, is the water layer) in each of the above test tubes. Record your results.

Add 1 mL of 6 M HCl to each of the above test tubes. **Thoroughly** mix each test tube. Concerning solubility, record your results. Again using pH indicator paper, test for pH in each water layer and record your results.

 Dispose of the residue in the organic waste container.

B. Amide Hydrolysis

1. Clean a test tube and add approximately 1 gram of urea. Now add about 5 mL of 3 M H_2SO_4 and thoroughly mix. Heat the mixture in a boiling water bath. After about five minutes in the boiling water bath, **carefully** note any odor coming from the test tube.

2. Clean a test tube and add approximately 1 gram of ethanamide (acetamide). Now add about 5 mL of 3 M H_2SO_4 and thoroughly mix. Heat the mixture in a boiling water bath. After about five minutes in the boiling water bath, **carefully** note any odor coming from the test tube. Add sufficient 3 M H_2SO_4 to acidify the mixture (use either pH indicator paper or litmus paper to verify that the mixture is acidic). Record what you observe upon addition of the acid.

 Dispose of the residues in the organic waste container.

C. Synthesis of a Nitrogen-Containing Dye

NOTE: The following reactions must be kept cold in an ice bath at all times.

Using a suitably sized plastic trough, fill the trough with ice/water for usage as an ice bath. Be sure there is sufficient water to allow the containers used in the reaction to be surrounded with ice/water, but not so much water as to permit the containers to tip and mix with the ice/water.

Prepare the following solutions and keep them in the ice bath.

Solution 1: In a 50 mL beaker place 0.3 gram of 1-naphthol. Add 15 mL of 1.0 M sodium hydroxide. Mix and keep cold.

Solution 2: In an ice bath cooled 50 mL flask place 0.5 gram of 4-nitroaniline. Add 10 mL of water. Add 1 mL of 6 M HCl and thoroughly mix. Keep cold.

Solution 3: In a test tube place 0.5 gram of sodium nitrite. Add 3 mL of water to the test tube and dissolve the sodium nitrite by stirring the mixture. Keep cold.

To the flask containing the 4-nitroaniline, add the sodium nitrite solution. While still in the ice bath, thoroughly stir the combined solutions to generate the diazonium salt.

Add the diazonium salt containing solution that you just made to the beaker containing the 1-naphthol. Keep in the ice bath and stir. Record your results.

 Dispose of the residue in the organic waste container.

NAME ______________________

SECTION ______ DATE ______________

REPORT FOR EXPERIMENT 32

INSTRUCTOR ______________________

Amines and Amides

A. Amine Solubility and Acidity Tests

SOLUBILITY *BEFORE* ADDITION OF ACID

Results for Test Tube 1 ______________________

Results for Test Tube 2 ______________________

Results for Test Tube 3 ______________________

Results for Test Tube 4 ______________________

Results for Test Tube 5 ______________________

SOLUBILITY *AFTER* ADDITION OF ACID

Results for Test Tube 1 ______________________

Results for Test Tube 2 ______________________

Results for Test Tube 3 ______________________

Results for Test Tube 4 ______________________

Results for Test Tube 5 ______________________

pH RESULTS

pH for Test Tube 1 *before* addition of HCl _____

pH for Test Tube 2 *before* addition of HCl _____

pH for Test Tube 3 *before* addition of HCl _____

pH for Test Tube 4 *before* addition of HCl _____

pH for Test Tube 5 *before* addition of HCl _____

pH for Test Tube 1 *after* addition of HCl _____

pH for Test Tube 2 *after* addition of HCl _____

pH for Test Tube 3 *after* addition of HCl _____

pH for Test Tube 4 *after* addition of HCl _____

pH for Test Tube 5 *after* addition of HCl _____

1. Are all of the amines equally soluble in water?

2. Did the addition of HCl change the solubilities and why?

3. Can you make any conclusions about amine structure and solubility? What are those conclusions?

4. Which amine is most acidic and why?

5. Which amine is most basic and why?

6. Did the addition of HCl change the pH of the water layers and why?

B. Amide Hydrolysis

1. For the urea reaction with H_2SO_4, describe the odor coming from the heated test tube. What chemical do you theorize as causing the odor?

2. What occurred upon addition of H_2SO_4 to the ethanamide solution?

3. Write a balanced equation for the addition of H_2SO_4 to the heated ethanamide solution.

C. Synthesis of a Nitrogen-Containing Dye

1. What did you observe when you added the diazonium salt to the 1-naphthol?

2. If the combined diazonium salt and 1-naphthol had been reacted at room temperature, would the colored product have been produced? Why?

QUESTIONS AND PROBLEMS

1. Write a balanced equation for the protonation of pyridine by HCl.

2. Using the *Handbook of Chemistry and Physics* and/or the *Merck Index* (or other appropriate reference book) give relevant solubility information about the compounds you used in this experiment.

ammonium hydroxide—

aniline—

pyridine—

diethylamine—

benzylamine—

3. Write a balanced equation for the acidic hydrolysis of propanamide by HCl.

4. Write a balanced equation for the basic hydrolysis of butanamide by KOH.

5. Again using the *Handbook of Chemistry and Physics* and/or the *Merck Index* (or other appropriate reference book) write the structural formula and use for pyridoxamine phosphate (one form of Vitamin B$_6$).

6. What would occur if you had mixed the diazonium salt with 2-naphthol instead of 1-naphthol? (A structural formula might be helpful in your explanation.)

EXPERIMENT 33

Polymers — Macromolecules

MATERIALS AND EQUIPMENT

Solids: Benzoyl peroxide $(C_6H_5COO)_2$. **Liquids:** methyl methacrylate, $[CH_2=C(CH_3)COOCH_3]$. **Solutions:** 0.40 M adipoyl chloride, $(C_6H_8O_2Cl_2)$, in cyclohexane, 0.40 M 1,6-diaminohexane, $[(CH_2)_6(NH_2)_2]$ in 0.40 M NaOH, 5% sodium bicarbonate $(NaHCO_3)$. A 4 inch wire with a hook at one end.

DISCUSSION

Polymers are high molar mass compounds, many of which have molar masses exceeding one million. Thus polymers are often referred to as macromolecules (very large molecules).

The process of forming very large, high molar mass molecules from smaller units is called polymerization. The large molecule is called the polymer, and the small unit, the monomer. The monomers may be alike or they may be different. When there are two or more monomers, the macromolecule is known as a copolymer.

Starch, glycogen, cellulose, and proteins are examples of naturally occurring polymers. Manmade polymers touch every phase of modern living. Examples of these are polyethylene, Nylon, Dacron, Bakelite, Lucite, polyvinyl chloride, etc.

An example of a naturally occurring polymer is starch, a sugar storage molecule found in plants. The monomer of starch is glucose and a portion of the starch is represented as follows:

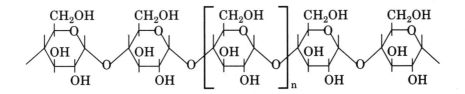

where n represents a large number of repeating units.

Alkenes, or substituted alkenes, are among the most common monomers for making synthetic polymers. For example, ethylene can be polymerized to polyethylene.

$$n\ CH_2{=}CH_2 \longrightarrow -CH_2CH_2\text{---}(CH_2CH_2)_n CH_2CH_2-$$

Ethylene   Polyethylene

Polymers that soften on heating, and therefore can be changed into different usable shapes, are known as thermoplastic polymers. Polymers that set to infusible solids and do not soften on reheating are known as thermosetting polymers.

All polymers may be classified as either addition polymers or condensation polymers. An addition polymer is one that is formed by the successive addition of repeating monomer molecules. A condensation polymer is one that is formed from monomers with the elimination of water or some other simple substance.

In the experiment, you will prepare a condensation polymer, Nylon (a polyamide) and an addition polymer, Lucite (polymethyl methacrylate).

Nylon 6-6 is a condensation polymer made from two different monomers, adipic acid and 1,6-diaminohexane. In the name Nylon 6-6, the first "6" refers to the number of carbons (methylene group carbons) in the diamine and the second "6" refers to the total number of carbons in the diacid or diacid chloride, including the carbonyl carbons and the methylene carbons. Thus, Nylon 5-10 may be synthesized from polymerizing 1,5-diaminopentane [$H_2N(CH_2)_5NH_2$] with sebacoyl chloride [$ClCO(CH_2)_8COCl$]).

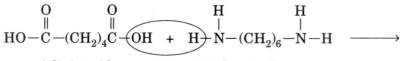

Adipic acid   1,6-diaminohexane

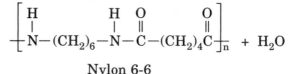

Nylon 6-6

Since the bonds linking the monomers are amide bonds ($-\overset{O}{\overset{\|}{C}}-\overset{H}{\overset{|}{N}}-$), Nylon is known as a polyamide. Polyamides can be made from diacids and diamines as shown above or from diacid chlorides and diamines. In this experiment, the diacid chloride is used, eliminating HCl to form the amide linkage.

$$Cl-\overset{O}{\overset{\|}{C}}-(CH_2)_4\overset{O}{\overset{\|}{C}}-Cl \quad + \quad H-\overset{H}{\overset{|}{N}}-(CH_2)_6-\overset{H}{\overset{|}{N}}-H \longrightarrow$$

Adipoyl chloride   1,6-diaminohexane

$$\left[-\overset{H}{\overset{|}{N}}-(CH_2)_6-\overset{H}{\overset{|}{N}}-\overset{O}{\overset{\|}{C}}-(CH_2)_4\overset{O}{\overset{\|}{C}}- \right]_n + HCl$$

Nylon 6-6

Nylon has good tensile strength, elasticity, and resistance to abrasion. It is used in fabrics, surgical sutures, thread, gears and bearings, rope, hosiery and clothing, tire cords, and so on.

Lucite, also known as Plexiglas, is an addition polymer made from the monomer methyl methacrylate. Lucite is one of the polyacrylics, the monomers of which are derivatives of acrylic acid.

$$CH_2{=}CHCOOH \qquad\qquad CH_2{=}\overset{\overset{\textstyle CH_3}{|}}{C}{-}COOCH_3$$

Acrylic acid   Methyl methacrylate

The polymerization of methyl methacrylate is a chain reaction catalyzed by benzoyl peroxide. The overall reaction is represented by this equation.

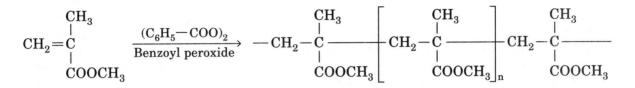

Methyl methacrylate   Polymethyl methacrylate (Lucite)

Polyacrylics are clear, colorless polymers. They take a high polish, are transparent to visible and ultraviolet light, and have excellent optical properties. Lucite is used for windshields in airplanes, for contact lenses, automobile finishes, molded ornamental objects, etc.

PROCEDURE

Wear protective glasses.

A. Synthesis of Nylon

⚠ CAUTION: The monomers used in this experiment are skin irritants, so be careful not to get them on your skin.

Place 5 mL of aqueous 0.40 M 1,6-diaminohexane solution into a test tube (to minimize adherence of the "rope" to the sides of the reaction container, a small, ~30 mL, wide-mouth vial or beaker may be used in place of a test tube). Obtain 5 mL of 0.40 M adipoyl chloride solution in cyclohexane and carefully pour this solution down the wall of the slightly tilted test tube containing the diamine. Do not allow the solutions to mix. Since the solution in cyclohexane is less dense it will float on the aqueous solution.

Observe that a thin film forms at the interface of the two solutions. This film is Nylon 6-6. Record your observations.

Using a piece of wire with a hook bent into one end, hook the polymer mass at the solution interface and slowly raise the wire so that a continuous "rope" of polymer is obtained. Wind the rope around another test tube as you pull it from the reaction mixture. If the Nylon rope breaks, start a new strand with the hooked wire. Do not touch the polymer with your hands since it may contain unreacted reagents.

Continue to wind the polymer around the test tube. When no more polymer is obtained transfer the Nylon to a small beaker, add 25 mL of 5% $NaHCO_3$ solution, stir, and decant the bicarbonate wash. Finally, wash the Nylon several times with water. Once the Nylon is washed it is safe to touch. Dry the polymer by pressing it between pieces of paper towel.

Dispose of the aqueous wash solutions down the drain.

Stir the remaining solutions to form additional Nylon and **dispose of this mixture in the container provided.**

B. Synthesis of Polymethyl Methacrylate (Lucite)

Measure 5 mL of water in a graduated cylinder and pour it into a test tube. Now pour an equal volume of methyl methacrylate from the reagent bottle into a second clean, dry test tube. Discard the 5 mL of water. Have your instructor add about 25 mg of the catalyst, benzoyl peroxide, to the methyl methacrylate. Dissolve the benzoyl peroxide by swirling the mixture.

Place the test tube in a beaker of gently boiling water. The volume of water in the beaker should be such that about one-half of the content of the test tube is submerged in the water. Watch the tube closely, and when gas bubbles start to form (5 to 10 minutes) remove the tube from the water and place it in the test tube rack. Continue to observe the tube as the polymerization proceeds to form a hard, glass-like polymer. While the polymerization is occurring, carefully feel the test tube to determine if the reaction is exothermic or endothermic.

 Dispose of the lucite according to the directions of your instructor.

REPORT FOR EXPERIMENT 33

Polymers—Macromolecules

A. Synthesis of Nylon

1. Describe what you observe at the interface of the two solutions.

2. Why did Nylon 6-6 polymer form only at the interface of the two liquid phases?

3. Nylon 6-8 has the formula

$$\left[\begin{array}{c} H \\ | \\ N-(CH_2)_6-N-C-(CH_2)_6-C \end{array}\right]_n$$

with H and O (double bond) substituents as drawn.

In the names, Nylon 6-6 and Nylon 6-8, to what do the numbers refer?

4. What would happen if you used $CH_3(CH_2)_4CH_2NH_2$ instead of 1,6-diaminohexane in the synthesis of Nylon 6-6?

5. Suppose you used citroyl chloride instead of adipoyl chloride. Draw enough of the polymer formula to show how the monomers would react with one another.

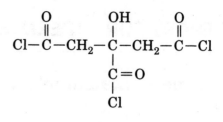

citroyl chloride

6. Another type of Nylon, Nylon 6, is made from caprolactam. In the reaction the ring opens between the $-C=O$ and the $-NH$ groups to give the monomer that polymerizes to the polyamide Nylon 6. Write the structure showing 3 units of Nylon 6.

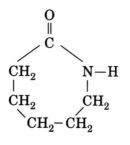

caprolactam

B. Synthesis of Lucite

1. Is this polymerization an exothermic or an endothermic reaction? What was your evidence?

2. Describe how the polymer compares with the monomer.

3. What change in the carbon-carbon bonding occurs in this polymerization?

4. Why is this polymerization reaction known as addition polymerization?

5. The polymerization of styrene to polystyrene can also be catalyzed by benzoyl peroxide. Write an equation for this reaction showing at least three units of the polymer.

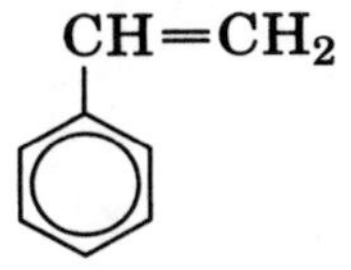

$CH=CH_2$

Styrene

EXPERIMENT 34

Carbohydrates

MATERIALS AND EQUIPMENT

Solids: Glucose, sucrose. **Liquids:** Concentrated sulfuric acid. **Solutions:** 1% solutions of arabinose, fructose, glucose, maltose, starch, sucrose (freshly prepared), and xylose; concentrated hydrochloric acid, 1% iodine in 2% potassium iodide (I_2–KI); 10% sodium hydroxide; reagent solutions for Barfoed, Benedict, Bial, Molisch, and Seliwanoff tests; fruit juices such as orange, lemon, lime, grapefruit, apple, etc.

DISCUSSION

Carbohydrates are one of the three principal classes of foods. They are major constituents of plants and are also found in animal tissues. They were so named because the carbon, hydrogen, and oxygen atom ratio in most carbohydrates approximates that of $C \cdot H_2O$. Carbohydrates, of course, are not hydrates of carbon but are relatively complex substances such as sugars, starches, and cellulose. Chemically, **carbohydrates** are polyhydroxy aldehydes or ketones or substances that, when hydrolyzed, yield polyhydroxyl aldehydes or ketones.

Carbohydrates are classified as monosaccharides, disaccharides, oligosaccharides, or polysaccharides based on the number of monosaccharide units present in the molecule. A **monosaccharide** is a carbohydrate that cannot be hydrolyzed to simpler carbohydrate molecules. A **disaccharide** yields 2 monosaccharide molecules, alike or different, when hydrolyzed. **Oligosaccharides,** on hydrolysis, yield 2 to 10 monosaccharide molecules. The monosaccharide molecule may be of only one kind, or they may be of two or more different kinds. **Polysaccharides,** when hydrolyzed, yield many monosaccharide molecules. These monosaccharide molecules are typically of only one kind.

The monosaccharides are further classified according to the length of the carbon chain, such as trioses ($C_3H_6O_3$), tetroses ($C_4H_8O_4$), pentoses ($C_5H_{10}O_5$), and hexoses ($C_6H_{12}O_6$). If they are aldehydes, they are called aldoses; if ketones, they are called ketoses. The most common monosaccharides are glucose, galactose, and fructose. Glucose and galactose are aldohexoses; fructose is a ketohexose. The three common disaccharides are sucrose (glucose and fructose), maltose (glucose and glucose), and lactose (galactose and glucose). All three have the formula $C_{12}H_{22}O_{11}$ and can be hydrolyzed to yield their respective monosaccharides by heating in a water solution containing a small amount of HCl or H_2SO_4. The three most common polysaccharides, starch, glycogen, and cellulose, have the formula $(C_6H_{10}O_5)_n$, where n ranges from about 200 to several thousand. All three are polymers of glucose.

Numerous tests have been devised for determination of the properties and for the differentiation of carbohydrates. A brief description of some of these tests follows:

Molisch Test

This is a very general test for carbohydrates. The test is based on the formation of furfural, or hydroxyfurfural when a carbohydrate reacts with concentrated sulfuric acid. The furfural reacts with the Molisch reagent, α-naphthol, to yield colored condensation products.

Seliwanoff Test

This test distinguishes fructose, a ketohexose, from aldohexoses and disaccharides. The reaction between fructose and the reagent (resorcinol in dilute HCl) occurs within one minute in boiling water. A reddish-colored product is formed; the color intensifies with further heating. Other carbohydrates subjected to this test will produce a faint red color if heating is prolonged. The color formation is attributable to transformation of glucose to fructose by the catalytic action of hydrochloric acid or by hydrolysis of sucrose to yield fructose.

The Benedict test and Barfoed test are reduction tests. Certain carbohydrates have a reducing ability because they have, or are able to form, a free aldehyde or ketone group in solution. In alkaline solution, copper(II) or silver ions are reduced to characteristic precipitates of Cu_2O or free Ag. All the common monosaccharides are reducing sugars. Some disaccharides and polysaccharides may initially be nonreducing but show reducing properties after heating in an acidic solution. During the heating, hydrolysis to monosaccharides occurs. Some widely used tests for sugars are based on this reducing ability of carbohydrates. An example is the Benedict test, which is used to detect sugar (glucose) in urine.

Benedict Test

In this test Cu^{2+} is reduced to Cu^+, forming Cu_2O, which is a brick-red colored precipitate. However, the color in a positive Benedict test may appear as green, yellow, orange, or red depending upon the amount of Cu_2O suspended in the dark blue reagent. The amount of Cu_2O formed depends on the concentration of sugar in the solution. Reducing sugars give a positive test.

Barfoed Test

The Barfoed reagent is used to distinguish between mono- and disaccharides. This test is also a copper reduction reaction but differs from the Benedict test in that the reagent is made in an acidic medium [copper(II) acetate and acetic acid]. Within the stated time interval, only monosaccharides will reduce the Cu^{2+} ions. If heated long enough, disaccharides will be hydrolyzed by the acid present and give a positive test. Barfoed reagent is not suitable for detection of sugar in urine.

Bial Test

This test is used to distinguish pentoses from hexoses. Pentoses occur in both plants and animals. The pentoses ribose and deoxyribose are universally found in the nucleic acid portion of nucleoproteins of the cells. Bial reagent contains orcinol (5-methylresorcinol) dissolved in concentrated HCl plus a small amount of $FeCl_3$. When mixed with the reagent, pentoses are converted to furfural, which reacts to yield a blue-green colored compound.

Iodine Test

Iodine reacts with starch to form a deep blue complex. When an acidified starch solution is boiled, it hydrolyzes to yield glucose ultimately. The blue color in the iodine test can be used to follow the course of this hydrolysis.

PROCEDURE

Wear protective glasses.

 Dispose of all materials in the containers provided.

A. Molisch Test

Run this test on each of the following 1% carbohydrate solutions and on water as a reference blank: (1) arabinose, (2) glucose, (3) fructose, (4) maltose, (5) sucrose, (6) starch, and (7) water (blank). To 5 mL of the carbohydrate solution in a test tube, add 3 drops of Molisch reagent and

 mix well. Now tilt the test tube at an angle of about 45 degrees and very carefully and slowly pour 2–3 mL of concentrated sulfuric acid from a 10 mL graduated cylinder down the side of the test tube so that the sulfuric acid forms a layer underneath the solution being tested.

 NOTE: It is very important that the lip of the graduated cylinder be touching the inner top of the test tube containing the carbohydrate and that the **acid be poured slowly.**

Set the test tubes in the rack and observe for evidence of reaction at the interface of the two liquid layers. Some reactions may take as long as 15 to 20 minutes, so proceed with the next part of the experiment.

B. Seliwanoff Test

Run this test on each of the following solutions: (1) arabinose, (2) glucose, (3) fructose, (4) maltose, (5) sucrose, and (6) water (blank). Mix together in a test tube 1 mL of the carbohydrate solution and 4 mL of Seliwanoff reagent. Place all six tubes in a beaker of boiling water for 2 minutes. (Do not overheat.) Observe and record the results.

C. Benedict Test

Run this test on each of the following solutions: (1) arabinose, (2) glucose, (3) fructose, (4) maltose, (5) sucrose, (6) starch, and (7) water (blank). Mix together in a test tube 1 mL of the carbohydrate solution and 5 mL of Benedict reagent. Place all seven tubes in a beaker of boiling water for 5 minutes. Observe and record the results.

D. Barfoed Test

Run this test on each of the following solutions: (1) arabinose, (2) glucose, (3) maltose, (4) sucrose, and (5) water (blank). Mix together in a test tube 1 mL of the carbohydrate solution and 5 mL of Barfoed reagent. Place all the tubes in a beaker of boiling water for 5 minutes. Observe and record the results.

E. Bial Test

Run this test on each of the following solutions: (1) arabinose, (2) xylose, (3) glucose, (4) fructose, and (5) water (blank). Mix together in a test tube 2 mL of the carbohydrate solution and 3 mL of Bial reagent. Carefully heat (with agitation) each tube over a burner flame until the mixture just begins to boil. Observe and record the results.

F. Dehydration

1. Place about 2 grams (no more) of sucrose in a test tube; add 1 mL of concentrated sulfuric acid. After about 30 seconds **very carefully** touch the test tube to feel the heat evolved. Observe and record the results. Allow the tube to cool for about ten minutes. Successively add and pour out small amounts of water to loosen the residue in the tube.

 Dispose of the black residue in the waste container provided.

2. Repeat this experiment using 2 grams of glucose and 1 mL concentrated sulfuric acid.

G. Hydrolysis of Disaccharides

Mix together 10 mL of sucrose solution and 5 drops of concentrated hydrochloric acid in one test tube and, in a second test tube, 10 mL of maltose solution and 5 drops of concentrated hydrochloric acid. Place both tubes in a beaker of boiling water for about 10 minutes. Cool the solutions and neutralize the acid with 10% NaOH solution (requires 18–20 drops of base). Use red litmus paper as an indicator. Now run Benedict and Seliwanoff tests using 2 mL samples of the hydrolyzed sugar solutions and 5 mL of each test reagent. Record your results.

H. Hydrolysis of Starch

Mix together in a 100 mL beaker 10 mL of 1% starch solution, 20 mL of water, and 10 drops of concentrated hydrochloric acid. Label five clean test tubes as (1) blank, (2) reference, (3) 5 minutes, (4) 10 minutes, and (5) 15 minutes. Add a medicine dropper full of distilled water to tube (1). Withdraw one medicine dropper full of the starch solution and put it in tube (2). Save these two samples for later reference.

Cover the beaker with a watch glass and gently boil the starch solution for 15 minutes. (Consider the zero time when the solution first starts to boil.) During the boiling period withdraw a medicine dropper full of the hot solution every 5 minutes and transfer the solution to the appropriate test tubes, (3), (4), and (5). Now add 1 drop of the I_2–KI solution to each of the five tubes and mix. Record your observations.

After the heating is completed, withdraw another medicine dropper full of the solution and transfer it to a test tube. Neutralize the acid present with 10% NaOH solution and test for the presence of a reducing sugar with Benedict reagent.

I. Fruit Juices

Test available fruit juices (preferably fresh), e.g., orange, lemon, lime, grapefruit, apple, etc., for the presence of reducing sugars and fructose using Benedict and Seliwanoff reagents. Use about 1 mL samples of juice for each test.

INSTRUCTOR DEMONSTRATION (OPTIONAL)

Colloidal Nature of Starch

Mix 5 mL of 1% starch suspension with 200 mL distilled water, divide into two equal portions and place them into 100 mL beakers. Fill a third 100 mL beaker with distilled water, align the three beakers with the water in the center and pass a narrow beam of light from a microscope illuminator (or other higher intensity source) through the beakers. Note the marked Tyndall effect apparent in even this very dilute starch. Observe the suspension from both the side and top views.

 Dispose of the starch solution down the sink drain.

REPORT FOR EXPERIMENT 34

Carbohydrates

A. Molisch Test

1. Describe the evidence for a positive test.

2. What is the chemical basis for a positive test?

3. Circle the names of the substances that gave a positive test.

   arabinose   glucose   fructose   maltose   sucrose   starch   water

B. Selinwanoff test

1. Describe the evidence for a positive test.

2. Circle the name(s) of the substances that gave a positive test.

   arabinose   glucose   fructose   maltose   sucrose   water

3. Upon prolonged heating, would you expect sucrose to give a positive Seliwanoff test? Explain.

C. Benedict Test

1. Describe the evidence for a positive test.

2. Circle the name(s) of the substances tested that gave a positive test.

arabinose   glucose   fructose   maltose   sucrose   starch   water

3. Circle the name(s) of the substances tested that are reducing carbohydrates.

arabinose   glucose   fructose   maltose   sucrose   starch

4. If the Cu^{2+} ions in Benedict reagent are reduced, what must have happened to the sugar molecules?

5. Complete and balance:

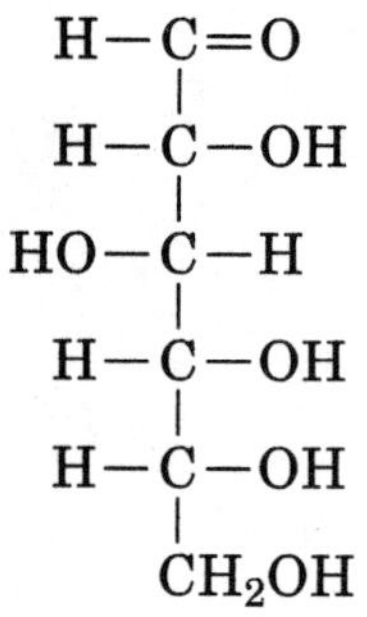

$$
\begin{array}{l}
H-C=O \\
\ \ \ \ | \\
H-C-OH \\
\ \ \ \ | \\
HO-C-H \\
\ \ \ \ | \\
H-C-OH \\
\ \ \ \ | \\
H-C-OH \\
\ \ \ \ | \\
\ \ \ \ CH_2OH
\end{array}
\ \ + \ \ Cu^{2+} \ \ + \ \ OH^- \longrightarrow
$$

D. Barfoed Test

1. What is the evidence for a positive test?

2. Circle the name(s) of the substances tested that gave a positive test.

arabinose   glucose   maltose   sucrose   water

3. How is Barfoed reagent able to distinguish a reducing monosaccharide from a reducing disaccharide?

4. Circle the name(s) of any of the following carbohydrates that will give a positive Barfoed test.

ribose   lactose   mannose   starch

E. Bial test

1. What is the evidence for a positive test?

2. Circle the name(s) of the substances that gave a positive test.

    arabinose   xylose   glucose   fructose   water

3. Write the structural formulas of two carbohydrates (other than those tested) that will give a positive Bial test.

F. Dehydration

1. Describe the effects that you noted after the addition of concentrated sulfuric acid to sucrose.

2. Write an equation for the reaction of sucrose with sulfuric acid. Assume complete dehydration and the formation of $H_2SO_4 \cdot H_2O$ as one of the products.

G. Hydrolysis of Disaccharides

1. Results after hydrolysis. Use + or – signs to indicate whether the test was positive or negative.

	Sucrose	Maltose
Benedict Test	_____	_____
Seliwanoff Test	_____	_____

2. What do the results of these tests indicate about the composition of sucrose and maltose?

H. Hydrolysis of Starch

1. What is the evidence for a positive iodine test for starch?

2. What evidence did you observe to indicate that the starch was hydrolyzing while being heated?

3. Did the Benedict tests prove that the starch actually was hydrolyzed? Explain.

I.   Fruit Juices

1. Tabulate your results using + or – signs to indicate whether a positive or a negative result was obtained with each fruit juice tested.

Fruit Juice	Benedict Test	Seliwanoff Test

2. Which carbohydrate(s) is proven to be present in each fruit juice tested?

QUESTIONS

1. The possible carbohydrates in the following problems are limited to these: arabinose, glucose, fructose, maltose, sucrose, and starch.

   (a) A carbohydrate solution gave a positive Molisch test and negative Benedict, Barfoed, Bial, and Seliwanoff tests. When treated with hydrochloric acid and boiled for several minutes, the solution showed positive Benedict, Barfoed, and Seliwanoff tests and a negative Bial test. Which carbohydrate was in the original solution?

   (b) A solution containing only one carbohydrate gave a Cu_2O precipitate with Benedict reagent. Which of the carbohydrates is present in the solution?

   (c) Another sample of the solution from (b) failed to give a Cu_2O precipitate with Barfoed reagent. Which carbohydrate is present in the solution?

2. Lactose is a reducing disaccharide. Describe how a solution of lactose would react toward these reagents.

   (a) Benedict

   (b) Barfoed

   (c) Seliwanoff

EXPERIMENT 35

Glucose Concentration in Aseptic or Simulated Blood

MATERIALS AND EQUIPMENT

Solutions: Aseptic or simulated blood, 0.2 M zinc sulfate ($ZnSO_4 \cdot 7H_2O$), 0.2 M barium hydroxide [$Ba(OH)_2 \cdot 8H_2O$], arsenomolybdate reagent, copper reagent, six standard glucose solutions ($C_6H_{12}O_6$) (see Table 35.1). Marbles (~20 mm diameter), 10 mL graduated pipet, protective gloves, spectrophotometer.

Do not use whole blood. Only aseptic, simulated, or synthetic blood that contains no harmful molecules or bloodborne disease organisms should be utilized in this experiment: It must be noted that more than 20 infectious diseases can be transmitted via whole blood or body-fluid contamination. These diseases include, but are not limited to, human immunodeficiency virus (HIV), hepatitis viruses (A, B, C, and D), cytomegalovirus, tetanus, tuberculosis, herpes simplex virus, malaria, rocky mountain spotted fever, and Creutzfeldt-Jacob disease. Therefore, **for critical safety reasons,** this experiment **must** be conducted with a disease-free blood or blood-like sample. An aseptic, simulated, or synthetic blood sample from a qualified supply source and guaranteed/verified to be disease-free is the **only** type of "blood" sample acceptable for this experiment.

DISCUSSION

Blood is a metabolic ocean of life-giving molecules. One of the most critical constituents of blood is glucose. Glucose is a source of energy and a building block for other cellular components. Normal adult levels of glucose fall between 80 mg and 120 mg per 100 mL of whole blood. Glucose is a reducing sugar capable of reducing Cu^{2+} to Cu^+. The Cu^+ ion is also a reducing agent. The reducing ability of Cu^+ enables it to form blue complexes with arsenomolybdate reagent. The color intensity of these blue complexes is proportional to the concentration of the original glucose.

To quantitatively determine the concentration of glucose in blood, you will use an instrument called a spectrophotometer to measure the amount of light absorbed by the blue complexes. The absorption depends primarily upon the concentration of the colored species. However, the distance the light travels through the solution (pathlength) and the wavelength of the light source also affect the amount of absorption. A spectrophotometer produces various wavelengths of light and can be set at a predetermined wavelength. In order to determine the concentration of an unknown sample, its absorption is compared to the absorption of standard solutions whose concentrations are known. In this comparison process, identical sample cuvettes (cells) are used. Since all samples have the same pathlength, pathlength becomes a constant. Using a pre-set wavelength of 495 nanometers (nm) for the blue complexes and fixing the pathlength, the absorption of light varies directly with the concentration of glucose.

To determine the concentration of glucose, a calibration curve is used. This curve is created by plotting the absorbance of each standard glucose sample versus its concentration. Figure 35.1 shows a sample calibration curve.

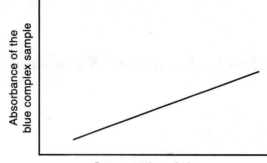

Figure 35.1   Calibration Curve for Glucose

This curve can then be used as a comparison tool. The solution derived from the blood sample is placed in a cuvette, and its absorbance measured. The corresponding concentration of glucose can then be determined from the calibration curve.

Glucose is only one of many components found in blood. Since we want to measure only the concentration of the blue complexes formed by the reducing sugar glucose, other substances that interfere, such as proteins, must be removed. Otherwise, values of glucose will be obtained that are higher than the true values. A method, known as the Somogyi-Nelson procedure, is used to remove both proteins and nonsugar-reducing substances from the blood.

In the Somogyi-Nelson procedure, barium hydroxide and zinc sulfate are added to precipitate both proteins and nonsugar-reducing substances. Proteins form insoluble zinc proteinates while the various reducing species are removed as either zinc or barium salts. These precipitates must be filtered or centrifuged from the mixture to obtain a clear solution for glucose analysis.

PROCEDURE

Wear protective glasses.

Wear protective gloves.

⚠ CAUTION: Use rubber suction bulb or water aspirator pump to draw liquids into your pipet. **Do not pipet by mouth.**

A.   Removal of Any Proteins and Nonsugar-reducing Substances

1. Pipet 2.0 mL of aseptic or simulated blood into a small Erlenmeyer flask.

2. Using a pipet, add 14.0 mL of water and mix.

3. Using a pipet, add 2.0 mL barium hydroxide solution and mix.

4. After a least one minute, pipet 2.0 mL zinc sulfate solution into the mixture. Mix thoroughly and let stand for five minutes.

5. Filter the mixture and collect the filtrate. The blood filtrate should be a water clear solution. Do not wet the filter paper before filtering.

 Discard the filter paper in the waste container provided.

B. Preparation of Samples

1. Thoroughly clean and dry nine test tubes and label them 1 to 9.

2. Fill one test tube with 24.0 mL of water and mark a line on the side to indicate this volume. Using this mark as a guide, mark all of the other tubes in a similar manner.

3. Pipet 3.0 mL of the appropriate liquid into each tube according to Table 35.1.

C. Production of Blue Complexes

1. Pipet 1.0 mL of the copper reagent into each test tube and swirl to mix.

2. Transfer all 9 test tubes (simultaneously, if possible) to a beaker containing boiling water. To prevent evaporation, place a marble on top of each test tube. Let the test tubes stand in this bath for at least 10 minutes.

Table 35.1 Material for Preparation of Samples

Tube No.	Material
1	Distilled water
2	2.0 mg/100 mL glucose standard
3	5.0 mg/100 mL glucose standard
4	8.0 mg/100 mL glucose standard
5	12.0 mg/100 mL glucose standard
6	15.0 mg 100 mL glucose standard
7	18.0 mg/100 mL glucose standard
8	Blood filtrate
9	Blood filtrate

3. Remove the marbles and place the test tubes in an ice-water bath for 5 minutes. Remove the test tubes from the ice-water bath and place them in the test tube rack.

4. Pipet 1.0 mL of the arsenomolybdate reagent into each test tube and gently mix. Let these solutions stand at room temperature for 5 minutes.

5. Dilute each solution with distilled water to the 24 mL mark. Using a rubber stopper to close the test tube, thoroughly mix the contents by repeated inversions.

D. Spectrophotometric Analysis

Plug in the spectrophotometer power line and switch the instrument on by turning the control knob clockwise past the click. Allow the instrument to warm up for 20 minutes before making any measurements.

1. Set the wavelength of your instrument to 495 nm. Since there are different types of spectrophotometers, your instructor will demonstrate the necessary procedures for adjusting your instrument.

2. **Calibrating the spectrophotometer:** Pour approximately 2 mL of solution from Tube No. 1 into a cuvette. This solution is used to zero the instrument and is called the "blank." No glucose is contained in this sample; therefore, set the spectrophotometer at 0.0 Å (no absorbance).

3. Remove the cuvette and thoroughly rinse it with the next solution to be tested (Tube No. 2). Make sure the cuvette is dry on the outside. Measure the absorbance of this solution and record your data. Repeat this procedure with Tubes 3–9. **Note:** rinse the cuvette twice with the solutions in Tubes 8 and 9. Record the absorbance on the data table in the Report form.

    Discard all solutions in the arsenic bottle provided.

REPORT FOR EXPERIMENT 35

Glucose Concentration in Aseptic or Simulated Blood

Data Table

Sample No.	Absorbance	Concentration (mg glucose/100 mL)
1		0.0
2		2.0
3		5.0
4		8.0
5		12.0
6		15.0
7		18.0
8		
9		

Plot the data obtained from Samples 2 through 7 on the graph paper provided. Draw the best straight line through the data points. Read the concentration values for Samples 8 and 9, and record them in the data table.

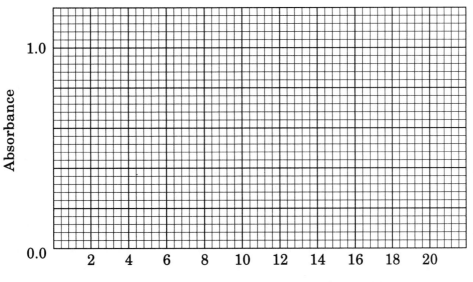

QUESTIONS AND PROBLEMS

1. Average the glucose concentration values for Samples 8 and 9. Calculate the concentration of glucose in the aseptic or simulated blood. (In your calculation consider that the aseptic or simulated blood has been diluted.)

2. The concentrations you have been working with have units of mg/100 mL. Calculate the molar concentration of glucose in a blood sample containing 115 mg/100 mL.

3. Did your aseptic or simulated blood glucose samples have exactly the same absorbance? List some reasons why they could vary in value.

EXPERIMENT 36

Amino Acids and Proteins

MATERIALS AND EQUIPMENT

Solids: Tyrosine [p-HOC$_6$H$_4$–CH$_2$CH(NH$_2$)COOH], urea [CO(NH$_2$)$_2$]. **Liquids:** Glacial acetic acid (HC$_2$H$_3$O$_2$), nonfat (skim) milk. **Solutions:** 2% albumin, 1% copper(II) sulfate (CuSO$_4$), 2% gelatin, 1% glycine (H$_2$NCH$_2$COOH), dilute (6 M) hydrochloric acid (HCl), 0.1 M lead(II) acetate [Pb(C$_2$H$_3$O$_2$)$_2$], 0.3% ninhydrin in acetone, dilute (6 M) nitric acid (HNO$_3$), concentrated nitric acid (HNO$_3$), 1-nitroso-2-naphthol (0.1% in acetone) (C$_{10}$H$_7$NO$_2$), 1% phenol (C$_6$H$_5$OH), 10% sodium hydroxide (NaOH).

DISCUSSION

Proteins are present in each cell in every living thing on earth. They function as structural material and as enzymes (catalysts) which regulate the multitude of chemical reactions that are necessary for life. Chemically proteins are polymers of α-amino acids (see your text for a list of the common α-amino acids and their structures).

$$
\begin{array}{c}
\quad\ \overset{\displaystyle H}{|}\ \ \overset{\displaystyle H}{|}\ \ \overset{\displaystyle O}{\|} \\
H-N-C-C-OH \\
\quad\quad\ \ \overset{|}{\ } \\
\quad\quad\ \ R
\end{array}
$$

An α-amino acid

Individual amino acids are joined together by bonds called **peptide linkages**. A peptide linkage in an amide structure is formed by splitting out a molecule of water between the carboxyl group of one amino acid and the amino group of another. In a typical protein, hundreds—sometimes thousands—of amino acids are linked together to form a polypeptide chain (Figure 36.1).

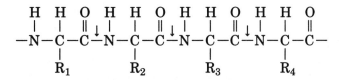

Figure 36.1   Segment of a polypeptide chain formed from α-amino acids.
The R's represent the remainders of the amino acids and the arrows point out the peptide linkages in the polypeptide chain.

The sequence of amino acids along the polypeptide chain is the primary structure of a protein. The polypeptide chains of proteins also have a secondary structure or configuration: the polypeptide chains are coiled into alpha helices or arranged side by side as pleated sheets.

Proteins are one of the three major classes of foods (proteins, carbohydrates, and fats). Proteins differ from carbohydrates and fats not only in their functions in the living organism, but also in elemental composition. In addition to carbon, hydrogen, and oxygen, which are present in carbohydrates and fats, proteins contain nitrogen. Most proteins also contain sulfur; sulfur is present in those that contain the amino acids cystine, cysteine, and methionine. Additional elements such as phosphorus, iron, copper, zinc, and iodine also occur in certain complex proteins. These elements are not part of the primary protein structure but are generally constituents of nonprotein substances combined with proteins.

Color reactions of proteins: Characteristic colors are produced when certain reagents react with one or more of the constituent groups in a protein molecule. The color produced with a given reagent will vary in intensity with different proteins. This is because all proteins either do not contain the same amino acids or do not contain the same amounts of a color-producing group. In this experiment, color-producing reactions will be used to obtain qualitative information about the composition of selected proteins. Various substances other than proteins or amino acids also give colors with some of the reagents. The colors produced with these substances will be compared to those produced with proteins and amino acids.

A. Separation of Casein from Milk

Nutritionally milk is an almost complete food containing proteins, fats, carbohydrates, many minerals, and a number of important vitamins. The proteins include casein, lactalbumin, and lactoglobin. Casein makes up about 80% of the protein in cow's milk and about 40% of that in human milk. The casein in milk is a phosphoprotein containing about 0.7% phosphorus. It is present as calcium caseinate in cow's milk and as potassium caseinate in human milk.

Casein is released from its salts and precipitated from nonfat milk by treating with dilute acetic or hydrochloric acids. However, care must be taken not to use too much acid because free casein acts as a base and redissolves in excess acid. In addition to being a food, casein is used industrially in the manufacture of adhesives, paints, paper coatings, and in printing textiles and wallpaper.

B. Biuret Test

This test derives its name from biuret, a compound formed by heating urea to 180°C.

$$\underset{\text{Urea}}{2 \; \overset{\displaystyle H}{\underset{\displaystyle H}{\text{H—N—}}} \overset{\displaystyle O}{\underset{\displaystyle \parallel}{\text{C}}} \overset{\displaystyle H}{\underset{\displaystyle H}{\text{—N—H}}}} \; \overset{\Delta}{\longrightarrow} \; \underset{\text{Biuret}}{\text{H—N—C—N—C—N—H}} \;\; + \;\; NH_3\,(g)$$

The biuret test involves heating a strongly alkaline solution of the test material with a little copper(II) sulfate and is very sensitive for proteins and polypeptides containing at least three peptide units. Biuret produces a violet color, but dipeptides do not give a positive biuret test. Colors ranging from pink to blue are produced both by polypeptides and proteins containing a structure with at least two peptide linkages (indicated by the arrows), thus

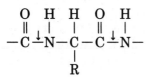

C. Tyrosine Test

This test is a very sensitive method for detecting the presence of the amino acid tyrosine, either by itself or in proteins.

D. Ninhydrin Test

Alpha-amino acids react with ninhydrin (triketohydrindene hydrate) to form a blue to purple colored complex. (Proline forms a yellow color.) This reaction is the basis for identification of amino acids by chromatography and also for the quantitative colorimetric determination of amino acids present in solution. During the reaction, ammonia is liberated which reacts with ninhydrin to give the colored complex.

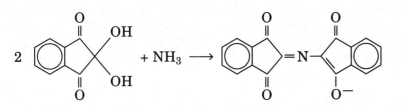

Ninhydrin   Colored Complex

E. Xanthoproteic Test

The xanthoproteic test (pronounced zan-tho-pro-teyic) is shown by most proteins and is due to the presence of the phenyl group, $-C_6H_5$, which reacts with nitric acid to form colored nitro compounds. (Phenylalanine generally does not give a positive test.)

F. Test for Cystine and Cysteine Sulfur in Proteins

When proteins containing cysteine are heated with a NaOH solution, the sulfur is converted into sulfide ions (S^{2-}). (Methionine usually does not react.) The presence of sulfide ion in solution is detected by reacting with lead(II) acetate to form a black precipitate of lead(II) sulfide.

PROCEDURE

Wear protective glasses.

WASTE
DISPOSE OF
PROPERLY
Dispose of all solids and liquids in the containers provided.

A. Separation and Testing of Casein

1. Separation of Casein: Mix 50 mL of nonfat milk with 50 mL of water in a 250 mL beaker. Add dilute acetic acid (1 vol. glacial acetic acid + 9 vol. H_2O) dropwise from a pipet or a medicine dropper, with vigorous stirring, until a flocculent precipitate forms. From 2 to 3 mL of the dilute acid is usually required. Avoid excess acid. Allow the precipitated casein to settle for a few minutes. Decant the supernant liquid through filter paper in a funnel. Then pour all of the precipitate into the funnel. Finally remove excess moisture from the casein by pressing the precipitate between absorbent paper.

The filtration process may take some time to complete, therefore, you should do all the other tests while the liquid drains from the funnel and then complete the experiment with casein.

2. Tests on Casein: For each of the following tests, mix a quantity of casein about the size of a small pea with the appropriate amount of water (3-5 mL). Perform the test and record whether the result is positive or negative.

(a) Biuret

(c) Ninhydrin

(b) Tyrosine

(d) Xanthoproteic

B. Biuret Test

1. Place 0.6 gram of urea in a dry test tube and carefully heat over a Bunsen flame. After the urea has melted, continue heating for about 30 seconds. Stop heating, and immediately note the odor (CAUTION) of the gas being emitted. Allow the tube to cool, add about 8 mL of water, and mix to dissolve some of the residue. Set aside for the Biuret test procedure.

Biuret Test Procedure: To a clean test tube add 4 mL of the solution to be tested, 1.0 mL 10% NaOH solution, and 4 drops 1% $CuSO_4$ solution. Mix and note the color that develops.

2. Do biuret tests on these solutions. Record your observations in the data table.

(a) Solution derived from heating urea in Part 1.

(b) 2% albumin

(c) 1% glycine

(d) Water (blank)

C. Tyrosine Test

Tyrosine Test Procedure: To a clean test tube add 0.5 to 1.0 mL of the solution to be tested and 2 drops of 0.1% 1-nitroso-2-naphthol (in acetone) and mix. At this stage of the tyrosine test, each tested solution should be a very light yellow. Add 3 drops dilute (6 M) HNO_3 and mix. Heat each tube in boiling water for approximately 1 minute. Note the color that develops (a positive tyrosine test will give a pink to red color for the heated solution).

Do the Tyrosine test on these solutions. Record your observations.

(a) 2% albumin

(b) 1% phenol

(c) 1% glycine

(d) 5 mL water to which a very small amount of tyrosine has been added (from the tip of a spatula)

(e) 2% gelatin

(f) Water (blank)

D. Ninhydrin Test

Ninhydrin Test Procedure: To a clean test tube add 5 mL of the solution to be tested and 10 drops of ninhydrin solution. Mix and heat the mixture in a boiling water bath for about 2 minutes. Note the color that develops.

Do the ninhydrin test on these solutions. Record your observations.

(a) 2% albumin

(b) 1% glycine

(c) 1% phenol

(d) Water (blank)

E. Xanthoproteic Test

Xanthoproteic Test Procedure: To a clean test tube add 3 mL of the solution to be tested and 10 drops of concentrated nitric acid. Mix and note any changes. Heat the tube 3 to 4 minutes in a beaker of boiling water and note any color changes. Cool and add 1% NaOH solution (3–4 mL) until the solution is alkaline (test with red litmus). Note the color change.

Do the xanthoproteic test on the following solutions. Record your observations.

(a) 2% albumin

(b) 1% glycine

(c) 1% phenol

(d) 2% gelatin

(e) Water (blank)

F. Test for Cysteine Sulfur in Proteins

Test Procedure: To a clean test tube add 3 mL of the solution to be tested and 2 mL of 10% NaOH solution. Heat the tube in a boiling water bath for 1 to 2 minutes. Remove the tube from the water bath and add 4 drops of lead(II) acetate solution and note the results. Add dilute (6M) HCl (2 to 3 mL) with slight warming until the dark color disappears. Carefully smell and note the odor of the solution.

Do the sulfur test on these solutions.

(a) 2% albumin

(b) 2% gelatin

(c) Water (blank)

NAME ______________________

SECTION ______ DATE ______________

REPORT FOR EXPERIMENT 36

INSTRUCTOR ______________________

Amino Acids and Proteins

A. Separation of Casein from Milk

Use (+) and (–) signs to indicate whether each test on casein was positive or negative.

Biuret	Tyrosine	Ninhydrin	Xanthoproteic

B. Biuret Test

Data Table

Solution	Color observation
(a) Heated urea	
(b) 2% albumin	
(c) 1% glycine	
(d) Water (blank)	

1. What is the odor of the gas emitted by heated urea? ______________________

2. How could you test for this gas other than by smelling it?

3. What is the evidence for a positive biuret test?

4. Write the formula of the organic substance responsible for the color reaction noted in tube (a).

5. Does albumin give a positive biuret test?

_____

6. Does glycine give a positive biuret test?

_____

7. Write the structure that must be present in a protein for a positive biuret test.

C. Tyrosine Test

Data Table

Solution	Observation
(a) albumin	
(b) 1% phenol	
(c) 1% glycine	
(d) Water + tyrosine	
(e) 2% gelatin	
(f) Water (blank)	

1. What is the evidence for a positive tyrosine test?

_____

2. Write the structural formula of the amino acid that will give a positive tyrosine test.

3. Which of the proteins tested contains this/these amino acids?

_____

4. Write the structural formula for a tripeptide that will give a positive tyrosine test.

D. Ninhydrin Test

Data Table

Solution	Observations
(a) 2% albumin	
(b) 1% glycine	
(c) 1% phenol	
(d) Water (blank)	

1. What is the evidence for a positive ninhydrin test?

2. What groups in the amino acid or protein are responsible for a positive ninhydrin test?

3. Write the name and structural formula of an amino acid (among those listed in your text) that will not give a positive ninhydrin test.

_____

4. Is it likely that there are proteins that will not give a positive ninhydrin test? Explain your answer.

E. Xanthoproteic Test

Data Table

Solution	Observations		
	After adding HNO_3	After heating	After adding NaOH
(a) 2% albumin			
(b) 1% glycine			
(c) 1% phenol			
(d) 2% gelatin			
(e) Water (blank)			

1. Which amino acids may be present in a protein showing a positive xanthoproteic test?

2. Why is the skin stained yellow when it comes in contact with nitric acid?

F.   Test for Cysteine Sulfur in Proteins

Data Table

Solutions	Observations	
	After adding lead(II) acetate	After adding hydrochloric acid
(a) 2% albumin		
(b) 2% gelatin		
(c) Water (blank)		

1. What is the evidence that sulfur is present?

2. Write the formula(s) of the amino acid(s) that may be present in a protein showing a positive sulfur test.

3. What is the dark colored substance that is formed when lead(II) acetate is added to the test solution? Write the equation for its formation.

4. What is the compound that you can smell after HCl is added to the dark-colored test mixture? Write the equation for its formation.

EXPERIMENT 37

Paper Chromatography

MATERIALS AND EQUIPMENT

Solutions: 0.2 M alanine, 0.2 M aspartic acid, 0.2 M leucine. 0.2 M lysine, a mixture of aspartic acid, alanine, leucine, and lysine all at 0.2 M; 0.3% ninhydrin in acetone; yellow, green, and blue food colors, isopropyl alcohol (C_3H_7OH)-water (2:1 by volume), and acetic acid (CH_3COOH)-1-butanol (C_4H_9OH)-water (1:3:1 by volume). A millimeter ruler, two 500 mL Erlenmeyer flasks, Whatman #1 filter paper (14 × 14 cm), micropipets, Pasteur pipets, aluminum foil (7 × 7 cm), hair dryer, a spray applicator for the ninhydrin solution, and unknown amino acid solutions.

DISCUSSION

Suppose that you have a solution containing two different types of molecules and you need to separate these two components from one another. How would you go about this task? Chemists face this problem often—if not daily. One possible separation technique is known as chromatography. In this experiment you will use one form of chromatography called paper chromatography. Mixtures can be separated because individual compounds (or components) are carried along the surface of paper at different rates by a solvent. This difference in mobility is due to the existence of two phases, stationary and mobile. Water absorbed on the paper fibers is the stationary phase and the free solvent is the mobile phase. The components which are the most water soluble are attracted to the stationary water phase and, therefore, will move along the paper surface more slowly, thus effecting a separation of the components in the mixture.

The ratio of the distance traveled by the component to that traveled by the solvent is the R_f value.

$$R_f = \frac{\text{The distance traveled by the component}}{\text{The distance traveled by the solvent front}}$$

This R_f value is a function of the characteristics of both the solvent and the component. The closer the R_f value is to 1.0, the more the component stayed in the mobile phase. Unknown components can be identified by comparing their R_f values to those of standard known compounds. The components must be compared with the standards under identical conditions.

You will investigate the separation of the components in (a) food colors and in (b) amino acid mixtures. To obtain a desired shade, most food colors contain more than one dye. In the first part of this experiment, three food colors are separated into their components. The second part of this experiment points out how chromatography can be used to separate and identify amino acids. By comparing the R_f value of an unknown amino acid, you can determine the identity of the unknown.

Literally, chromatography means "writing with colors." However, unlike the food colors, amino acids have no color. Therefore, to identify the positions of the amino acids, the dried paper is sprayed with ninhydrin, which reacts with amines to yield a purple to reddish-brown spot at each amino acid location.

These are the structures for the amino acids that will be used in this experiment.

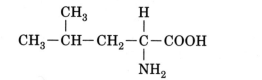

Alanine (ala)

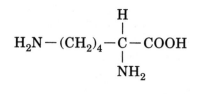

Aspartic acid (asp)

$$CH_3-CH-CH_2-\overset{\overset{\displaystyle H}{|}}{\underset{\underset{\displaystyle NH_2}{|}}{C}}-COOH$$

Leucine (leu)

$$H_2N-(CH_2)_4-\overset{\overset{\displaystyle H}{|}}{\underset{\underset{\displaystyle NH_2}{|}}{C}}-COOH$$

Lysine (lys)

PROCEDURE

Wear protective glasses.

NOTE: After starting Part A, begin Part B as soon as possible. The two parts should overlap. **Dispose of all material in the container provided.**

A. Food Color Separation

1. Cut a piece of Whitman #1 filter paper into a 14 cm square. Using a pencil, not a pen, draw a thin line across the sheet 2 cm from one edge. Measure in 3 cm, 7 cm, and 11 cm along this line and make a small x at each point. Label them 1, 2, and 3 as shown in Figure 37.1.

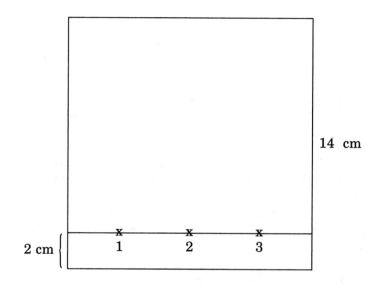

Figure 37.1

2. At the first x apply a small drop of yellow food color, at the second x, a drop of green, and at the third x a drop of blue. Each spot should be about 0.5 to 1 cm in diameter. Allow all the spots to dry for 5 minutes.

3. Pour 50 mL of isopropyl alcohol-water solvent into a 500 mL Erlenmeyer flask.

4. Roll the sheet of filter paper into a cylinder small enough to pass through the mouth of the flask. The spots should be on the outside surface at the end of the cylinder. The cylinder is then carefully inserted into the flask so that the spots are about 1 cm from the surface of the solvent. Just the bottom edge of the paper should touch the solvent. Tightly cover the mouth of the flask with a piece of aluminum foil. **Be careful not to splash the solvent on the paper.**

5. The solvent will begin to move up the paper. Allow the solvent front to travel up the paper about 7 cm above the pencil line.

NOTE: Start Part B while the solvent in Part A is moving up the paper.

6. Remove the paper and **immediately mark** the solvent front with a light pencil line. Place an x at the leading edge of each colored spot or smear.

7. Measure the distance, to the nearest 0.1 cm, between the original line at the bottom of the paper and the solvent front. Also, measure the distances from the bottom line to each x. These are the measurements used to calculate R_f values. Record your data.

B. Amino Acid Separation and Unknown Identification

In this part of the experiment, the R_f values for four known amino acids are determined. By using these R_f values, the amino acids in an unknown sample may be determined.

1. Cut a piece of Whatman #1 filter paper into a 14 cm square. Using a pencil, not a pen, mark a line across the sheet 2 cm up from one side. At 2 cm intervals draw crosses on this line. Label the crosses ala, asp, leu, lys, mixture, and unknown as shown in Figure 37.2. **Be sure you handle the paper only at the extreme edges.**

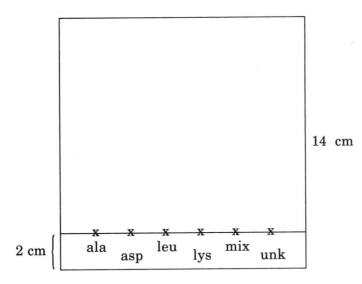

Figure 37.2

2. Obtain an unknown sample (test tube) from your instructor. The unknown will be one of the other five samples used in this experiment.

3. Fill a micropipet with one of the samples (the four amino acids, the known mixture of the four amino acids, or the unknown) and touch it to the appropriate x. Use a different micropipet for each sample. Repeat the application process to each of the six spots twice more to increase the amount of sample on each spot.

4. Pour 50 mL of the acetic acid-butanol-water solvent into a 500 mL Erlenmeyer flask.

5. Roll the sheet of filter paper into a cylinder small enough to pass through the mouth of the flask. The spots should be on the outside surface at the end of the cylinder. The cylinder is then carefully inserted into the flask so that the spots are about 1 cm from the surface of the solvent. Just the bottom edge of the paper should touch the solvent. Tightly cover the mouth of the flask with a piece of aluminum foil. **Be careful not to splash the solvent on the paper.**

6. The solvent will begin to move up the paper. Allow the solvent front to travel up the paper about 7 cm above the pencil line.

7. Remove the paper and **immediately mark** the solvent front with a light pencil line. Allow the paper to dry thoroughly. If available, use a hair dryer to speed the drying process.

8. Do this part in a hood.

Spray the paper with a fine mist of ninhydrin solution. Colored spots should appear within 15 minutes. The application of a small amount of heat will make these spots appear more quickly. Now mark each spot at its leading edge. Measure the distance, to the nearest 0.1 cm, between the original line at the bottom of the paper and the solvent front. Also, measure the distances from the bottom line to each of the spots. These are the measurements used to calculate the R_f values. Record your data.

REPORT FOR EXPERIMENT 37

Paper Chromatography

Attach both chromatographs here.

A. Data from your food color chromatogram.

Solvent front distance _____

Yellow spot(s) distance(s) _____

Yellow R_f value(s) _____

Green spot(s) distance(s) _____

Green R_f value(s) _____

Blue spot(s) distance(s) _____

Blue R_f value(s) _____

B. Data from your amino acid chromatogram.

Solvent front distance _____

Alanine distance _____

Alanine R_f _____

Aspartic acid distance _____

Aspartic acid R_f _____

Leucine distance _____

Leucine R_f _____

Lysine distance _____

Lysine R_f _____

Unknown distances _____

Unknown R_f value(s) _____

Unknown number is _____

Which amino acid(s) is in your unknown? _____

_____

_____

_____

QUESTIONS AND PROBLEMS

1. Why did you use pencil to mark the paper and not a pen?

2. In each part, A and B, what was the mobile phase and what was the stationary phase?

3. If two samples have identical R_f values does this mean that they are necessarily identical molecules? Explain.

4. For the green food color, how would it be possible to show that the component spots you separated do produce a green food color?

5. Why did you chromatograph a mixture containing all four of the amino acids?

6. Why is it important in the amino acid chromatogram not to touch the paper with your fingers or hands? Does this matter in the food color experiment?

EXPERIMENT 38

Ion-Exchange Chromatography of Amino Acids

MATERIALS AND EQUIPMENT

Solutions: 0.1% 1-nitroso-2-naphthol (in acetone) $(C_{10}H_7NO_2)$, 0.2% ninhydrin in 1-butanol, 3 M nitric acid (HNO_3), 0.2 M phosphate buffer, 0.1% tyrosine-0.1% arginine mixture, Dowex-50 slurry. A screw clamp, rubber tubing (about 3 cm long), glass wool (Pyrex), 30 test tubes, marker capable of writing on glass, a wire (about 20 cm long), a 600 mL beaker, and either a chromatography column (polypropylene column with funnel from Kontes) or a 10 mL graduated pipet.

DISCUSSION

In many areas of chemistry the separation of components in a mixture is a major challenge. One very powerful separation technique is chromatography. Many types of chromatographic techniques are used in the modern laboratory. This experiment involves ion-exchange chromatography.

Ions present in a mixture can be separated by using ion-exchange chromatography. Separation occurs because the ions that are being selectively removed bind electrostatically to ion-attracting groups having an opposite charge. The ion-attracting groups are part of an insoluble resin. Positively charged ions will bind to this resin, hence the name cation-exchange resin. Anion-exchange resins have positively charged groups capable of binding to negative ions (anions).

Amino acids exist in ionized forms (see the Discussion in Experiment 39). Amino acids which have more than one positive group will bind to the cation-exchange resin more tightly than one with only one positive charge. In this experiment you will use the amino acids tyrosine and arginine.

HO—⟨benzene ring⟩—CH$_2$—$\overset{\overset{\displaystyle H}{|}}{\underset{\overset{\displaystyle |}{\underset{+}{NH_3}}}{C}}$—COO$^-$

$\overset{+}{H_2N}$=C—NH—(CH$_2$)$_3$—$\overset{\overset{\displaystyle H}{|}}{\underset{\underset{+}{NH_3}}{C}}$—COO$^-$
  |
  NH$_2$

Tyrosine at pH 6   Arginine at pH 6

At a pH of 6 arginine has one more positive charge than tyrosine and will bind more tightly (with greater affinity) to Dowex-50 than tyrosine. This equilibrium is involved:

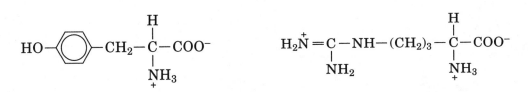

$$\text{Resin}-O-\overset{\overset{\displaystyle O}{||}}{\underset{\underset{\displaystyle O}{||}}{S}}-O^- H^+ + \text{Amino acid}^+ \rightleftarrows \text{Resin}-\overset{\overset{\displaystyle O}{||}}{\underset{\underset{\displaystyle O}{||}}{S}}-O^- \text{Amino acid}^+ + H^+$$

In this experiment, a column filled with Dowex-50 is used. The Dowex-50 is suspended in phosphate buffer at pH 6. The buffer keeps the pH constant and thus maintains the charged states of the amino acids. A mixture of 0.1% tyrosine and 0.1% arginine in phosphate buffer at pH 6 is added to the column. Once the tyrosine-arginine sample is on the resin, additional buffer is allowed to flow through the resin, and fifteen 5 mL fractions are collected. Separation is achieved since both amino acids are continuously leaving and returning to the resin (the equilibrium process). Since the arginine has a more positive charge, it has a higher affinity for the Dowex-50 and stays on the column longer than does the tyrosine. The early fractions will contain the tyrosine, and the arginine will be found in the later fractions.

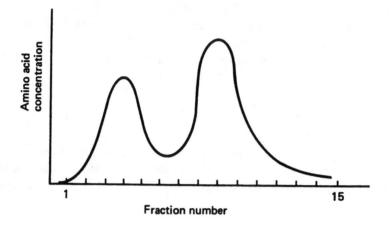

Figure 38.1   Elution profile for tyrosine and arginine

A typical "elution" profile is shown in Figure 38.1. In this particular example, Fraction 4 would contain the highest concentration of one amino acid and Fraction 9 the highest concentration of the other amino acid.

Since amino acids are not colored there is no visible evidence of separation when the fractions are collected. We can detect the amino acids by using ninhydrin, which reacts to give a blue to purple colored complex. Tyrosine can be specifically identified by using 1-nitroso-2-naphthol reagent. A positive test is indicated by a red-colored complex. With each of these reagents, the intensity of color reflects the relative amount of amino acid present in any fraction.

PROCEDURE

Wear protective glasses.

1. Label 15 test tubes with numbers from 1 to 15. Fill one test tube with 5 mL of water and mark a line on the side to indicate this volume. Use this as a guide and mark all of the others in a similar manner. This will allow you to know immediately when you have collected a 5 mL fraction from the column.

2. Mount a chromatography column (or 10 mL graduated pipet) on a ringstand with the tip down (see Figure 38.2). Force a piece of rubber tubing, about 3 cm long, onto the tip and close the tubing with a screw clamp. Place a small amount of Pyrex glass wool inside the column. Using a wire, push the wool down the column to form a loose mat that covers the hole. When the clamp is open, this mat of glass wool prevents the Dowex-50 from running out the opening

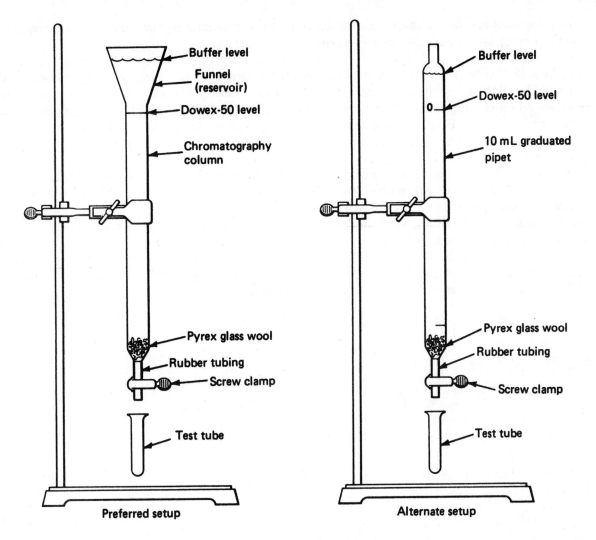

Figure 38.2 Setup of chromatographic column

but allows the buffer to flow freely. If the glass wool is packed too tightly it will prevent or significantly impede the flow of liquid through the column.

3. Obtain 25 mL of Dowex-50 slurry in a 50 mL beaker. Be sure to stir the slurry while pouring it into the beaker. Pour the slurry into the column until the resin is about 1 cm from the top of the column (not to the top of the funnel). You will need to drain the column slowly to pack the resin into place. **It is very important never to allow the buffer to drop below the resin level.** There must always be a layer of liquid above the resin.

4. Drain the buffer until only 2 mm of liquid remain above the resin, and close the clamp.

5. Using a thin dropper, carefully "load" the column by adding 1 mL of a solution containing 0.1% tyrosine and 0.1% arginine. Do not disturb the surface of the resin.

6. Open the clamp and slowly (about 1 mL/min) drip the solution into the first test tube. Stop just before the surface of the sample touches the resin. Carefully fill the funnel (reservoir) with buffer solution. Finish collecting the first 5 mL fraction. As soon as you collect the first fraction, switch to the second test tube, and so on, until you have collected all fifteen fractions. **Fill the reservoir with buffer solution after you collect each fraction.**

7. Number a second set of fifteen test tubes. Pour approximately half of each 5 mL fraction into the empty test tube with the corresponding number.

8. Test the contents of the first set of tubes for the presence of an amino acid by adding 10 drops of the ninhydrin solution to each test tube with mixing. Simultaneously place all of the test tubes in a 600 mL beaker containing 100 mL of boiling water, and heat the solutions for about three minutes. Remove the test tubes from the beaker, set aside, and record your observations.

9. Test the contents of the second set of tubes for the presence of tyrosine by adding 5 drops of 0.1% 1-nitroso-2-naphthol to each test tube with mixing. Add 5 drops of 3 M HNO_3 and mix. Simultaneously place all of the test tubes in a 600 mL beaker containing 100 mL of boiling water and heat the solutions for about three minutes. Remove the tubes from the beaker, and record your observations.

WASTE
DISPOSE OF
PROPERLY
Dispose of all liquids in the container provided. Dispose of the Dowex resin in the used resin container. The resin can be used again.

REPORT FOR EXPERIMENT 38

Ion-Exchange Chromatography of Amino Acids

Data Table: In recording your observations estimate the color intensity of each tube. Place a 0 in the result column if no color is detected. Use one + to indicate a light color; two +'s for medium color; and three +'s for the most intense color.

Data Table

Tube No.	Ninhydrin Results	1-nitroso-2-naphthol Results
1		
2		
3		
4		
5		
6		
7		
8		
9		
10		
11		
12		
13		
14		
15		

Plot all of the data on the graph provided. Draw a separate curve for each reagent.

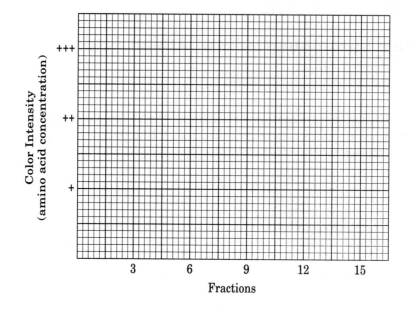

QUESTIONS AND PROBLEMS

1. Could you use ion-exchange chromatography to separate aspartic acid from glutamic acid? Why?

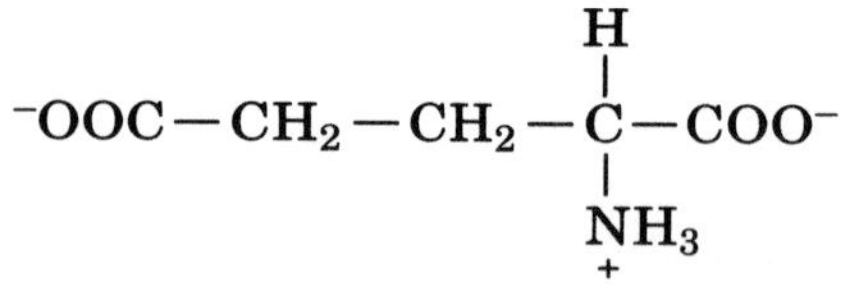

$$^-OOC-CH_2-CH_2-\overset{\displaystyle H}{\underset{\displaystyle \underset{+}{NH_3}}{C}}-COO^-$$

Glutamic acid at pH 6

$$^-OOC-CH_2-\overset{\displaystyle H}{\underset{\displaystyle \underset{+}{NH_3}}{C}}-COO^-$$

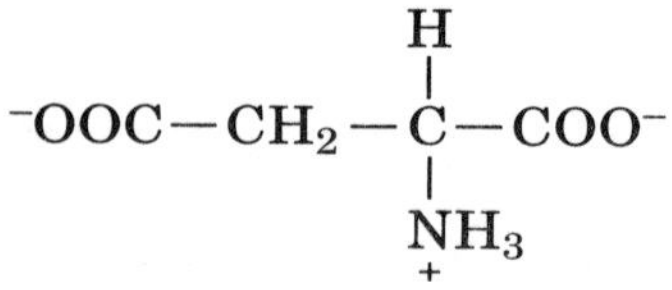

Aspartic acid at pH 6

2. How would you separate aspartic acid from tyrosine using ion-exchange chromatography at pH 6?

3. From the results of your experiment, are you guaranteed that arginine has been separated from the tyrosine? Explain.

4. Suppose you were careless for a moment and let the buffer drop 4 to 5 cm below the top of the resin during the collection of Fraction 2. What results do you imagine this would have on the separation process?

EXPERIMENT 39

Identification of an Unknown Amino Acid by Titration

MATERIALS AND EQUIPMENT

Solutions: 0.10 M glycine · HCl, 0.10 M sodium hydroxide (NaOH), unknowns of 0.10 M alanine · HCl, 0.10 M glutamic acid · HCl, 0.10 M histidine · HCl, 0.10 M lysine · HCl, and/or 0.10 M cysteine · HCl. Standard pH 7.0 buffer. A pH meter, one buret, a magnetic stirrer with stirring bar, and a 10 mL pipet.

DISCUSSION

Proteins are organic polymers formed from amino acid monomers. There are 20 common amino acids found in proteins. The identification of amino acids is an important part of biochemistry. Knowledge of the amino acid composition of proteins is of value in (1) planning adequate diets, (2) detecting certain diseases that are characterized by abnormal concentrations of amino acids, and (3) synthesizing proteins, such as human insulin. Many techniques have been developed to identify amino acids. The method used in this experiment is one of the first developed and is still an excellent technique.

The R-group shown in Figure 39.1 differs for each of the 20 common amino acids.

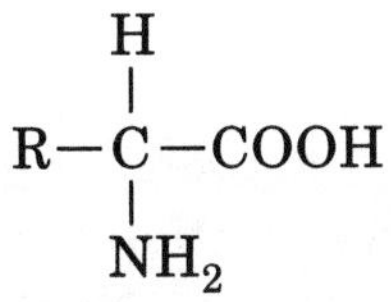

$$\begin{array}{c} H \\ | \\ R-C-COOH \\ | \\ NH_2 \end{array}$$

Figure 39.1   General structure for an amino acid

Most organic molecules have relatively low melting points, but amino acids are exceptions. For example alanine has a melting point of 297°C and phenylalanine a melting point of 285°C. These unusually high values are attributable to the ionic nature of amino acids. Both the carboxyl group (–COOH) and the amino group (–NH$_2$) undergo ionization.

$$-COOH \rightleftharpoons -COO^- + H^+ \qquad \text{(Equation 1)}$$

Acid   Base   Acid

$$-NH_2 + H^+ \rightleftharpoons -NH_3^+ \qquad \text{(Equation 2)}$$

Base   Acid

Remember that an acid is a proton donor and a base is a proton acceptor. Equations 1 and 2 show that these functional groups in amino acids are either acids or bases, depending upon whether they donate or accept protons.

At the pH of blood, approximately 7.4, amino acids exist as di-ions or Zwitterions (Figure 39.2). These di-ion charges are the attractive forces that account for the high melting points of amino acids.

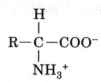

$$R-\underset{\underset{NH_3^+}{|}}{\overset{\overset{H}{|}}{C}}-COO^-$$

Figure 39.2   An amino acid in Zwitterion form

Remember that pH is defined as the negative logarithm of the hydrogen ion concentration:

$$pH = -\log [K^+] \qquad \text{(Equation 3)}$$

Every acid and every base has a characteristic ionization constant, K_a, that reflects the strength of that acid or base. The pK_a is the negative logarithm of this ionization constant.

$$pK_a = -\log K_a \qquad \text{(Equation 4)}$$

In addition to the amino and carboxyl groups found in every amino acid, the R-groups of some amino acids contain another ionizable group. For example the R-group in lysine contains an amino group.

$$H^+ \quad + \quad H_2N-(CH_2)_4-\underset{\underset{NH_2}{|}}{\overset{\overset{H}{|}}{C}}-COO^- \quad \rightleftharpoons \quad H_3\overset{+}{N}-(CH_2)_4-\underset{\underset{NH_2}{|}}{\overset{\overset{H}{|}}{C}}-COO^- \qquad \text{(Equation 5)}$$

In this experiment, you will titrate the amino acid glycine and an unknown amino acid and plot their respective titration curves. The titration curve is a plot of the pH versus the volume of base added. An example of a typical titration curve is shown in Figure 39.3. This curve is for aspartic acid which has three ionizable groups. As the base (NaOH) concentration increases, each group will ionize at a specific and characteristic pH.

The mid-points of the flat portions of the curve labeled *1, 2,* and *3* in Figure 39.3 represent the pK_a's for aspartic acid (i.e., 2.1, 3.9, and 9.8). Amino acids with only two ionizable groups (e.g., glycine) will have only two such points, one for the carboxyl group and one for the amino group.

The ionization equilibrium equation for the first ionizable group of aspartic acid is shown in Equation 6.

$$HOOC-CH_2-\underset{\underset{NH_3^+}{|}}{\overset{\overset{H}{|}}{C}}-COOH \quad \rightleftharpoons \quad HOOC-CH_2-\underset{\underset{NH_3^+}{|}}{\overset{\overset{H}{|}}{C}}-COO^- \quad + \quad H^+ \qquad \text{(Equation 6)}$$

Equations similar to Equation 6 can be written for each of the other two ionizable groups of aspartic acid.

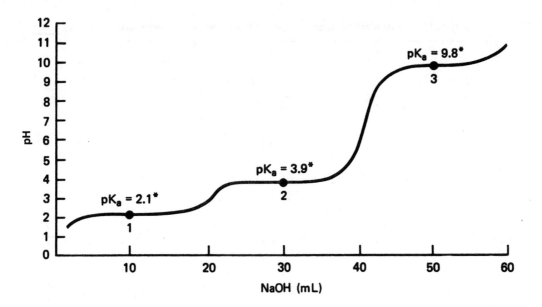

Figure 39.3 Titration of 20 mL of 0.10 M aspartic acid with 0.10 M NaOH
(*Even though this represents pH at this point it is also equal to the pK$_a$.)

In this experiment, small volume increments of standard base are added from a buret to a glycine solution. The pH is determined during this titration by means of a pH meter. After the addition of each increment of base, the pH and total volume of base are recorded. This titration is repeated using an unknown amino acid. The data collected are then used to plot the titration curves for both amino acids. The pK$_a$'s are then determined from the titration curves. By comparing the pK$_a$'s of your unknown to the values in Table 39.1 you should be able to identify your unknown amino acid.

PROCEDURE

Wear protective glasses. Do not pipet by mouth.

A. Standardization of the pH Meter

Your laboratory instructor will demonstrate how to use and how to standardize a pH meter.

B. Titration of Glycine

1. Pipet 10.0 mL of 0.10 M glycine into a 100 mL beaker, and add 30 mL of distilled water.

2. Continuous mixing is necessary. Place the beaker on a magnetic stirrer and put a stirring bar into the solution; lower the pH meter electrodes into the solution.

    CAUTION: Position the electrodes so that the bar will not strike them when it spins.

3. Start the magnetic stirrer and adjust it to a moderate speed.

Table 39.1 Formulas, names and pKa's for selected amino acids.

Formula	Name	pK_a –COOH	pK_a –NH_3^+	pK_a R-Group
H—C(H)(NH₂)—COOH	Glycine	2.3	9.6	—
CH₃—C(H)(NH₂)—COOH	Alanine	2.3	9.7	—
HS—CH₂—C(H)(NH₂)—COOH	Cysteine	1.7	10.8	8.3
HO—⟨C₆H₄⟩—CH₂—C(H)(NH₂)—COOH	Tyrosine	2.2	9.1	10.1
HOOC—CH₂—C(H)(NH₂)—COOH	Aspartic acid	2.1	9.8	3.9
HOOC—CH₂—CH₂—C(H)(NH₂)—COOH	Glutamic acid	2.2	9.7	4.3
(imidazole)—CH₂—C(H)(NH₂)—COOH	Histidine	1.8	9.2	6.0
H₂N—(CH₂)₄—C(H)(NH₂)—COOH	Lysine	2.2	9.0	10.5

4. Record the initial pH, and begin the titration by adding 1.0 mL of 0.10 M NaOH. Allow to mix for a few seconds, and record the pH and volume of added NaOH.

Continue the addition of 1.0 mL increments of NaOH, recording the pH and total volume added after the addition of each increment. When the pH begins to change rapidly, use 0.5 mL increments of base during the rapid change. Continue the titration until a pH of 11.5 has been reached or 30 mL of base has been added.

C. Titration of an Unknown Amino Acid

Pipet 10.0 mL of 0.10 M unknown into a second 100 mL beaker. Add 30 mL of water, and repeat the titration procedure.

WASTE
DISPOSE OF
PROPERLY
Dispose of all solutions down the sink drain.

D. Plotting and Reading the Data

1. Plot the data on the graphs provided.

2. Draw smooth curves through the plotted points to obtain the titration curves.

3. Read the pH corresponding to the middle of each horizontal plateau. These pH values are equal to the pK_a's of the amino acids.

NAME ______________________

SECTION ______ DATE ______________

INSTRUCTOR ______________________

REPORT FOR EXPERIMENT 39

Identification of an Unknown Amino Acid by Titration

DATA TABLES

B. Titration of Glycine

NaOH (mL)	pH	NaOH (mL)	pH
0.00			

Titration of Glycine (table continues on page 340)

C. Titration of Unknown

NaOH (mL)	pH	NaOH (mL)	pH
0.00			

Titration of Unknown (table continues on page 340)

B. Titration of Glycine (continued)

NaOH (mL)	pH	NaOH (mL)	pH

C. Titration of Unknown (continued)

NaOH (mL)	pH	NaOH (mL)	pH

TITRATION CURVES

B. Titration Curve for Glycine

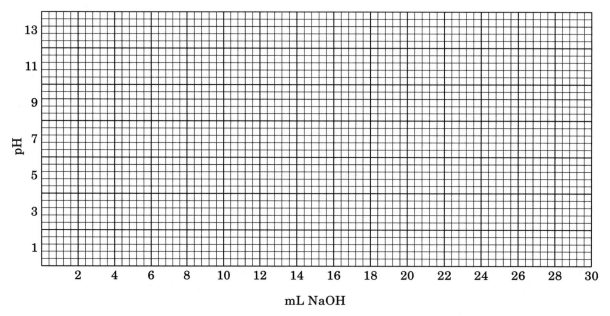

mL NaOH

C. Titration Curve for the Unknown

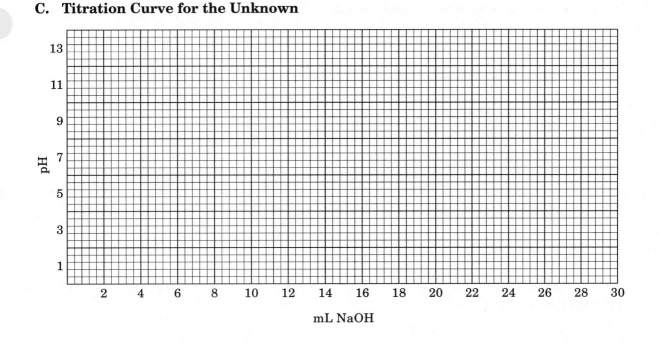

mL NaOH

QUESTIONS AND PROBLEMS

1. Based on your graph, what are the pK_a values for glycine at the two plateaus? Explain why your values may differ from the reported ones.

2. (a) What are the pK_a's for your unknown amino acid?   Unknown No. _____

    (b) What is your unknown? (Your choices are glutamic acid, histidine, alanine, lysine, or cysteine.)

3. On the basis of titration data, why might it be difficult to distinguish between lysine and tyrosine?

4. What is the pH of each of these solutions?

   (a) 10^{-3} M HNO_3

   (b) 6.21×10^{-4} M HCl

5. Write the appropriate equilibria (see Equation 6) for each pK_a of glycine and of your unknown.

6. Suppose that you titrated 0.10 M tyrosine solution with 0.10 M NaOH. Draw the titration curve that you would expect to obtain. (Assume that you have 10.0 mL of the tyrosine solution.)

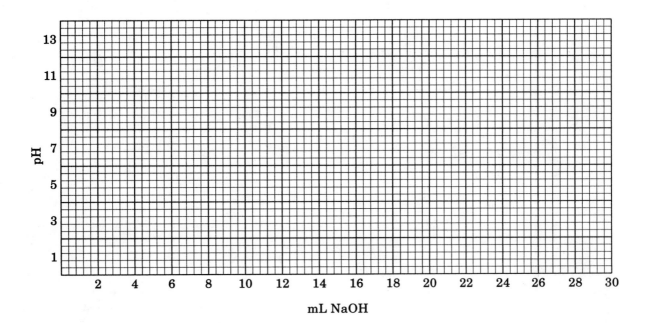

EXPERIMENT 40

Enzymatic Catalysis—Catalase

MATERIALS AND EQUIPMENT

Solutions: 6% hydrogen peroxide (H_2O_2), 0.1 M copper(II) sulfate ($CuSO_4$), 1% sodium hydroxide (NaOH), 0.01 M hydrochloric acid (HCl) (gas collection apparatus, 2 mL and 3 mL volumetric pipets), or a 5 mL graduated pipet, ice, glass wool, and a potato.

DISCUSSION

Almost all reactions in the body are controlled by enzymes. Many different enzymes are required to break down food (carbohydrates, fats, and proteins) into forms usable by the cells. Other enzymes are necessary for the cells to synthesize their own carbohydrates, fats, and proteins. Enzymes are special types of protein whose function is to control cellular reactions. These reactions occur very quickly. Speed is necessary to maintain constant cellular conditions. Enzymes are used over and over again by the cells to catalyze life-sustaining reaction.

Enzymes are very sensitive to their local surroundings. Changes in pH, temperature, ion concentration, and types of ions or molecules in solution drastically alter the enzyme's ability to catalyze a reaction. Enzymes are also very specific in catalyzing only one reaction out of the millions that are possible.

An enzyme is a protein and is built of amino acids. Each enzyme has a specific three-dimensional structure. This structure contains a pocket called the **active site** where the reaction occurs. The enzyme and substrate (reactant) combine to form the enzyme-substrate complex. The substrate, activated by the enzyme in the complex, reacts to form the products, regenerating the enzyme. (See Figure 40.1.)

Various forces (hydrogen-bonding, salt bridges, disulfide linkages, etc.) hold the enzyme in its specific shape. These forces can be overcome by denaturants, which inactivate the enzyme. One such denaturant is heat. A heated enzyme (boiled) loses its three-dimensional shape and becomes inactive or **denatured**.

Another method of inactivating an enzyme is to introduce a chemical **inhibitor** capable of specifically binding to the active site of the enzyme. These inhibitors prevent normal substrate-enzyme interactions. (See Figure 40.1.)

In this experiment we will investigate an enzyme called catalase. Catalase catalyzes the reaction:

$$2\,H_2O_2 \longrightarrow 2\,H_2O + O_2\,(g)$$

Hydrogen peroxide, H_2O_2, is found in many cellular locations and is toxic above a certain concentration. Therefore, the removal of hydrogen peroxide is important to the cell's survival.

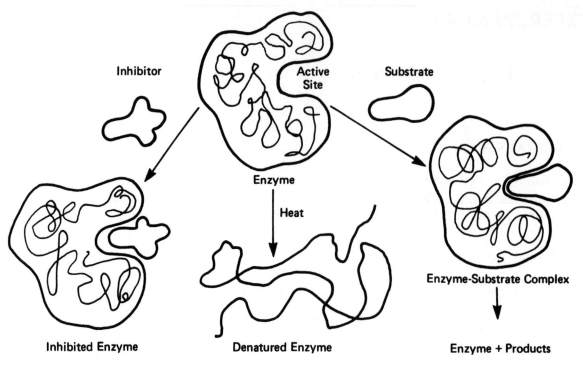

Inhibitor　Active Site　Substrate

Enzyme

Heat

Inhibited Enzyme　Denatured Enzyme

Enzyme-Substrate Complex

Enzyme + Products

Figure 40.1　Enzyme reactions

Catalase is found in high concentrations in red blood cells and in many plants. Our source of this enzyme will be the potato. We will study the effect of variations of pH and temperature on catalase activity. We will also see the effect of inhibition and denaturation in enzymatic catalysis.

In order to determine the enzyme's activity (assay procedure), we must measure either the loss of a reactant or the increase of a product. In this experiment, we will measure the amount of oxygen produced during the reaction.

PROCEDURE

Wear protective glasses. Do not pipet by mouth.

A.　Enzyme Preparation—By Students

Catalase can be extracted from a potato. Using clean equipment, cut approximately 30 grams of peeled potato into small pieces and mash them to form a paste. Transfer the paste to an Erlenmeyer flask and add 25 mL of distilled water and swirl for 15 minutes. Filter this suspension through glass wool to obtain a clear liquid containing the catalase. Filtration can be speeded up with a suction filter (see Experiment 31).

WASTE
DISPOSE OF
PROPERLY
　Dispose of filter paper and contents in the wastebasket.

A-1.　Enzyme Preparation—By Instructor

As an alternative "fast" method to the Enzyme Preparation described above, the instructor may prepare a bulk quantity catalase solution. This procedure may be performed before the laboratory class to minimize the time required to complete the experiment or at the beginning

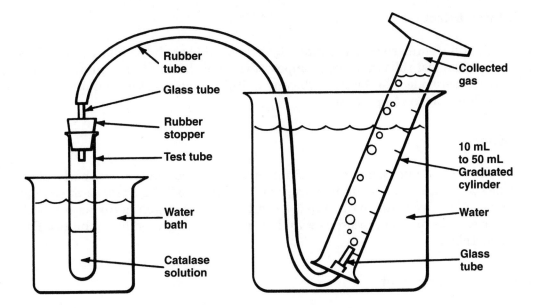

**Figure 40.2 Apparatus for measuring the volume of oxygen
produced by decomposition of hydrogen peroxide**

of the laboratory as a demonstration. Pulverize two peeled potatoes in a food processor. The "potato slurry or paste" generated by the food processor is mixed with approximately 1 liter of chilled distilled water to extract the catalase. Filter the catalase mixture through a funnel packed loosely with glass wool to obtain a reasonably clear solution of catalase (a brownish to reddish color is normal). Store the solution of catalase in an ice container until used.

B. Assays

The volume of oxygen produced from hydrogen peroxide can be determined by measuring the amount of water displaced (see Figure 40.2). The oxygen produced from the hydrogen peroxide decomposition is bubbled into the cylinder and displaces water. The volume of gas collected is measured by the graduations on the cylinder.

WASTE
DISPOSE OF
PROPERLY
Dispose of all liquids in Procedures B1–4 down the sink drain.

1. Temperature Effect

(a) Place a test tube in an ice-water bath and pipet 3.0 mL of 6% hydrogen peroxide into the test tube. Let this stay in the ice-water for at least 5 minutes. In the meantime, set up the apparatus shown in Figure 40.2. Rinse the pipet and pipet 2.0 mL of the catalase solution into the cooled hydrogen peroxide and mix quickly. With the test tube in the ice-water, replace the stopper so that the gas will bubble into the graduated cylinder. After 3 minutes (more or less time may be needed depending on the activity of the enzyme) measure and record the volume of oxygen produced.

(b) Repeat the foregoing procedure using a room temperature water bath. Measure and record the volume of oxygen produced.

(c) Repeat, using a 37°C warm water bath. Measure and record the volume of oxygen produced.

2. Inhibitor Effect

Pipet 2.0 mL of the catalase solution into a test tube, add one drop of 0.01 M $CuSO_4$ solution, and mix. Place the test tube in a room temperature water bath. After 5 minutes, pipet 3.0 mL of 6% hydrogen peroxide into the test tube and quickly stopper. After 3 minutes, measure and record the volume of oxygen produced.

3. pH Effect

Label two test tubes A and B, and pipet 2.0 mL of the catalase solution into each tube. Add five drops of 0.01 M HCl to Tube A and five drops of 1% NaOH to Tube B. Place the two test tubes in a room temperature water bath. Pipet 3.0 mL of 6% hydrogen peroxide into Tube A and quickly stopper. Measure and record the volume of oxygen produced in 3 minutes. Repeat this process with Tube B.

4. Denaturation by Heat

Pipet 2.0 mL of the catalase solution into a test tube and place it in a boiling water bath for 5 minutes. Remove the test tube, and place it in the room temperature water bath to cool. After 10 minutes, pipet 3.0 mL of 6% hydrogen peroxide into the tube and quickly stopper. Measure and record the volume of oxygen produced in 3 minutes.

REPORT FOR EXPERIMENT 40

Enzymatic Catalysis—Catalase

Data Table

Experiment	Temperature	Volume of Oxygen Produced
1. Temperature Effect (a)	0°C	
(b)	Room Temp.	
(c)	37°C	
2. Inhibitor Effect	Room Temp.	
3. pH Effect HCl	Room Temp.	
NaOH	Room Temp.	
4. Denaturation Effect	Room Temp.	

QUESTIONS AND PROBLEMS

1. (a) What effect does changing from room temperature to 0°C have on the activity of the enzyme?

    (b) What effect does changing from room temperature to 37°C have on the activity of the enzyme?

2. (a) Does the addition of acid increase or decrease the activity of catalase?

   (b) Does the addition of base increase or decrease the activity of catalase?

3. What does copper(II) sulfate do to the enzyme? Explain.

4. Why did the heat-treated enzyme behave differently than the non-heated enzyme? Explain.

EXPERIMENT 41

Lipids

MATERIALS AND EQUIPMENT

Solids: Cholesterol ($C_{27}H_{45}OH$), potassium bisulfate ($KHSO_4$), stearic acid [$CH_3(CH_2)_{16}COOH$], vegetable shortening (Spry, Crisco, etc.). **Liquids:** Acetic anhydride [$(CH_3CO)_2O$], chloroform ($CHCl_3$), glycerol ($CH_2(OH)CH(OH)CH_2OH$), hexane (C_6H_{14}), oleic acid [$CH_3(CH_2)_7CH=CH(CH_2)_7COOH$], concentrated sulfuric acid (H_2SO_4), vegetable oils (cottonseed, peanut, corn, soybean, etc.). **Solutions:** 5% bromine (Br_2) in 1,1,1-trichloroethane (CCl_3CH_3), 10% sodium hydroxide (NaOH).

DISCUSSION

Lipids are naturally occurring substances that are arbitrarily grouped together on the basis of their solubility in fat solvents, such as ether, benzene, chloroform, and carbon tetrachloride, and their insolubility in water. Lipids are subdivided into classes based on structural similarities. Two important classes are the simple lipids and the steroids.

Simple Lipids. (a) Fats and oils; esters of fatty acids and glycerol. (b) Waxes: esters of high molar mass fatty acids and high molar mass alcohols.

Steroids. These are substances possessing a 17-carbon unit structure containing four fused rings known as the steroid nucleus. Cholesterol and several hormones are in this class. Steroids having an –OH (alcohol) group attached to the ring are known as sterols. Cholesterol is an example of such a substance.

Fats (and oils) are one of the three general classes of foodstuffs—carbohydrates, proteins, and fats. The distinction between a fat and an oil is that a fat is a solid at room temperature while an oil is a liquid. Both have similar molecular structure. It is a triester (triacylglycerol) that may be considered as being derived from a glycerol molecule and three fatty acid molecules; thus

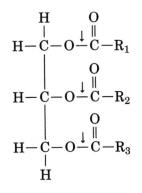

In this generalized formula, the part derived from glycerol is at the left, and the ester linkages are marked by small arrows. The portions to the right of the arrows are derived from three fatty acids. The fatty acids usually contain even numbers of carbon atoms ranging from 4 to 20 or more. The number of carbon-carbon double bonds in the carbon chains (represented by R_1, R_2, or R_3 in the generalized formula) usually varies from 0 to 4. Oleic acid, an 18-carbon acid with one double bond, is the most common unsaturated fatty acid.

A fat with no carbon-carbon double bonds is saturated. If carbon-carbon double bonds are present, the fat is unsaturated. The term polyunsaturated means that the molecules of a particular product each contain several double bonds.

Halogens add readily to carbon-carbon double bonds.

$$X_2 + \quad -CH=CH- \quad \longrightarrow \quad -CHX-CHX-$$

A solution of bromine in 1,1,1-trichloroethane is used to detect and estimate the degree of unsaturation in fats. The reddish-brown color disappears when bromine adds to double bonds.

Fats are saponified when heated with a strong base, such as sodium hydroxide, yielding sodium salts of fatty acids (soaps) and glycerol. Fats are hydrolyzed to fatty acids and glycerol in the presence of fat splitting enzymes (lipases) or when heated with a strong acid.

Saponification: Fat + Sodium hydroxide $\xrightarrow{\Delta}$ Sodium salts of fatty acids + Glycerol
(Soaps)

Hydrolysis: Fat + Water $\xrightarrow{\text{Enzymes or H}^+}$ Fatty acids + Glycerol

Potassium bisulfate ($KHSO_4$) is used to distinguish glycerol esters from other lipids. It is both a strong acid and a powerful dehydrating agent. When $KHSO_4$ is heated with a fat, hydrolysis occurs and the glycerol produced is dehydrated to acrolein. Acrolein ($CH_2=CHCHO$) is an unsaturated aldehyde with a characteristic irritating odor.

Steroids are widely distributed in plants and animals and have a variety of functions. Examples of steroids are cholesterol, the sex hormones, ergosterol (which is a precursor of vitamin D), bile salts (which aid in the digestion of fats) and cortisone (a potent hormone useful in the treatment of allergies). Steroids have a 17-carbon skeletal structure consisting of four fused rings, numbered as shown.

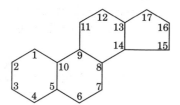

This hydrocarbon skeleton may have varying amounts of unsaturation and be substituted at various points, especially at positions 3 and 17.

Cholesterol is the most abundant steroid in the body and occurs in foods from animal sources such as eggs, butter, meat, and cheese. Cholesterol is not found in plant tissues, but closely related steroids are present in plants.

Minute amounts of cholesterol and related steroids can be detected by the Lieberman-Burchard test. This test involves treating a chloroform solution of the steroid with acetic anhydride and concentrated sulfuric acid. The formation of a blue-green color is a positive test.

PROCEDURE

Wear protective glasses.

NOTES:

1. Clean, dry test tubes are needed in this experiment and may be prepared as follows:

    (a) Clean each tube by brushing with hot water containing detergent.

    (b) Rinse thoroughly with water. For spotless tubes, rinse once with distilled water.

    (c) Wipe the outside of the tube, shake out excess water, grasp the tube with a test tube holder and heat over a nonluminous burner flame (without an inner cone) until dry.

    (d) Allow the tube to cool.

2. The tubes containing the reaction mixture from the acrolein test should be filled with hot water and allowed to soak for several minutes before cleaning by the method given in Note 1, above.

 Dispose of all substances in the container provided.

A. Acrolein Test

Place about 0.4 g (no more than 0.5 g) potassium bisulfate in a clean, dry test tube. Add 2 drops (or 0.05 to 0.1 g) of the material to be tested, and make sure that it is in contact with the potassium bisulfate. Now, using a test tube holder, heat the bottom of the tube over a hot burner flame. Incline the mouth of the tube at an angle of about 45° and constantly agitate the mixture by shaking the tube while heating, moving it in and out of the flame to control the rate of heating.

 CAUTION: Keep the mouth of the tube pointed away from yourself and away from others while it is being heated; there is danger of spattering, and hot potassium bisulfate is an extremely disagreeable substance.

Continue heating until the potassium bisulfate is melted and a very slight darkening of the tube contents is visible. Stop heating and cautiously note the odor of the vapor coming from the tube by fanning it toward your nose from a distance of 3 to 6 inches. Acrolein produces a characteristic sharp irritating odor.

Do the acrolein test on the following materials in the order given. Describe and record qualitative differences, as well as intensity differences, among the odors produced.

1. Vegetable oil (cottonseed, corn, peanut, or safflower)

2. Oleic acid (or stearic acid)

3. Glycerol

B. Reaction with Bromine

Place 7 clean, dry test tubes in a rack, and add 3 mL of hexane to each. Hexane is used here as a solvent. Add the following substances to the hexane in the test tubes.

Tube No.	Material
1	Nothing (Control)
2	3 drops vegetable oil (Peanut, cottonseed, or corn)
3	3 drops of glycerol
4	0.1 g stearic acid
5	3 drops oleic acid
6	0.1 g cholesterol
7	3 drops melted vegetable shortening

Mix the contents of each tube thoroughly for several seconds by firmly holding the top with one hand and tapping the side of the tube near the bottom with the forefinger of the other hand. **Do the following in the hood.** Add 2 drops of 5% bromine in 1,1,1-trichloroethane to Tube 1. Mix and note whether the bromine color has faded at 10 seconds and at 30 seconds following the bromine addition. Record your observations on the Report Sheet using this code: 0 = No fading, or barely noticeable fading, of color. 1 = Definitely noticeable color fading, but not complete. 2 = Complete, or nearly complete, disappearance of color.

Repeat the addition of bromine and the 10 and 30 second observations for each of the 6 remaining tubes in succession, and record your observations on the Report Sheet.

WASTE
DISPOSE OF
PROPERLY **Dispose of liquids in the waste organic solvent bottle.**

C. Lieberman-Burchard Test

Place 5 clean, dry test tubes in a rack and add the materials to be tested as follows:

Tube No.	Material
1	Stearic acid, pinhead sized quantity
2	Cholesterol, pinhead sized quantity
3	Glycerol, 1 drop
4	Vegetable oil, 1 drop (Cottonseed, peanut, soybean, or corn)
5	Melted vegetable shortening, 1 drop

Do the following in the hood. Add 2 mL chloroform, 10 drops acetic anhydride, and 2 drops concentrated sulfuric acid to each tube, and mix thoroughly. After 5 minutes note the color present and its relative intensity in each solution. Record your data on the Report Form.

This is a sensitive test for the presence of sterols. Colors changing from pinkish, to blue, to green develop in the course of the reaction. The intensity of the color is roughly proportional to the amount of sterol present.

 Dispose of liquids in the waste organic solvent bottle.

D. Saponification

Put 4 drops of vegetable oil or melted shortening into a test tube, and add 10 drops of 10% NaOH solution. Now using a test tube holder, incline the tube at an angle of about 45° and heat the contents over a low burner flame.

 CAUTION: Be sure to wear your safety glasses and keep the tube pointed away from others, and from yourself, while it is being heated.

Shake the tube constantly while heating so that the liquid is continually sloshed about in the bottom third of the tube. Control the rate of heating by moving the tube in and out of the flame. Continue heating with constant agitation until the mixture in the tube is almost dry or until a solid begins to form. Allow the tube to cool for at least 2 minutes; add 5 mL of distilled water, stopper, and shake the tube vigorously for several seconds. Note and record the results.

To another test tube, add 4 drops of the vegetable oil and 10 drops of 10% NaOH solution. Do not heat this tube, but add 5 mL distilled water; stopper, and shake the tube vigorously for several seconds. Compare the results with those obtained with the first tube.

 Dispose of residue down the sink drain.

REPORT FOR EXPERIMENT 41

Lipids

A. Acrolein Test

1. Describe the odors produced by (a) vegetable oil, (b) oleic acid (or stearic acid), and (c) glycerol.

2. Which substance(s) gave a positive acrolein test?

B. Reaction with Bromine

Tube No.	Material tested	Color 10 sec. (0, 1, or 2)	Color 30 sec. (0, 1, or 2)	Conclusions (Saturated or unsaturated)
1				
2				
3				
4				
5				
6				
7				

C. Lieberman-Burchard Test

Tube No.	Material tested	Color and relative intensity	Conclusions Positive or negative
1			
2			
3			
4			
5			

D. Saponification

Observations in:
  First test tube:

  Second test tube:

QUESTIONS AND PROBLEMS

1. Write the structural formula of a triacylglycerol molecule that is made from glycerol and three molecules of stearic acid. Draw a box around the portion(s) of the molecule that make it soluble in solvents such as benzene or hexane.

2. Glycerol is a saturated trihydroxy alcohol. Acrolein is an unsaturated aldehyde having the formula, $CH_2=CHCHO$. Hot potassium bisulfate is a powerful dehydrating agent. Using structural formulas for organic substances, write an equation to show how hot $KHSO_4$ converts glycerol to acrolein.

3. Explain why the fat, glyceryl tripalmitate, and cholesterol would be expected to give different results with the Acrolein Test.

4. Write an equation showing the addition of bromine to oleic acid.

5. Does the vegetable oil (or oils) tested contain substances other than glycerides? Cite experimental results which justify your answer.

6. What was the evidence in Part D that the vegetable oil was saponified?

7. Write condensed structural formulas for the soaps produced in Part D if vegetable oil contained glycerides of lauric, palmitic, and oleic acids.

EXPERIMENT 42

Cholesterol Levels in Aseptic or Simulated Blood

MATERIALS AND EQUIPMENT

Liquids: Isopropyl alcohol [$CH_3CH(OH)CH_3$], acetone (CH_3COCH_3), glacial acetic acid (CH_3COOH), concentrated sulfuric acid (H_2SO_4). **Solutions:** Aseptic or simulated blood, 1:1 ethanol (C_2H_5OH)–acetone solution, cholesterol standard solution ($C_{27}H_{45}OH$), digitonin solution ($C_{56}H_{92}O_{29}$), and iron reagent ($FeCl_3 \cdot 6H_2O$). Spectrophotometer, centrifuge and centrifuge tubes, 1 mL and 5 mL graduated pipets, protective gloves.

> ⚠ **Do not use whole blood. Only aseptic, simulated, or synthetic blood that contains no harmful molecules or bloodborne disease organisms should be utilized in this experiment.** It must be noted that more than 20 infectious diseases can be transmitted via whole blood or body-fluid contamination. These diseases include, but are not limited to, human immunodeficiency virus (HIV), hepatitis viruses (A, B, C, and D), cytomegalovirus, tetanus, tuberculosis, herpes simplex virus, malaria, rocky mountain spotted fever, and Creutzfeldt-Jacob disease. Therefore, **for critical safety reasons,** this experiment **must** be conducted with a disease-free blood or blood-like sample. An aseptic, simulated, or synthetic blood sample from a qualified supply source and guaranteed/verified to be disease-free is the **only** type of "blood" sample acceptable for this experiment.

DISCUSSION

Listen to the radio, watch television, or read the newspaper for a day and you will probably encounter some question or statement about cholesterol. The Office of the Surgeon General of the United States believes cholesterol can either directly or indirectly cause heart disease. Only animals produce and require cholesterol. Humans need a certain level of cholesterol to survive. Cholesterol is the principal steroid in vertebrates and is a starting molecule for the biosynthesis of some steroid hormones and bile acids. Cholesterol occurs in both blood and cells. A normal blood cholesterol level is in the range of 150–200 mg/100 mL. Problems may arise when cholesterol levels go significantly above this range. High levels of cholesterol indicate possible coronary disease, nephritis, or diabetes (among other possibilities). Since cholesterol is an alcohol, it can be esterified. A major portion of blood cholesterol exists as esters of fatty acids (see page 362).

In this experiment, you will use a spectrophotometer to measure the concentration of both total cholesterol and free cholesterol by the Leffler method. By subtracting the free cholesterol from the total cholesterol, you obtain the amount of esterified cholesterol.

Whole blood consists primarily of plasma and red blood cells (RBC). Usually, the RBCs are removed from whole blood by centrifugation. Then the cholesterols are extracted into an organic solvent. The cholesterols (both free and esterified) are then reacted with iron (III) chloride to produce a colored complex. This complex absorbs light at 560 nm. The concentration of total cholesterol in the blood will be determined by comparing its absorbance with that of a standard cholesterol solution.

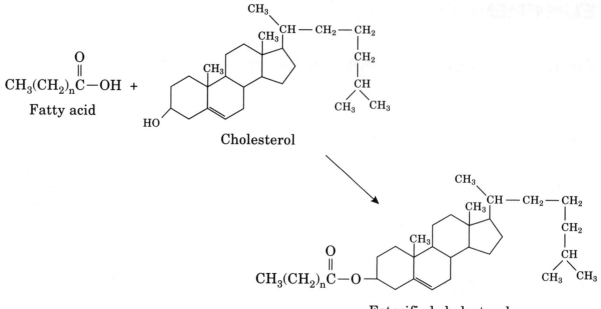

Fatty acid + Cholesterol → Esterified cholesterol

Free cholesterol is selectively precipitated from solution by digitonin. Once the free cholesterol is separated, it is redissolved and analyzed.

PROCEDURE

Wear protective glasses. Wear protective gloves.

 Dispose of all materials in the containers provided.

 CAUTION: Use rubber suction bulb or water aspirator to draw liquids into your pipet. **Do not pipet by mouth.**

NOTE: To keep track of the various samples in this experiment, label four centrifuge tubes C-1 through C-4 and six test tubes T-1 through T-6. Use clean and dry tubes.

A. Separation of Plasma from Aseptic or Simulated Blood

1. Put 1.5 to 2.0 mL of aseptic or simulated blood in centrifuge tube C-1 and centrifuge for 10 minutes or until the red blood cells (or simulated red blood cell materials) have sedimented to the bottom of the tube.

2. Cool test tube T-1 in an ice bath.

3. Carefully pipet the plasma (upper layer) in centrifuge tube C-1 and transfer it to the cooled test tube T-1; keep this plasma cold. (Do not pick up any solids.)

 4. **Discard the RBCs in tube C-1 in the container provided.**

B. Determination of Total and Free Cholesterol

1. Pipet 0.20 mL of plasma from test tube T-1 into centrifuge tube C-2.

2. Pipet 0.20 mL of the cholesterol standard (2.0 mg/mL) into tubes C-2 and T-2.

3. Pipet 5.0 mL of isopropyl alcohol into C-2 and T-2 and **thoroughly** mix.

4. Centrifuge the plasma sample in C-2 for about 5 minutes to remove the precipitate.

5. Pipet 1.0 mL of the isopropyl-cholesterol supernatant from C-2 into T-3 and set aside for later use.

6. Pipet 1.0 mL of the isopropyl-cholesterol standard from T-2 into T-4 and set it aside for later use.

7. Pipet 0.20 mL of plasma from T-1 in the ice bath into C-3. Now pipet 3.0 mL of 1:1 ethanol-acetone solution into C-3. Thoroughly mix and centrifuge for about 5 minutes.

8. Pipet 2.0 mL of the supernatant from C-3 into C-4. Pipet 2.0 mL of the digitonin solution into C-4 and thoroughly mix.

9. Heat the mixture in C-4 in a water bath at approximately 55°C for 15 minutes and then cool in an ice bath for about one minute. Centrifuge, decant and discard the supernatant; **save** the precipitate.

10. Add 4 mL of acetone to the precipitate in C-4 and thoroughly mix; centrifuge for 15 minutes and decant the supernatant; **save** the precipitate.

11. Pipet 1.0 mL of isopropyl alcohol onto the precipitate in C-4; thoroughly mix and set aside for later use.

12. Pipet 1.0 mL of isopropyl alcohol into T-6. This is the blank required to zero the spectrophotometer. Save this blank.

13. Pipet 3.0 mL of glacial acetic acid into each of tubes C-4, T-3, T-4, and T-6 and **thoroughly** mix. Be sure that no precipitate in C-4 remains undissolved. Pour the solution from C-4 into T-5.

14. Pipet 0.30 mL of the iron reagent into each of the four tubes, T-3 through T-6, and thoroughly mix.

⚠ 15. **USING EXTREME CAUTION,** pipet 3.0 mL of concentrated sulfuric acid into each of the four tubes, T-3 through T-6, and thoroughly mix.

16. Cool the four samples in No. 15 to room temperature in a water bath.

17. Absorbency measurements: Warm up the spectrophotometer for at least 20 minutes. Set the wavelength of the spectrophotometer to 560 nm. Transfer the solution from tube T-6 (blank) to a cuvette and zero the instrument. Using another cuvette measure the absorbencies of the solutions from tubes T-3, T-4, and T-5. Make three measurements for each solution (zero the instrument between each measurement). Record your data on the report form.

C. Calculations

To obtain the cholesterol concentrations, use the following formulas:

$$\text{Total cholesterol} = \left(\frac{\text{Total cholesterol absorbance}}{\text{Std. cholesterol absorbance}}\right) \times \left(\begin{array}{c}\text{Std. cholesterol}\\\text{concentration}\end{array}\right)$$

$$\text{Free cholesterol} = \left(\frac{\text{Free cholesterol absorbance}}{\text{Std. cholesterol absorbance}}\right) \times \left(\begin{array}{c}\text{Std. cholesterol}\\\text{concentration}\end{array}\right) \times \left(0.31\right)$$

The concentration of the cholesterol standard is 2.0 mg/mL.

The factor 0.31 is a correction factor and is the ratio of the amount of the total cholesterol to the amount of free cholesterol measured, i.e., (1.0/5.2)/(2.0/3.2).

Total cholesterol is measured in Tube T-3 and free cholesterol in Tube T-5. Tube T-4 contains the cholesterol standard.

REPORT FOR EXPERIMENT 42

Cholesterol Levels in Aseptic or Simulated Blood

Data Table

Tube No.	Absorbance	Average Absorbance
T-3 T-3 T-3		
T-4 T-4 T-4		
T-5 T-5 T-5		

Calculated concentrations (show your calculations).

Concentration of total cholesterol in Tube T-3 = _____

Concentration of free cholesterol in Tube T-5 = _____

Concentration of esterified cholesterol = _____

QUESTIONS AND PROBLEMS

1. Cholesterol will react with stearic acid in the presence of a small amount of mineral acid. Name the product and draw its structure.

2. Does the cholesterol level in your aseptic or simulated blood sample indicate the donor (real or simulated) is possibly subject to cholesterol related diseases? Explain.

3. In the calculation for free cholesterol, we used a 0.31 factor. Show where this originates. Remember (1.0/5.2)/(2.0/3.2).

STUDY AID 1

Significant Figures

Every measurement that we make has some inherent error due to the limitations of the measuring instrument and the experimenter. The numerical value recorded for a measurement should give some indication of the reliability (precision) of that measurement. In measuring a temperature using a thermometer calibrated at one-degree intervals we can easily read the thermometer to the nearest one degree, but we normally estimate and record the temperature to the nearest tenth of a degree (0.1°C). For example, a temperature falling between 23°C and 24°C might be estimated at 23.4°C. There is some uncertainly about the last digit, 4, but an estimate of it is better information than simply reporting 23°C or 24°C. If we read the thermometer as "exactly" twenty-three degrees, the temperature should be reported as 23.0°C, not 23°C, because 23.0°C indicates our estimate to the nearest 0.1°C. Thus in recording any measurement, we retain one uncertain digit. The digits retained in a physical measurement are said to be significant, and are called **significant figures.**

Some numbers are exact (have no uncertain digits) and therefore have an infinite number of significant figures. Exact numbers occur in simple counting operations, such as 5 bricks, and in defined relationships, such as 100 cm = 1 meter, 24 hours = 1 day, etc. Because of their infinite number of significant figures, exact numbers do not limit the number of significant figures in a calculation.

Counting Significant Figures. Digits other than zero are always significant. Depending on their position in the number, zeros may or may not be significant. There are several possible situations:

1. All zeros between other digits in a number are significant; for example: 3.076, 4002, 790.2. Each of these numbers has four significant figures.

2. Zeros to the left of the first nonzero digit are used to locate the decimal point and are not significant. Thus 0.013 has only two significant figures (1 and 3).

3. Zeros to the right of the last nonzero digit, and to the right of the decimal point, are significant, for they would not have been included except to express precision. For example, 3.070 has four significant figures; 0.070 has two significant figures.

4. Zeros to the right of the last nonzero digit, but to the left of the decimal, as in the numbers 100, 580, 37000, etc., may not be significant. For example, in 37000 the measurement might be good to the nearest 1000, 100, 10, or 1. There are two conventions which may be used to show the intended precision. If all the zeros are significant, then an expressed decimal may be added, as 580., or 37000. But a better system, and one which is applicable to the case when some but not all of the zeros are significant, is to express the number in exponential notation, including only the significant zeros. Thus for 300, if the zero following 3 is significant, we would write 3.0×10^2. For 17000, if two zeros are significant, we would write 1.700×10^4. The number we correctly expressed as 580. can also be expressed as 5.80×10^2. With exponential notation there is no doubt as to the number of significant figures.

Addition or Subtraction. The result of an addition or subtraction should contain no more digits to the right of the decimal point than are in that quantity which has the least number of digits to the right of the decimal point. Perform the operation indicated and then round off the number to the proper number of significant figures.

Example: 24.372
            72.21
            6.1488
            102.7308 (102.73)

Example: 242.112
          − 84.9
            157.212 (157.2)

Since the digit 1 in 72.21 is uncertain, the sum can have no digits beyond this point, so the sum should be rounded off to 102.73. In the subtraction, the answer should contain four significant figures.

Multiplication or Division. In multiplication or division, the answer can have no more significant figures than the factor with the least number of significant figures. In multiplication or division, the position of the decimal point has nothing to do with the number of significant figures in the answer.

Example: $3.1416 \times 7.5 \times 252 = 5937.624 \ (5.9 \times 10^3)$

The operations of arithmetic supply all the digits shown, but this does not make the answer precise to seven significant figures. Most of these digits are not realistic because of the limited precision of the number 7.5. So the answer must be rounded to two significant figures, 5900 or 5.9×10^3. It should be emphasized that in rounding-off the number you are not sacrificing precision, since the digits discarded are not really meaningful.

Example: $\dfrac{189}{24} = 7.875 \ (7.9)$

The answer contains two significant figures.

STUDY AID 2

Formulas and Chemical Equations

The atomic compositions of substances are expressed by formulas. In order to write correct chemical equations, it is necessary to know how to write correct formulas.

Ions are atoms or groups of atoms that have either a positive or a negative electrical charge. The relative electrical charge on an ion is indicated by a plus (+) or minus (–) sign together with a number. The number and the plus or minus sign are written as a superscript at the upper right corner of the symbol for the ion. Examples: Na^+ represents a sodium ion with a +1 charge (the number 1 is omitted but understood to be present). SO_4^{2-} represents a sulfate ion which has a –2 charge and is composed of one sulfur atom and four oxygen atoms.

In the examples given below, the ions shown are used to illustrate the formation of possible compounds. The first series of ions is hypothetical but will be used to illustrate the principle of combining positive and negative ions to write formulas of compounds. A list of common ions is provided at the end of this manual.

Ion	Name	Ion	Name	Ion	Name
X^+	(Hypothetical)	Na^+	Sodium	Br^-	Bromide
Y^{2+}	(Hypothetical)	NH_4^+	Ammonium	Cl^-	Chloride
Z^{3+}	(Hypothetical)	K^+	Potassium	NO_3^-	Nitrate
E^-	(Hypothetical)	Ca^{2+}	Calcium	O^{2-}	Oxide
G^{2-}	(Hypothetical)	Pb^{2+}	Lead(II)	S^{2-}	Sulfide
		Mg^{2+}	Magnesium	CrO_4^{2-}	Chromate
		Al^{3+}	Aluminum	SO_2^{2-}	Sulfate
		Fe^{3+}	Iron(III)	PO_4^{3-}	Phosphate

Examples of compounds formed by combining positive and negative ions:

Ions	Hypothetical Compounds	Specific Compound with Corresponding Ion Ratios
X^+, E^-	XE	$NaCl$, NH_4Cl, KBr, $NaNO_3$
Y^{2+}, E^-	YE_2	$CaCl_2$, $Mg(NO_3)_2$, $Pb(NO_3)_2$
X^+, G^{2-}	X_2G	Na_2O, $(NH_4)_2SO_4$, K_2CrO_4
Y^{2+}, G^{2-}	YG	MgO, $PbCrO_4$, $CaSO_4$
Z^{3+}, E^-	ZE_3	$AlCl_3$, $Fe(NO_3)_3$
Z^{3+}, G^{2-}	Z_2G_3	Al_2O_3, $Fe_2(SO_4)_3$

When more than one atom of an element is used in the formula of a compound, it is indicated by a numerical subscript written to the right of the element; e.g., $CaCl_2$. The formula $CaCl_2$ indicates one Ca^{2+} ion and two Cl^- ions. If an ion contains more than one element, e.g., NO_3^-, and two or more of these ions are needed in the formula of a compound, the ion is enclosed in parentheses and the subscript is written after the parenthesis; e.g., $Cu(NO_3)_2$. This formula indicates one Cu^{2+} ion and two NO_3^- ions.

A chemical change or reaction results in the formation of products whose compositions are different from the starting substances (reactants). A chemical equation is a shorthand expression for a chemical reaction. Substances in the equation are represented by their formulas. The equation indicates both the reactants and their products. The reactants are written on the left side and the products on the right side of the equation. An arrow ($\longrightarrow$) pointing to the products separates the reactants from the products. A plus sign (+) is used to separate one reactant (or product) from another.

Reactant $\longrightarrow$ Products

Example 1. The combustion of carbon in oxygen or air

Word equation: Carbon + Oxygen $\longrightarrow$ Carbon dioxide

Formula equation: C + O_2 $\longrightarrow$ CO_2

Example 2. The combustion of magnesium in oxygen or air

Word equation: Magnesium + Oxygen $\longrightarrow$ Magnesium oxide

Formula equation: $2\ Mg$ + O_2 $\longrightarrow$ $2\ MgO$

Balancing equations. There is no detectable change in mass resulting from a chemical reaction. Therefore, the mass of the products must equal the mass of the reactants before the chemical change occurred. In representing the chemical change by an equation, this conservation of mass is attained by "balancing the equation." Because each atom has a particular mass, we balance an equation by adjusting the number of atoms of each kind of element to be the same on each side of the equation. This adjustment should not be made by changing subscripts in correct formulas. In example 1 there are two oxygen atoms and one carbon atom on each side of the equation; therefore, the mass of the reactants and the products must be equal. The equation is balanced.

If the equation in Example 2 is written and left as

$Mg + O_2 \longrightarrow MgO$ (Unbalanced)

it is not balanced since there are two oxygen atoms on the left side and only one oxygen atom on the right side. We can balance the oxygen atoms by placing a 2 in front of MgO, which gives us two oxygen atoms on each side of the equation.

$Mg + O_2 \longrightarrow 2\ MgO$ (Unbalanced)

Now the magnesium atoms are unbalanced. We can balance the magnesium atoms and the whole equation by placing a 2 in front of Mg.

$2\ Mg + O_2 \longrightarrow 2\ MgO$ (Balanced)

Remember! A correct formula of a substance may not be changed for the convenience of balancing an equation. Small whole numbers, as needed, are placed in front of the formulas to balance the equation. However, it is important to notice that when a number is placed in front of one formula to balance a particular element, as in 2 MgO above, it may unbalance another element in the equation.

A number placed in front of a formula multiplies every atom in the formula by that number. Thus 2 MgO means 2 Mg atoms and 2 O atoms; 3 $CaCl_2$ means 3 Ca atoms and 6 Cl atoms; 4 H_2SO_4 means 8 H atoms, 4 S atoms, and 16 O atoms; 3 $Cu(NO_3)_2$ means 3 Cu atoms, 6 N atoms, and 18 O atoms or 3 Cu atoms and 6 NO_3^- ions.

Many equations may be balanced by this "trial and error" or inspection method.

STUDY AID 3

Preparing and Reading a Graph

A graph is often the most convenient way to present or display a set of data. Various kinds of graphs have been devised, but the most common type uses a set of horizontal and vertical coordinates, x and y, to show the relationship between two variables, the independent and dependent variables. The dependent variable is a measurement that changes as a result of changes in the independent variable. The independent variable either changes itself (like time) or is controlled by the experimenter. Usually the independent variable is plotted on the x-axis (abscissa) and the dependent variable is plotted on the y-axis (ordinate). See Figure S3.1.

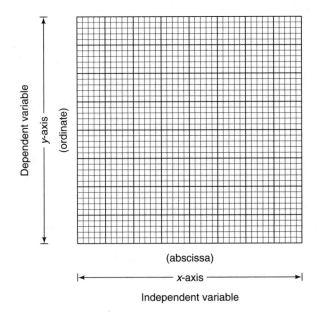

Figure S3.1   Rectangular coordinate graph paper

The values for each variable are called data and listed in a data table to facilitate the construction of a graph. As a specific example of how a graph is constructed, let us graph the relationship between the volume of a liquid and its mass. A chemist measured increasing volumes of a liquid and determined the mass of each volume. The data are recorded in Table S3.1. In this study aid, we will use this data to illustrate the steps for making graphs by hand (Part A) and by computer (Part C).

A.   STEPS IN PREPARING A GRAPH

Most scientists today use computers to help them make graphs from their data. Before we show you how to work with a computer to do this, it is important to learn how to make a graph with pencil, ruler, and graph paper. Use the following step-by-step procedure to plot the data in Table S3.1 on the graph paper provided in Figure S3.2. Your completed graph should resemble the graph in Figure S3.3 very closely. After you complete this first graph, practice your graphing skills by making another graph using the data in Table S3.2 and the grid provided in Figure S3.4.

PROCEDURE

1. Examine the graph paper in Figure S3.2 and count how many blocks are available along each axis: This paper has 40 blocks along the x-axis and 40 blocks along the y-axis.

Table S3.1   Volume vs. Mass Data

Volume, mL	Mass, g
21.0	19.1
30.0	27.3
37.5	34.1
44.0	40.0
47.0	42.8
50.0	45.5

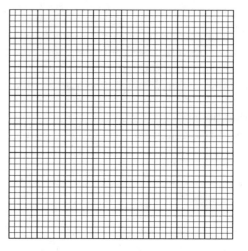

Figure S3.2   Graph paper sample

2. Examine the data in Table S3.1 and determine the independent and dependent variables: The amount of liquid in each sample was varied by the experimenter, so volume is the independent variable and will be plotted along the x-axis. The mass of each sample changed as the volume was changed, so mass is the dependent variable and will be plotted along the y-axis. Usually the independent-versus-dependent-variable decision can be reasoned out like this example. If not, then the placement of the variables on the axes can be arbitrary.

3. Determine the range for each variable: The independent variable ranges from 21.0 mL to 50.0 mL. This is a range of 29.0 mL. The dependent variable ranges from 19.1 g to 45.5 g. This is a range of 26.4 g.

4. Determine the scale for each axis; that is, how many units each block will represent. The calculation for the independent variable using this particular piece of graph paper is:

independent variable scale = 29.0 mL/40 blocks = 0.73 mL/block

But, if we adopted this scale, the graph would be extremely awkward to plot and read. So, we round **up** (never round down) this preliminary scale to a more convenient value per block. The most convenient scales to use are generally 0.5, 1, 2, 5, or 10 units per block. The scale is never rounded up to more than double its preliminary value. For this sample data, 0.73 mL/block is rounded up to 1.0 mL/block because it is convenient and less than 1.46 (double 0.73).

Now, we do the same calculations for the dependent variable on the y-axis.

dependent variable scale = 26.4 g/40 blocks = 0.66 g/block. It is not very convenient to count by units of 0.66, so this value is rounded **up** to 1.0 g per block.

5. Determine the starting values for each coordinate: Although it is common for the axes to be numbered starting with zero at the origin (lower left corner), it is not required and some-

times it is a poor choice. For instance, in our example, all of the data for the independent variable are greater than 20.0, so from 0–20, there would be no data plotted. Therefore, we start numbering the x-axis at 20.0 mL and the y-axis at 15.0 g.

6. Determine the major and minor increments for each axis: We never number every block. Instead we number in major increments of several blocks with minor, unnumbered increments (blocks) in between. Because of our choice of scales, we will label both the x-axis and the y-axis every 5 blocks. The axes do not have to be numbered every 5 divisions, but often the graph paper has darker lines every five blocks and it is convenient to number at these heavier lines. The numbered increments must be on lines.

7. Label each axis so it is clear what each one represents: In our example, we label the x-axis as Volume, mL, and the y-axis as Mass, g. Labels and units on the coordinates are absolutely essential.

8. Plot the data points: Here is how a point is located on the graph: Using the 44.0 mL and 40.0 g data as an example, trace a vertical line up from 44.0 mL on the x-axis and a horizontal line across from 40.0 g on the y-axis and mark the point where the two lines intersect. This process is called plotting. The remaining five points are plotted on the graph in the same way. It is often helpful if each data point is neatly circled so it will be more visible. Then, if more than one set of data is plotted on the same graph, another symbol (an open triangle or square, for example) can be used.

9. Draw a smooth line through the plotted points: In our example, if the six points have been plotted correctly, they lie on a straight line so that a straight edge can be used to draw the smooth line connecting the six points. When plotting data collected in the laboratory, the best smooth line will not necessarily touch each of the plotted points. Thus, some judgment must be exercised in locating the best smooth line, whether it be straight or curved.

10. Title the graph: Every graph should have a title that clearly expresses what the graph represents. Titles may be placed above the graph or on the upper part of the graph. The latter choice, which is illustrated in Figure S3.3, is the most common for student laboratory reports. Of course, the title must be placed so as not to interfere with the plot on the graph. A completed graph of the data in Table S3.1 is shown in Figure S3.3.

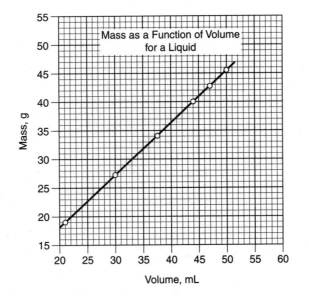

Figure S3.3   Sample graph of volume vs. mass in Table S3.1

Practice Plotting: Sample Data

Table S3.2 is a set of data for you to practice plotting a graph with the steps just described. Use the graph paper in Figure S3.4. Plot °C on the *x*-axis and °F on the *y*-axis.

Table S3.2
Temperature Scales

Temperature, °C	Temperature, °F
0	32
20	68
37	98.6
50	122
100	212

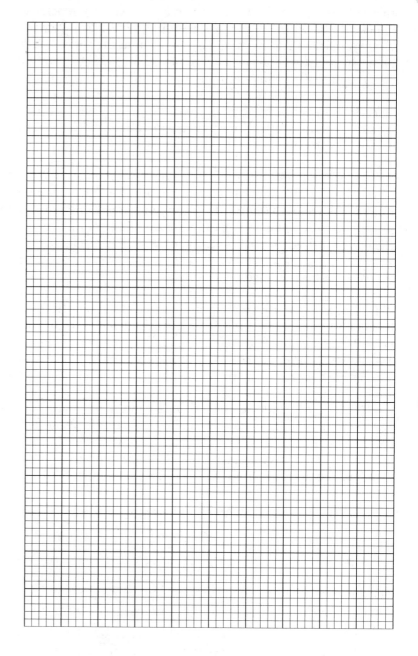

Figure S3.4   Grid for practice plotting of Table S3.2 temperature data

B.   READING A GRAPH

Although graphs are prepared from a limited number of data points (the graph in Figure S3.5 was prepared from six data points), it is possible to extract reliable data for points between the experimental data points and to infer information beyond the range of the plotted data. These skills require that you understand how to read a graph.

Table S3.3

Temperature °C	Solubility, g KClO₃/100g water
10	5.0
20	7.4
30	10.5
50	19.3
60	24.5
80	38.5

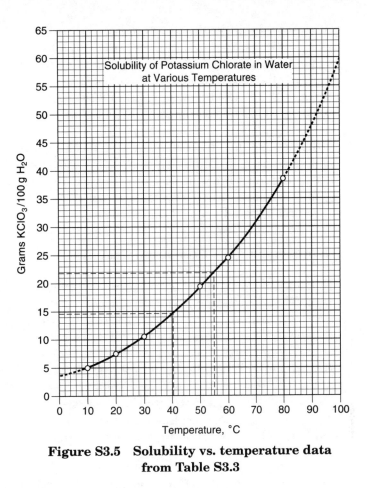

Figure S3.5   Solubility vs. temperature data from Table S3.3

Figure S3.5 is a graph showing the solubility of potassium chlorate in water at various temperatures. The solubility curve on this graph was plotted from experimentally determined solubilities at six temperatures shown in Table S3.3.

These experimentally determined solubilities are all located on the smooth curve traced by the solid line portion of the graph. We are therefore confident that the solid line represents a very good approximation of the solubility data for potassium chlorate covering the temperature range from 10°C to 80°C. All points on the plotted curve represent the composition of saturated solutions. Any point below the curve represents an unsaturated solution.

The dashed line portions of the curve are extrapolations; that is, they extend the curve above and below the temperature range actually covered by the plotted data. Curves such as this are often extrapolated a short distance beyond the range of the known data although the extrapolated portions may not be highly accurate. Extrapolation is justified only in the absence of more reliable information.

The graph in Figure S3.5 can be used with confidence to obtain the solubility of $KClO_3$ at any temperature between 10°C and 80°C but the solubilities between 0°C and 10°C and between 80°C and 100°C are less reliable. For example, what is the solubility of $KClO_3$ at 40°C, at 55°C, and at 100°C? First, draw a vertical line from each temperature to the plotted solubility curve. Now from each of these points on the curve, draw a horizontal line to the solubility axis and read the corresponding solubility. The values that we read from the graph are

40°C	14.7 g KClO₃/100 g water
55°C	21.9 g KClO₃/100 g water
100°C	60.0 g KClO₃/100 g water

Of these solubilities, the one corresponding to 55°C is probably the most reliable because experimental points are plotted at 50°C and 60°C. The 40°C solubility value is probably a bit less reliable because the nearest plotted points are at 30°C and 50°C. The 100°C solubility is the least reliable of the three values because it was taken from the extrapolated part of the curve, and the nearest plotted point is 80°C. Actual handbook solubility values are 14.0 g and 57.0 g of $KClO_3$/100 g water at 40°C and 100°C, respectively.

Although making and reading graphs by hand in this way gets easier with practice, most scientists now use computers to graph their experimental data. If you have access to a computer either in the laboratory, library, or at home, we encourage you to learn computer graphing after you have mastered all the skills of graphing data by hand. Instructions for computer graphing are provided in Part C of this study aid.

Requirements for Computer Graphing

There are several software programs that can be used to generate graphs of scientific data. We have chosen **Microsoft Graph**, a feature of *Microsoft Word 97* (or *Microsoft Word 98 Macintosh Edition*). It requires PC Windows 95 or Macintosh OS 7.5 or later with at least 16 MB or more RAM (32 MB is recommended). You should be sitting in front of a PC computer with Microsoft Word 97 active as you begin these instructions (or a Power Mac computer with Microsoft Word 98 for the Macintosh active). These instructions assume that you have some knowledge of Microsoft Word and that you know how to write graphs manually as instructed in Part A of this study aid. Unless you are comfortable with your computer and have used this software, there should be someone around that can help you out occasionally when you try this the first time.

C. STEPS IN PREPARING A COMPUTER GRAPH

1. If a new document is not already open in Microsoft Word 97, open one now by clicking in the new document button in the upper left of the Word tool bar.

2. Go to **Table** in the menu bar and choose **Insert Table.** A dialogue box will appear. Adjust, using the scroll arrows, to get 2 columns and 7 rows. Then adjust the width of the columns to 1.5 inch with the arrow beside the column width option. Choose **OK** when this is done, and an empty table should appear on the left of the screen. The blocks within the table are called **cells.**

3. Go back to the menu bar and choose **Table** again. Drag down to **Cell Height and Width.** Another dialogue box will appear. Under **Row,** choose **Alignment** and select **center** and **OK.** The table should move to the center of the screen.

4. Type in the data exactly as it appears in the Table S3.1 (copied at the top of the following page for your convenience). The top row of the table will contain labels (sometimes called *categories*) and the rest of the cells will have real numbers (sometimes called *values*). Move up and down or right and left from cell to cell using the arrow keys.

5. OPTIONAL (if the table does not have a border around each cell): Highlight the whole table and click in the outside border button at the end of the tool bar. ▢ This will put a border around the whole table and will also open the menu for border options. Choose ⊞ to put lines around all the individual cells.

Volume, mL	Mass, g
21.0	19.1
30.0	27.3
37.5	34.1
44.0	40.0
47.0	42.8
50.0	45.5

6. Move down the page a few lines by using the return key so the cursor is about an inch below the table you just inserted. Highlight the **entire** table, the cells with the labels for the variables (top row) and the cells with the data.

7. Click the **Insert** button on the menu bar. This will open a menu list of options. Drag the arrow to **Picture,** which will open another list of options. Drag again to **chart** and release. This will open up the Microsoft Graph window. A different data table and a strange-looking graph surrounded by a thick frame will appear like Figure S3.6. (If the recommended default settings have not been made for you the screen may be a little different. Continue anyway.) The Graph 97 menu bar and tool bar will appear across the top.

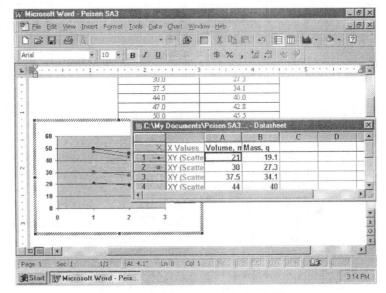

Figure S3.6 Opening screen when Microsoft Graph is opened

8. Choose **Chart** from this tool bar and drag down to <u>Chart type</u> on the menu. Choose XY Scatter and drag over to <u>Subtypes</u> and select the <u>Smoothed line</u> button shown on the right. This selection should bring up a description of this subtype: <u>Scatter with data points connected by smoothed lines.</u> (This step may have already been done for you when the default settings were made.) Click **OK.**

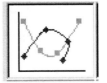

9. Return to the graphing toolbar and select **Data.** You will now tell the computer how the data are presented in the table so it can format the graph for you.

    a. Since the various data for each variable are arranged in columns, select <u>Series in Columns.</u>

b. The independent variable data are in the first column (titled **volume, mL**). Highlight this column and select **Data** again, followed by <u>Plot on x axis</u>.

c. Finally, since the titles in each of these columns are not data points (values), highlight the top row of cells, which contain **volume, mL** and **mass, g**. Select **Data** again, followed by <u>Exclude row/column</u>. From the dialogue box which appears next, choose **Rows.** Click **OK.**

10. Close the datasheet by using the Close box. (If necessary, it can be opened at any time by clicking on the View datasheet button.) ▦ On the screen you will have Document page with the Table that you typed when all this started, and a primitive graph that looks something like Figure S3.7A.

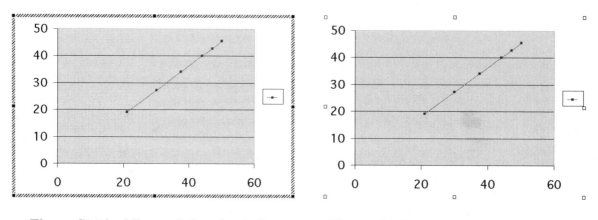

Figure S3.7A   Microsoft Graph window   **Figure S3.7B   Microsoft Word window**

11. Now, you are going to spread out the primitive graph on the page. If at any time from this point forward, the border around the graph should change from S3.7A (Microsoft Graph) to S3.7B (Microsoft Word), you should double-click inside the graph, and Microsoft Graph will be reactivated. Put the cursor on one of the pull points of the graph frame. Hold down on the mouse and stretch the graph so it fills the page, then Scroll the graph up so the table disappears and the screen is filled with the graph.

12. Click in the middle of the plot area. This presents a **Format Plot Area** menu of <u>Pattern</u> options for formatting the <u>Border</u> and the <u>Plot area</u>.

a. **Border:**   <u>Automatic</u>   Does not need to be changed unless you dislike the default setting.

b. **Area:**   <u>Color</u>   Select a block filled with white if you don't have a color printer.

<u>Fill effects</u>   No change for now. Experiment with this at some other time.

Click **OK.**

13. To format the y-axis, put the arrow on the y-axis of the graph and click once. (Double-click with the Macintosh for a shortcut to opening the dialogue boxes in the following sections.) Select **Format** on the tool bar and drag to <u>Format selected axis</u>. This will present a Format Axis dialogue box. Working from left to right across the top of this screen, the following selections are recommended for the data in Table S3.1. When your choices are complete, click **OK** and go to step 15.

a. **Patterns:** Major tick, select Cross

Minor tick, select Inside

Tick mark labels, select Next to axis

b. **Scale:** Minimum = 15. This should be the minimum of the range of data rounded down to a sensible number, often a multiple of 5. For the data in this example, a good minimum is 15 because the lowest data point for the y-axis is 19.1.

Maximum = 50. This should be the maximum of the range of data rounded up to a sensible number, often a multiple of 5. For the data in this example, a good maximum would be 50 because that is the highest data point.

Major unit = 5. This is the unit for numbering ticks on the grid. If the axis is going to go up in multiples of 5, the major unit will be 5 (a good choice for these sample data).

Minor unit = 1. This is the value of each tick mark between the numbered ticks. If you choose to have the minor unit = 1, with the major unit = 5, then the computer will put 4 minor tick marks between every numbered tick. (This is similar to the units/block step on the hand-drawn graph.) For these sample data, the minor unit = 1 is a good choice.

Value (X) crosses at = 15. Usually the x-axis crosses the y-axis at the minimum value of the y-axis. Later you may choose something else, but for these sample data, choose 15.

c. **Font:** Geneva. This is a common font used for graphing and scientific presentations. Arial is another common choice. The font is your choice.

Style = regular. This is a personal choice. You may prefer bold.

Size = 9. For the numbers labeling the major ticks, a small font is best. You can experiment with other fonts in the future.

d. **Number:** General.

e. **Alignment:** Orientation. This orients the numbers either vertically or horizontally as shown. The default choices are usually appropriate.

Click **OK.**

14. Point the cursor arrow on the x-axis and repeat formatting instructions in step 13 for the independent variable as follows: Select **Format** on the tool bar and then choose Selected axis as you did for the y-axis. The following choices are recommended.

a. **Patterns:** Major tick, select Cross

Minor tick, select Inside

Tick mark labels, select Next to axis

b. **Scale:**   Minimum = 20

                    Maximum = 50

                    Major = 5

                    Minor = minor unit could be 1 again, but the x-axis is much longer; it would look better if the minor tick marks were 0.5. You can try it both ways and see what this means. (A graph looks better if the grid is made up of squares rather than rectangles.)

                    Value at which the y-axis will cross this x-axis = 20 (the minimum)

c. **Font:**   Geneva or Arial, 9, Regular (should match the y-axis choices)

d. **Number:**   General

e. **Alignment:**   Automatic

15. To add titles to the graph and axes or add gridlines and labels, select **Chart** from the menu bar and drag down to **Chart options.** This brings up a menu of options as follows:

a. **Titles:**   Chart title: Type in the title you have chosen for this graph. If the title is too long for the space provided, keep typing anyway. The spacing and placement of the title will be adjusted later. For these data, a good title is

                   Mass of an Unknown Liquid as a Function of Volume

                   Value X title: Type in the label and units for the x-axis, **Volume, mL.**

                   Value Y title: Type in the label and units for the y-axis, **Mass, g.**

b. **Axes:**   Primary axes are Value (X) and Value (Y) and both should have ✓ beside them.

c. **Gridlines:**   This step allows you to put gridlines on your graph if you want to make it look like it is drawn on graph paper. This step is a matter of personal taste. Gridlines extend tick marks into lines across and up/down the plot area. To insert gridlines, put a ✓ beside the ticks that you want to extend. For these data, we suggest that you ✓ Major and Minor ticks on both the Value (X) and Value (Y) axes.

d. **Legend:**   This is a table to show which symbol and line corresponds to each set of data points. Since there is only one kind of symbol and one line in this graph, the legend is not needed. Click on the ✓ to remove the legend.

e. **Data labels:** Usually the choice here is None. If you choose Show Value, then the actual data value for the y-axis will be printed beside the symbol. The computer will take the value from the data table automatically. You don't have to do anything once you make this choice.

Click **OK** when finished.

16. To format the line and data points, put the arrow on the line or any data point and double-click. A dialogue box will appear for **Format Data Series.** Make choices as follows:

a. **Patterns:** Line. Choose Custom followed by the style, color, and weight that you prefer. The appearance of the line you design will appear under sample. For this graph, choose a solid line in a dark color.

b. **Marker:** This will format the symbol for each data series. Choose Custom then click and drag on the arrow beside Marker and choose the shape of the symbol that you like best. If you choose the same color for the Foreground and Background the symbol will be filled/solid. If you choose a dark color for the foreground with light color (white) for the background the result is an open symbol.

c. **Y Error Bars:** Does not apply to this example.

d. **X Error Bars:** Does not apply to this example.

e. **Data Labels:** Usually the choice here is NONE. If you choose SHOW VALUE then the actual data value for the y-axis will be printed beside the symbol. The computer will take the value from the data table automatically.

f. **Options:** Vary colors by point does not apply unless you have a color printer.

Click **OK** when finished.

17. Now look at the title and axis labels. If they need to be moved, click on them. This will put a border around the words (shown on the right) that can then be dragged to a new location. To adjust the font (while the border "selects" the title being modified), select **Format** from the menu bar and drag to format Selected axis title. This will bring up a Format dialogue box that you can use to change pattern, font, style, size, and alignment.

The chart title can be changed and moved in the same way.

18. Now you are done. Click twice on the page outside the graph. This will deactivate Microsoft Graph and put you back in Microsoft Word. Put the cursor below the graph and type your name and the name of anyone else who worked on this graph with you.

19. To make sure the size of the graph and placement on the page is appropriate, select **View** from the menu bar and drag down to **Page Layout.** Adjust the view to 50% using the tool bar and click **OK.** This will bring up a screen that shows more of the entire page so you can see if the graph fills the space available and is not too large or too small. To adjust the size of the graph, click once in the middle of the graph to get the Microsoft Word frame with pull points (shown in Figure S3.7B). Put the cursor arrow in the pull points and adjust the size of the frame in which the graph is located so it fits within the margins and the page is balanced. When you are satisfied with the appearance of the page, choose **Print Preview** from the **File** menu and make sure you are satisfied. Sometimes it is necessary to position the paper horizontally by choosing **Page Setup** from the **File** menu. After all the adjustments are final, choose **Print** from the **File** menu and print copies for each person on the team helping with the graph.

An example of the graph that results from this computer graphing exercise is shown on the following page.

volume, mL	mass, g
21.0	19.1
30.0	27.3
37.5	34.1
44.0	40.0
47.0	42.8
50.0	45.5

Mass of an Unknown Liquid as a Function of Volume

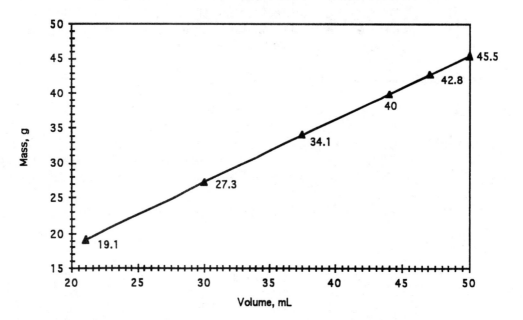

Figure S3.8   Sample computer graph (with data labels and without gridlines)

Study Aid 4

Using a Scientific Calculator

A calculator is useful for most calculations in this book. You should obtain a scientific calculator; that is, one that has at least the following function keys on its keyboard.

Addition $\boxed{+}$		Second function $\boxed{F}$ or $\boxed{INV}$ or $\boxed{Shift}$	
Subtraction $\boxed{-}$		Change sign $\boxed{+/-}$	
Multiplication $\boxed{\times}$		Exponential number $\boxed{EXP}$	
Division $\boxed{\div}$		Logarithm $\boxed{LOG}$	
Equals $\boxed{=}$		Antilogarithm $\boxed{10^x}$	

All calculators do not use the same symbolism for these function keys. Not all calculators work the same way. Save the instruction manual that comes with your calculator. It is very useful for determining how to do special operations on your particular model. Refer to your instruction manual for variations from the function symbols shown above and for the use of other function keys.

Some keys have two functions, upper and lower. In order to use the upper (second) function, the second function key $\boxed{F}$ must be pressed in order to activate the desired upper function.

The display area of the calculator shows the numbers entered and often shows more digits in the answer than should be used. Therefore, the final answer should be rounded to reflect the proper number of significant figures of the calculations.

Addition and Subtraction

To add numbers:

1. Enter the first number to be added followed by the plus key $\boxed{+}$.

2. Enter the second number to be added followed by the plus key $\boxed{+}$.

3. Repeat Step 2 for each additional number to be added, except the last number.

4. After the last number is entered, press the equal key $\boxed{=}$. You should now have the answer in the display area.

5. When a number is to be subtracted, use the minus key $\boxed{-}$ instead of the plus key.

As an example, to calculate 16.0 + 1.223 + 8.45, enter 16.0 followed by the $\boxed{+}$ key; then enter 1.223 followed by the $\boxed{+}$ key; then enter 8.45 followed by the $\boxed{=}$ key. The display shows 25.673, which is rounded to the answer 25.7.

Examples of Addition and Subtraction

Calculation	Enter in Sequence	Display	Rounded Answer
a. 12.0 + 16.2 + 122.3	12.0 [+] 16.2 [+] 122.3 [=]	150.5	150.5
b. 132 − 62 + 141	132 [−] 62 [+] 141 [=]	211	211
c. 46.23 + 13.2	46.23 [+] 13.2 [=]	59.43	59.4
d. 129.06 + 49.1 − 18.3	129.06 [+] 49.1 [−] 18.3 [=]	159.86	159.9

Multiplication

To multiply numbers using your calculator

1. Enter the first number to be multiplied followed by the multiplication key [×].

2. Enter the second number to be multiplied followed by the multiplication key [×].

3. Repeat Step 2 for all other numbers to be multiplied except the last number.

4. Enter the last number to be multiplied followed by the equal key [=]. You now have the answer in the display area. Round off to the proper number of significant figures. As an example, to calculate (3.25)(4.184)(22.2) enter 3.25 followed by the [×] key; then enter 4.184 followed by the [×] key; then enter 22.2 followed by the [=] key. The display shows 301.8756, which is rounded to the answer 302.

Examples of Multiplication

Calculation	Enter in Sequence	Display	Rounded Answer
a. (12)(14)(18)	12 [×] 14 [×] 18 [=]	3024	3.0×10^3
b. (122)(3.4)(60.)	122 [×] 3.4 [×] 60. [=]	24888	2.5×10^4
c. (0.522)(49.4)(6.33)	0.522 [×] 49.4 [×] 6.33 [=]	163.23044	163

Division

To divide numbers using your calculator:

1. Enter the numerator followed by the division key [÷].

2. Enter the denominator followed by the equal key $\boxed{=}$ to give the answer in the display area. Round off to the proper number of significant figures.

3. If there is more than one denominator, enter each denominator followed by the division key except for the last number, which is followed by the equal key. As an example, to calculate $\dfrac{126}{12}$, enter 126 followed by the $\boxed{\div}$ key; then enter 12 followed by the $\boxed{=}$ key. The display shows 10.5, which is rounded to the answer 11.

Examples of Division

Calculation	Enter in Sequence	Display	Rounded Answer
a. $\dfrac{142}{25}$	142 $\boxed{\div}$ 25 $\boxed{=}$	5.68	5.7
b. $\dfrac{0.422}{5.00}$	0.422 $\boxed{\div}$ 5.00 $\boxed{=}$	0.0844	0.0844
c. $\dfrac{124}{(0.022)(3.00)}$	124 $\boxed{\div}$ 0.022 $\boxed{\div}$ 3.00 $\boxed{=}$	1878.7878	1.9×10^3

Exponents

In scientific measurements and calculations we often encounter very large and very small numbers. A convenient method of expressing these large and small numbers is by using exponents or powers of 10. A number in exponential form is treated like any other number, that is, it can be added, subtracted, multiplied, or divided.

To enter an exponential number into your calculator first enter the non-exponential part of the number, then press the exponent key $\boxed{\text{EXP}}$, followed by the exponent. For example, to enter 4.94×10^3, enter 4.94, then press $\boxed{\text{EXP}}$, then press 3. When the exponent of 10 is a minus number, press the Change of Sign key $\boxed{+/-}$ after entering the exponent. For example, to enter 4.94×10^{-3}, enter in sequence 4.94 $\boxed{\text{EXP}}$ 3 $\boxed{+/-}$. In most calculators the exponent will appear in the display a couple of spaces after the non-exponent part of the number; for example, 4.94   03 or 4.94   −03.

Examples Using Exponential Numbers

Calculation	Enter in Sequence	Display	Rounded Answer
a. $(4.94 \times 10^3)(21.4)$	4.94 $\boxed{\text{EXP}}$ 3 $\boxed{\times}$ 21.4 $\boxed{=}$	105716	1.06×10^5
b. $(1.42 \times 10^4)(2.88 \times 10^{-5})$	1.42 $\boxed{\text{EXP}}$ 4 $\boxed{\times}$ 2.88 $\boxed{\text{EXP}}$ 5 $\boxed{+/-}$ $\boxed{=}$	0.40896	0.409
c. $\dfrac{8.22 \times 10^{-5}}{5.00 \times 10^7}$	8.22 $\boxed{\text{EXP}}$ 5 $\boxed{+/-}$ $\boxed{\div}$ 5.00 $\boxed{\text{EXP}}$ 7 $\boxed{=}$	1.644 −12	1.64×10^{-12}

Logarithms

The logarithm of a number to the base 10 is the power (exponent) to which 10 must be raised to give that number. For example, the log of 100 is 2.0 (log $100 = 10^{2.0}$). The log of 200 is 2.3 (log $200 = 10^{2.3}$). Logarithms are used in chemistry to calculate the pH of an aqueous solution. The answer (log) should contain the same number of significant figures to the right of the decimal as there are significant figures in the original number. Thus the log $100 = 2.0$ but the log 100. is 2.000.

The log key on many calculators is a function key. To determine the log using your calculator, enter the number, then press the log key [**LOG**]. For example to determine the log of 125, enter 125, then press the log key [**LOG**]. The display shows 2.09691, which is rounded to the answer 2.097.

Examples. Determine the log of the following:

Calculation	Enter in Sequence	Display	Rounded Answer
a. log 42	42 [LOG]	1.6232492	1.62
b. log 1.62×10^5	1.62 [EXP] 5 [LOG]	5.209515	5.210
c. log 6.4×10^{-6}	6.4 [EXP] 6 [+/−] [LOG]	−5.19382	−5.19
d. log 2.5	2.5 [LOG]	0.39794	0.40

Antilogarithms (Inverse Logarithms)

An antilogarithm is the number from which the logarithm has been calculated. It is calculated using the [10^x] key on your calculator. Many calculators use the [**F**] or [**INV**] or [**Shift**] key to access this function. To use a function that appears above a key you must first press the [**F**] key. For example, to determine the antilogarithm of 2.891, enter 2.891 into your calculator, then press the [**F**] key followed by the [10^x] key. The display shows 778.03655, which is rounded to the answer 778. The answer should contain the same number of figures as there are to the right of the decimal in the antilog.

Examples. Determine the antilogarithm of the following:

Calculation	Enter in Sequence	Display	Rounded Answer
a. antilog 1.628	1.628 [F] [10^x]	42.461956	42.5
b. antilog 7.086	7.086 [F] [10^x]	12189896	1.22×10^7
c. antilog −6.33	6.33 [+/−] [F] [10^x]	4.6773514 −07	4.7×10^{-7}

Additional Practice Problems

Only the problem, the display, and the answers are given.

	Problem	*Display*	*Answer*
1.	$143.5 + 14.02 + 1.202$	158.722	158.7
2.	$72.06 - 26.92 - 49.66$	−4.52	−4.52
3.	$2.168 + 4.288 - 1.62$	4.836	4.84
4.	$(12.3)(22.8)(1.235)$	346.3434	346
5.	$(2.42 \times 10^6)(6.08 \times 10^{-4})(0.623)$	916.65728	917
6.	$\dfrac{(46.0)(82.3)}{19.2}$	197.17708	197
7.	$\dfrac{0.0298}{243}$	1.2263374 −04	1.23×10^{-4}
8.	$\dfrac{(5.4)(298)(760)}{(273)(1042)}$	4.2992554	4.3
9.	$(6.22 \times 10^6)(1.45 \times 10^3)(9.00)$	8.1171 10	8.12×10^{10}
10.	$\dfrac{(1.49 \times 10^6)(1.88 \times 10^6)}{6.02 \times 10^{23}}$	4.6531561 −12	4.65×10^{-12}
11.	$\log 245$	2.389166	2.389
12.	$\log 6.5 \times 10^{-6}$	−5.1870866	−5.19
13.	$(\log 24)(\log 34)$	2.1137644	2.11
14.	antilog 6.34	2187761.6	2.2×10^6
15.	antilog −6.34	4.5708819 −07	4.6×10^{-7}

STUDY AID 5

Use of the Mole in Chemical Calculations: Stoichiometry

It is often necessary to calculate the amount of product that can be obtained from a given amount of reactant or, conversely, to determine how much reactant is required to produce a stated amount of product. Calculations of this kind, based upon balanced chemical equations, are called **stoichiometry** (from Greek, meaning element measure).

In solving stoichiometric problems, the concept of the mole is very useful. In its broadest sense a mole is Avogadro's Number (6.022×10^{23}) of any chemical species. Even though the word "mole" is used as a short expression for "molar mass," it is quite permissible to refer to moles of chemical species that are not really molecular in character. Reference may be made to moles of such diverse species as sulfur **atoms** (S), oxygen **atoms** (O), oxygen **molecules** (O_2), sulfuric acid **molecules** (H_2SO_4), sodium chloride **formula units** (NaCl), ammonium **ions** (NH_4^+), nitrate **ions** (NO_3^-)—or even to moles of electrons or protons.

Since a mole, by definition, consists of Avogadro's Number of particles, it represents a mass in grams numerically equal to the molar mass of the particular species under consideration.

Consideration of Avogadro's famous hypothesis that "equal volumes of all gases, at the same temperature and pressure, contain the same number of molecules" leads to the conclusion that one mole (6.022×10^{23} molecules) of any gas will have the same volume as one mole of any other gas at the same temperature and pressure. The volume of one mole of any gas is known to be 22.4 liters at $0°C$ ($273°K$) and one atmosphere (760 torr) pressure; i.e., at standard temperature and pressure (STP).

These three facets of the mole concept are summarized below:

1. One mole is 6.022×10^{23} particles of the species under consideration.

2. One mole is a quantity in grams numerically equal to the molar mass of the substance under consideration.

3. If the substance is a gas, one mole is 22.4 liters of that gas at STP.

To use the mole concept to advantage in dealing with a particular problem, these four steps are generally needed:

Step 1. Write the balanced chemical equation for the reaction involved or check the equation given to make sure that it is balanced.

Step 2. Examine the problem statement and determine what is the **given substance** upon which to base the calculations. Then set up the calculation needed to convert the specified quantity of the substance to **moles of given substance.**

Step 3. Obtain the chemically equivalent **moles of desired substance** by multiplying the moles of given substance (from Step 2) by the **mole ratio** obtained from the balanced equation:

$$\text{Mole ratio} = \frac{\text{Moles of desired substance (in the equation)}}{\text{Moles of given substance (in the equation)}}$$

The number of moles in this mole ratio are the numbers (coefficients) in front of the respective substances in the balanced equation.

Step 4. Convert the moles of desired substance (obtained in Step 3) to whatever units are required in the problem by multiplying by the appropriate factor—i.e., molecules/mole, grams/mole, or liters/mole.

Consider the following solutions of problems related to the heating of potassium chlorate to produce potassium chloride and oxygen.

Problem 1. How Many Grams of Oxygen Can Be Obtained from 15.0 g of KClO₃?

SOLUTION

Step 1. Write the balanced equation for the reaction.

$$2\ KClO_3 \longrightarrow 2\ KCl + 3\ O_2$$

Molar masses: $KClO_3$ = 122.6 g/mol KCl = 74.55 g/mol O_2 = 32.00 g/mol

Step 2. $KClO_3$ is the given substance; convert the given quantity of $KClO_3$ to moles of $KClO_3$.

$$15.0\ \text{g}\ KClO_3 \times \frac{1\ \text{mol}\ KClO_3}{122.6\ \text{g}\ KClO_3} = \text{mol}\ KClO_3$$

Step 3. Convert moles of $KClO_3$ to the chemically equivalent moles of O_2 by multiplying by the mole ratio taken from the equation.

$$15.0\ \text{g}\ KClO_3 \times \frac{1\ \text{mol}\ KClO_3}{122.6\ \text{g}\ KClO_3} \times \frac{3\ \text{mol}\ O_2}{2\ \text{mol}\ KClO_3} = \text{mol}\ O_2$$

Step 4. Convert the moles of O_2 to grams O_2.

$$15.0\ \text{g}\ KClO_3 \times \frac{1\ \text{mol}\ KClO_3}{122.6\ \text{g}\ KClO_3} \times \frac{3\ \text{mol}\ O_2}{2\ \text{mol}\ KClO_3} \times \frac{32.00\ \text{g}\ O_2}{\text{mol}\ O_2} = 5.87\ \text{g}\ O_2 \quad \text{(Answer)}$$

The answer was obtained by completing the indicated arithmetic, $\dfrac{15.0 \times 3 \times 32.00}{122.6 \times 2}$.

Problem 2. How Many Oxygen Molecules Can Be Obtained from 15.0 g of KClO$_3$?

SOLUTION

The first three steps are exactly the same as in Problem 1, but the fourth step converts moles of O$_2$ to molecules of O$_2$.

$$15.0 \text{ g KClO}_3 \times \frac{1 \text{ mol KClO}_3}{122.6 \text{ g KClO}_3} \times \frac{3 \text{ mol O}_2}{2 \text{ mol KClO}_3} \times \frac{6.022 \times 10^{23} \text{ molecules O}_2}{\text{mol O}_2}$$

$$= 1.11 \times 10^{23} \text{ molecules O}_2 \quad \text{(Answer)}$$

Problem 3. How Many Liters of Oxygen Gas, Measured at STP, Can Be Obtained from 15.0 g of KClO$_3$?

SOLUTION

Here again the first three steps are identical to those of Problems 1 and 2. The final step converts moles of O$_2$ to liters of O$_2$.

$$15.0 \text{ g KClO}_3 \times \frac{1 \text{ mol KClO}_3}{122.6 \text{ g KClO}_3} \times \frac{3 \text{ mol O}_2}{2 \text{ mol KClO}_3} \times \frac{22.4 \text{ L O}_2}{\text{mol O}_2} = 4.11 \text{ L O}_2 \quad \text{(Answer)}$$

Problems 1, 2, and 3 have dealt with questions relating to the determination of the amount of product obtainable from a given quantity of reactant. The following problems illustrate how the same general method may be used to predict the amount of reactant required to obtain a specified amount of product.

Problem 4. How Many Grams of KClO$_3$ Must Be Decomposed to Produce 25.0 g of KCl?

SOLUTION

Step 1. Balance the equation for the reaction as shown in Problem 1.

Step 2. Consider KCl to be the given substance; convert to moles of KCl.

$$25.0 \text{ g KCl} \times \frac{1 \text{ mol KCl}}{74.55 \text{ g KCl}} = \text{ mol KCl}$$

Step 3. Convert moles of KCl to chemically equivalent moles of KClO$_3$ by use of the mole ratio taken from the balanced equation.

$$25.0 \text{ g KCl} \times \frac{1 \text{ mol KCl}}{74.55 \text{ g KCl}} \times \frac{2 \text{ mol KClO}_3}{2 \text{ mol KCl}} = \text{ mol KClO}_3$$

Step 4. Convert moles of KClO$_3$ to grams of KClO$_3$.

$$25.0 \text{ g KCl} \times \frac{1 \text{ mol KCl}}{74.55 \text{ g KCl}} \times \frac{2 \text{ mol KClO}_3}{2 \text{ mol KCl}} \times \frac{122.6 \text{ g KClO}_3}{\text{mol KClO}_3} = 41.1 \text{ g KClO}_3 \quad \text{(Answer)}$$

Problem 5. What Mass of KClO$_3$ Must Be Decomposed to Obtain 7.50 Liters of O$_2$ Measured at STP?

SOLUTION

Oxygen is the given substance. Liters of O$_2$ are converted to grams of KClO$_3$ in the following manner:

$$\text{Liters O}_2 \longrightarrow \text{moles O}_2 \longrightarrow \text{moles KClO}_3 \longrightarrow \text{grams KClO}_3$$

$$7.50 \text{ L O}_2 \times \frac{1 \text{ mol O}_2}{22.4 \text{ L O}_2} \times \frac{2 \text{ mol KClO}_3}{3 \text{ mol O}_2} \times \frac{122.6 \text{ g KClO}_3}{\text{mol KClO}_3} = 27.4 \text{ g KClO}_3 \quad \text{(Answer)}$$

STUDY AID 6

Organic Chemistry—An Introduction

Organic chemistry is known as the chemistry of the carbon compounds. All compounds that are classified as organic contain carbon. Organic compounds are found in all living matter, food stuffs (fats, proteins, and carbohydrates), fuels of all kinds, plastics, fabrics, wood and paper products, paints and varnishes, dyes, soaps and detergents, cosmetics, medicinals, insecticides, refrigerants, etc. There are over eleven million known organic compounds.

To help study their properties, organic compounds are grouped into classes or series according to the similarity of their chemical makeup or structure. Some of the common classes are hydrocarbons, alcohols, aldehydes, ketones, acids, esters, ethers, amines, and amides.

The major reason for the large number of organic compounds is that carbon atoms have the ability to bond together, forming long chains and rings. Carbon atoms share electrons, forming covalent bonds. Between two carbon atoms, single, double, or triple covalent bonds may be formed by sharing one, two, or three pairs of electrons, respectively. These types of bonds are illustrated below:

C : C	C :: C	C ::: C
C—C	C=C	C≡C
Single bond	Double bond	Triple bond

A dash between carbon atoms indicates a covalent bond and represents one pair of electrons.

The names of the alkane series hydrocarbons are important to organic chemistry because they represent the basis for the systematic nomenclature of organic compounds. The first 10 members of this series and their molecular formulas are listed below:

CH_4	Methane	C_4H_{10}	Butane	C_6H_{14}	Hexane	C_8H_{18}	Octane
C_2H_6	Ethane	C_5H_{12}	Pentane	C_7H_{16}	Heptane	C_9H_{20}	Nonane
C_3H_8	Propane					$C_{10}H_{22}$	Decane

Structural Formulas

A great many organic compounds are composed of carbon, hydrogen, and oxygen atoms. In these compounds we find these atoms bonded to each other in the following ways:

Carbon to carbon	(C—C), (C=C), (C≡C)
Carbon to carbon to carbon	(C—C—C)
Carbon to hydrogen	(C—H)
Carbon to oxygen	(C=O)
Carbon to oxygen to hydrogen	(C—O—H)
Carbon to oxygen to carbon	(C—O—C)

In chemical compounds, with some rare exceptions, a carbon atom will always have four bonds; a hydrogen atom, one bond; and an oxygen atom, two bonds. A bond consists of a pair of electrons shared between any two atoms.

Because of the different arrangements in which carbon atoms bond with each other, a single written molecular formula may present more than one arrangement of the atoms, giving rise to more than one compound. For example, there are 2 different butanes (C_4H_{10}), 3 pentanes (C_5H_{12}), and 75 decanes ($C_{10}H_{22}$). To illustrate different compounds with the same molecular formula, we use structural formulas. Structural formulas show the order in which the atoms are bonded to each other, while molecular formulas show only the number and kind of each atom in a molecule. A few examples will illustrate.

Methane, CH_4

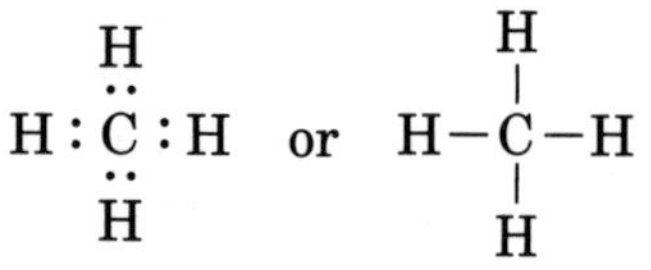

Structural formula

CH_4

Condensed structural formula

Ethane, C_2H_6

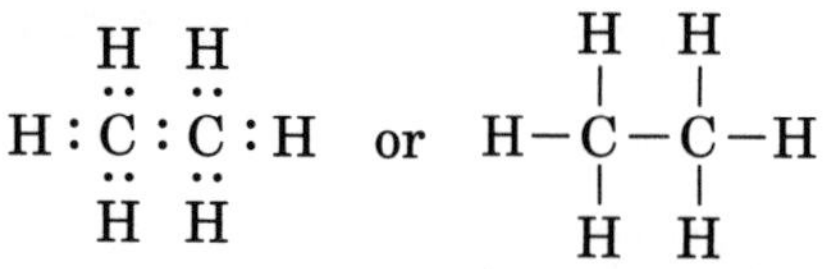

Structural formula

CH_3-CH_3 or CH_3CH_3

Condensed structural formula

Note that in the condensed structural formula, which is a convenient simplification of the structural formula, all of the atoms or groups attached to each carbon atom are written to the right of it.

Propane, C_3H_8

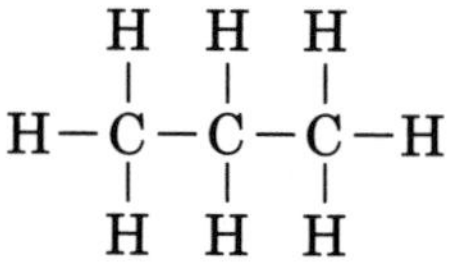

Structural formula

$CH_3CH_2CH_3$

Condensed structural formula

Methyl alcohol, CH_3OH

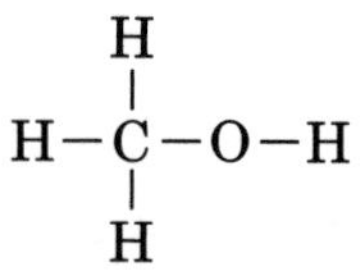

Structural formula

CH_3OH

Condensed structural formula

Butane, C_4H_{10}

(a) n-Butane (n = normal)

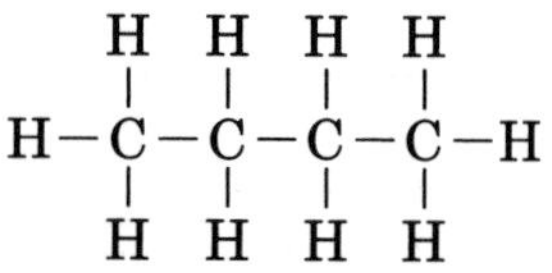

Structural formula

$CH_3CH_2CH_2CH_3$

Condensed structural formula

(b) Isobutane (2-methylpropane)

$$\begin{array}{c}
\text{H} \\
| \\
\text{H}-\text{C}-\text{H} \\
\quad | \\
\text{H} \quad | \quad \text{H} \\
| \quad | \quad | \\
\text{H}-\text{C}-\text{C}-\text{C}-\text{H} \\
| \quad | \quad | \\
\text{H} \ \ \text{H} \ \ \text{H}
\end{array}$$

Structural formula

$$\begin{array}{c}
\text{CH}_3 \\
| \\
\text{CH}_3\text{CHCH}_3 \quad \text{or}
\end{array}$$

$CH_3CH(CH_3)CH_3$ or

$CH_3CH(CH_3)_2$

Condensed structural formula

Isomerism

We have shown that there are two possible ways to bond 4 carbon atoms and 10 hydrogen atoms to form the structures of the two butanes. These butanes are indeed different compounds, each with its own physical and chemical properties. For example, n-butane boils at $-0.5°C$ and isobutane boils at $-11.7°C$.

The phenomenon of two or more compounds having the same molecular formula is known as **isomerism.** The individual compounds are called **isomers.** Thus there are 2 isomers of butane, 3 of pentane, 18 of octane, and 75 of decane.

Alkyl Groups

The nomenclature of organic chemistry is sprinkled with such terms as methyl, ethyl, and isopropyl. These terms represent **alkyl groups** derived from alkane hydrocarbons. An alkyl group is formed by removing one hydrogen atom from an alkane. For example, the methyl group, CH_3-, is formed by removing one hydrogen from methane, CH_4. The name methyl is formed by dropping the *ane* from the name *methane* and adding the letters *yl* to the remaining stem *meth*. Other alkyl groups are derived in a similar manner—ethyl from ethane, etc. A few of the more common alkyl groups are listed below. The dash in the formula indicates the carbon atom from which the hydrogen atom has been removed.

CH_3-	methyl
CH_3CH_2-	ethyl
$CH_3CH_2CH_2-$	n-propyl (n = normal)
CH_3CHCH_3- (with $\vert$ below)	isopropyl
$CH_3CH_2CH_2CH_2-$	n-butyl

Note that two different propyl groups are formed, depending on whether the hydrogen atom removed is from an end carbon or from a middle carbon atom. In a like manner, four different butyl groups are formed (only one of which is shown). Alkyl groups are often designated by the letter R—. Thus RH represents an alkane hydrocarbon.

Functional Groups

The formulas for many classes of organic compounds may be derived from the formulas of the alkane hydrocarbons by substituting a different group for one more of the hydrogen atoms in the hydrocarbon chain. The groups are known as **functional groups** and characterize the classes of compounds that they represent.

The functional group of the alcohols is –OH. Two examples are methyl alcohol (CH_3OH) and ethyl alcohol (CH_3CH_2OH or C_2H_5OH). In a like manner, the formulas of an entire series of alcohols may be written by substituting an –OH group for a hydrogen atom on the alkane chain or by combining the –OH group with the alkyl groups given above. Thus n-propyl alcohol and isopropyl alcohol are $CH_3CH_2CH_2OH$ and CH_3CHCH_2, respectively.
$$\qquad\qquad\qquad\qquad\qquad\qquad\qquad\qquad\quad |$$
$$\qquad\qquad\qquad\qquad\qquad\qquad\qquad\qquad\ OH$$

Other common functional groups, together with their classes of compounds, are given below.

Functional Group	Class of Compound	Examples*	
–OH	Alcohol	CH_3OH	Methanol (Methyl alcohol)
		CH_3CH_2OH	Ethanol (Ethyl alcohol)
$-\overset{\overset{O}{\|\|}}{C}-OH$ or $-COOH$	Acid	HCOOH	Methanoic acid (Formic acid)
		CH_3COOH	Ethanoic acid (Acetic acid)
$-\overset{\overset{H}{\|}}{C}=O$	Aldehyde	$H-\overset{\overset{H}{\|}}{C}=O$	Methanal (Formaldehyde)
		$CH_3\overset{\overset{H}{\|}}{C}=O$	Ethanal (Acetaldhyde)
$R-\underset{\underset{O}{\|\|}}{C}-R$	Ketone	$CH_3\underset{\underset{O}{\|\|}}{C}CH_3$	Propanone (Acetone)
$-\overset{\overset{O}{\|\|}}{C}-OR$	Ester	$H-\overset{\overset{O}{\|\|}}{C}-OCH_3$	Methyl methanoate (Methyl formate)
		$CH_3-\overset{\overset{O}{\|\|}}{C}-OCH_2CH_3$	Ethyl ethanoate (Ethyl acetate)

*IUPAC name
Common name in parentheses

EXERCISE 1

Significant Figures and Exponential Notation

1. How many significant figures are in each of the following numbers?

   (a) 7.42 _____ (b) 4.6_____ (c) 3.40 _____ (d) 26,000 _____

   (e) 0.088 _____ (f) 0.0034_____ (g) 0.0230 _____ (h) 0.3080 _____

2. Write each of the following numbers in proper exponential notation:

   (a) 423   (a) _____

   (b) 0.032   (b) _____

   (c) 8,300   (c) _____

   (d) 302.0   (d) _____

   (e) 12,400,000   (e) _____

   (f) 0.0007   (f) _____

3. How many significant figures should be in the answer to each of the following calculations?

   (a) 17.10   (b) 57.826   (a) _____
       + 0.77   − 9.4
                                                              (b) _____

   (c) 12.4 × 2.82 =   (d) 6.4 × 3.1416 =   (c) _____

                                                              (d) _____

   (e) $\dfrac{0.5172}{0.2742} =$ (f) $\dfrac{0.0172}{4.36} =$ (e) _____

                                                              (f)_____

   (g) $\dfrac{5.82 \times 760. \times 425}{723 \times 273} =$ (h) $\dfrac{0.92 \times 454 \times 5.620}{22.4} =$ (g) _____

                                                              (h) _____

4. For each of these problems, complete the answer with a 10 raised to the proper power. Note that each answer is expressed to the correct number of significant figures.

(a) $2.71 \times 10^4 \times 2.0 \times 10^2 = 5.4 \times$ _____   (a) _____

(b) $\dfrac{4.523 \times 10^4}{2.71 \times 10^2} = 1.67 \times$ _____   (b) _____

(c) $4.8 \times 10^4 \times 3.5 \times 10^4 = 1.7 \times$ _____   (c) _____

(d) $\dfrac{1.64 \times 10^{-4}}{1.2 \times 10^2} = 1.4 \times$ _____   (d) _____

(e) $\dfrac{4.70 \times 10^2}{8.42 \times 10^5} = 5.58 \times$ _____   (e) _____

5. Solve each of the following problems, expressing each answer to the proper number of significant figures. Use exponential notation for (c), (d), and (e).

(a)   1.842
      45.21
   + 37.55
   _____

(b)   714.3
   − 28.52
   _____

(a) _____

(b) _____

(c) $2.83 \times 10^3 \times 7.55 \times 10^7 =$ (c) _____

(d) $4.4 \times 5{,}280 =$ (d) _____

(e) $\dfrac{7.07 \times 10^{-4} \times 6.51 \times 10^{-2}}{2.92 \times 10^4} =$ (e) _____

Answers

1. (a) 3, (b) 2, (c) 3, (d) 2, (e) 2, (f) 2, (g) 3, (h) 4.

2. (a) 4.23×10^2, (b) 3.2×10^{-2}, (c) 8.3×10^3, (d) 3.020×10^2, (e) 1.24×10^7, (f) 7×10^{-4}.

3. (a) 4, (b) 3, (c) 3, (d) 2, (e) 4, (f) 3, (g) 3, (h) 2.

4. (a) 10^6, (b) 10^2, (c) 10^9, (d) 10^{-6}, (e) 10^{-4}.

5. (a) 84.60, (b) 685.8, (c) 2.14×10^{11}, (d) 2.3×10^4, (e) 1.58×10^{-9}.

NAME ____________

SECTION ______ DATE ____________

EXERCISE 2

INSTRUCTOR ____________

Measurements

For each of the following problems, show your calculation setup. In both your setup and answer, show units and follow the rules of significant figures. See Experiment 2 and the appendixes for any needed formulas or conversion factors.

1. Convert 78°F to degrees Celsius.

   ____

2. Convert –13°C to degrees Fahrenheit.

   ____

3. An object weighs 8.22 lbs. What is the mass in grams?

   ____

4. A stick is 12.0 cm long. What is the length in inches?

   ____

5. The water in a flask measures 423 mL. How many quarts is this?

   ____

6. A piece of lumber measures 98.4 cm long. What is its length in:

   (a) Millimeters?

   ____

   (b) Feet?

   ____

7. A block is found to have a volume of 35.3 cm^3. Its mass is 31.7 g. Calculate the density of the block.

_____

8. A graduated cylinder was filled to 25.0 mL with liquid. A solid object weighing 73.5 g was immersed in the liquid, raising the liquid level to 43.9 mL. Calculate the density of the solid object.

_____

9. The density of the liquid in Problem 8 is 0.874 g/mL. What is the mass of the liquid in the graduated cylinder?

_____

10. How many joules of heat are absorbed by 500.0 g of water when its temperature increases from 20.0°C to 80.0°C? (sp. ht. water = 1.00 cal/g°C)

_____

11. A beaker contains 421 mL of water. The density of the water is 1.00 g/mL. Calculate:

    (a) The volume of the water in liters.

_____

    (b) The mass of the water in grams.

_____

12. The density of carbon tetrachloride, CCl_4, is 1.59 g/mL. Calculate the volume of 100.0 g of CCl_4.

_____

NAME ____________________

SECTION ______ DATE ____________

EXERCISE 3

INSTRUCTOR ____________________

Names and Formulas I

Give the names of the following compounds:

1. $NaCl$ ____________________
2. $AgNO_3$ ____________________
3. $BaCrO_4$ ____________________
4. $Ca(OH)_2$ ____________________
5. $ZnCO_3$ ____________________
6. Na_2SO_4 ____________________
7. Al_2O_3 ____________________
8. $CdBr_2$ ____________________
9. KNO_2 ____________________
10. $Fe(NO_3)_3$ (a) ____________________

    (b) ____________________
11. $(NH_4)_3PO_4$ ____________________
12. $KClO_3$ ____________________
13. MgS ____________________
14. $Cu_2C_2O_4$ ____________________

Give the formulas of the following compounds:

1. Barium chloride   1. _____

2. Zinc fluoride   2. _____

3. Lead(II) iodide   3. _____

4. Ammonium hydroxide   4. _____

5. Potassium chromate   5. _____

6. Bismuth(III) chloride   6. _____

7. Magnesium perchlorate   7. _____

8. Copper(II) sulfate   8. _____

9. Iron(III) chloride   9. _____

10. Calcium cyanide   10. _____

11. Copper(I) sulfide   11. _____

12. Silver carbonate   12. _____

13. Cadmium hypochlorite   13. _____

14. Sodium bicarbonate   14. _____

15. Aluminum acetate   15. _____

16. Nickel(II) phosphate   16. _____

17. Sodium sulfite   17. _____

18. Tin(IV) oxide   18. _____

NAME ____________________

SECTION ______ DATE ____________

EXERCISE 4

INSTRUCTOR ____________________

Names and Formulas II

Give the names of the following compounds:

1. $(NH_4)_2S$ ____________________
2. NiF_2 ____________________
3. $Sb(ClO_3)_3$ ____________________
4. $HgCl_2$ ____________________
5. $H_2SO_4(aq)$ ____________________
6. $CrBr_3$ ____________________
7. Cu_2CO_3 (a) ____________________

   (b) ____________________
8. $K_2Cr_2O_7$ ____________________
9. $FeSO_4$ (a) ____________________

   (b) ____________________
10. $AgC_2H_3O_2$ ____________________
11. HCl ____________________
12. HCl(*aq*) ____________________
13. $KBrO_3$ ____________________
14. $Cd(ClO_2)_2$ ____________________

Give the formulas of the following compounds:

1. Sodium oxalate

1. _____

2. Manganese(II) iodate

2. _____

3. Zinc nitrite

3. _____

4. Potassium permanganate

4. _____

5. Titanium(IV) bromide

5. _____

6. Sodium arsenate

6. _____

7. Manganese(IV) sulfide

7. _____

8. Bismuth(III) arsenate

8. _____

9. Sodium peroxide

9. _____

10. Magnesium bicarbonate

10. _____

11. Lead(II) acetate

11 _____

12. Phosphoric acid

12. _____

13. Nitric acid

13. _____

14. Acetic acid

14. _____

15. Arsenic(III) iodide

15. _____

16. Ammonium thiocyanate

16. _____

17. Cobalt(II) chlorite

17. _____

18. Stannous fluoride

18. _____

NAME ______________________

SECTION ______ DATE ______________

EXERCISE 5

INSTRUCTOR ______________________

Names and Formulas III

Give the names of the following compounds:

1. CO_2 ______________________
2. H_2O_2 ______________________
3. $Ni(MnO_4)_2$ ______________________
4. $Co_3(AsO_4)_2$ ______________________
5. KCN ______________________
6. Sb_2O_5 ______________________
7. BaH_2 ______________________
8. $NaHSO_3$ ______________________
9. $As(NO_2)_5$ ______________________
10. KSCN ______________________
11. Ag_2CO_3 ______________________
12. CrF_3 ______________________
13. SnS_2 (a) ______________________

    (b) ______________________
14. $H_2SO_3(aq)$ ______________________
15. HgC_2O_4 ______________________
16. $Pb(HCO_3)_2$ ______________________
17. $Cu(OH)_2$ ______________________

Give the formulas of the following substances:

1. Ammonium hydrogen carbonate   1. _____

2. Hydrogen sulfide   2. _____

3. Barium hydroxide   3. _____

4. Carbon tetrachloride   4. _____

5. Nickel(II) perchlorate   5. _____

6. Lead(II) nitrate   6. _____

7. Sulfur dioxide   7. _____

8. Carbonic acid   8. _____

9. Copper(I) carbonate   9. _____

10. Calcium cyanide   10. _____

11. Arsenic(III) oxide   11. _____

12. Silver dichromate   12. _____

13. Nitrous acid   13. _____

14. Copper(II) bromide   14. _____

15. Ammonia   15. _____

16. Chlorine   16. _____

17. Chromium(III) sulfite   17. _____

18. Chloric acid   18. _____

19. Barium arsenate   19. _____

20. Manganese(IV) chloride   20. _____

21. Carbon disulfide   21. _____

22. Aluminum fluoride   22. _____

EXERCISE 6

Equation Writing and Balancing I

Balance the following equations:

1. $Mg + O_2 \xrightarrow{\Delta} MgO$

2. $KClO_3 \xrightarrow{\Delta} KCl + O_2$

3. $Fe + O_2 \xrightarrow{\Delta} Fe_3O_4$

4. $Mg + HCl \longrightarrow MgCl_2 + H_2$

5. $Na + H_2O \longrightarrow NaOH + H_2$

Beneath each word equation write the formula equation and balance it. Remember that oxygen and hydrogen are diatomic molecules.

1. Sulfur + Oxygen $\xrightarrow{\Delta}$ Sulfur dioxide

2. Zinc + Sulfuric acid $\longrightarrow$ Zinc sulfate + Hydrogen

3. Carbon + Oxygen $\xrightarrow{\Delta}$ Carbon dioxide

4. Hydrogen + Oxygen $\xrightarrow{\Delta}$ Water

5. Aluminum + Hydrochloric acid $\longrightarrow$ Aluminum chloride + Hydrogen

Balance the following equations:

1. $N_2 +$ $H_2 \xrightarrow{\Delta}$ NH_3

2. $CoCl_2 \cdot 6\,H_2O \xrightarrow{\Delta}$ $CoCl_2 +$ H_2O

3. $Fe +$ $H_2O \xrightarrow{\Delta}$ $Fe_3O_4 +$ H_2

4. F_2 $+$ $H_2O \xrightarrow{\Delta}$ $HF +$ O_2

5. $Pb(NO_3)_2 \xrightarrow{\Delta}$ $PbO +$ $NO +$ O_2

Beneath each word equation write and balance the formula equation. Oxygen, hydrogen, and bromine are diatomic molecules.

1. Aluminum + Oxygen $\xrightarrow{\Delta}$ Aluminum oxide

2. Potassium + Water $\longrightarrow$ Potassium hydroxide + Hydrogen

3. Arsenic(III) oxide + Hydrochloric acid $\longrightarrow$ Arsenic(III) chloride + Water

4. Phosphorus + Bromine $\longrightarrow$ Phosphorus tribromide

5. Sodium bicarbonate + Nitric acid $\longrightarrow$ Sodium nitrate + Water + Carbon dioxide

EXERCISE 7

Equation Writing and Balancing II

Complete and balance the following double displacement reaction equations (assume all reactions will go):

1. $NaCl + AgNO_3 \longrightarrow$

2. $BaCl_2 + H_2SO_4 \longrightarrow$

3. $NaOH + HCl \longrightarrow$

4. $Na_2CO_3 + HCl \longrightarrow$

5. $H_2SO_4 + NH_4OH \longrightarrow$

6. $FeCl_3 + NH_4OH \longrightarrow$

7. $Na_2SO_3 + HCl \longrightarrow$

8. $K_2CrO_4 + Pb(NO_3)_2 \longrightarrow$

9. $NaC_2H_3O_2 + HCl \longrightarrow$

10. $NaOH + NH_4NO_3 \longrightarrow$

11. $BiCl_3 + H_2S \longrightarrow$

12. $K_2C_2O_4 + HCl \longrightarrow$

13. $H_3PO_4 + Ca(OH)_2 \longrightarrow$

14. $(NH_4)_2CO_3 + HNO_3 \longrightarrow$

15. $K_2CO_3 + NiBr_2 \longrightarrow$

Complete and balance the following equations. (Combination, 1–4; Decomposition, 5–8; Single displacement, 9–12; Double displacement, 13–16.)

1. $K + Cl_2 \longrightarrow$

2. $Zn + O_2 \longrightarrow$

3. $BaO + H_2O \longrightarrow$

4. $SO_3 + H_2O \longrightarrow$

5. $MgCO_3 \xrightarrow{\Delta}$

6. $NH_4OH \xrightarrow{\Delta}$

7. $Mn(ClO_3)_2 \xrightarrow{\Delta}$

8. $HgO \xrightarrow{\Delta}$

9. $Ni + HCl \longrightarrow$

10. $Pb + AgNO_3 \longrightarrow$

11. $Cl_2 + NaI \longrightarrow$

12. $Al + CuSO_4 \longrightarrow$

13. $KOH + H_3PO_4 \longrightarrow$

14. $Na_2C_2O_4 + CaCl_2 \longrightarrow$

15. $(NH_4)_2SO_4 + KOH \longrightarrow$

16. $ZnCl_2 + (NH_4)_2S \longrightarrow$

EXERCISE 8

Equation Writing and Balancing III

For each of the following situations, write and balance the formula equation for the reaction that occurs.

1. A strip of zinc is dropped into a test tube of hydrochloric acid.

2. Hydrogen peroxide decomposes in the presence of manganese dioxide.

3. Copper(II) sulfate pentahydrate is heated to drive off the water of hydration.

4. A piece of sodium is dropped into a beaker of water.

5. A piece of limestone (calcium carbonate) is heated in a Bunsen burner flame.

6. A piece of zinc is dropped into a solution of silver nitrate.

7. Hydrochloric acid is added to a sodium carbonate solution.

8. Potassium chlorate is heated in the presence of manganese dioxide.

9. Hydrogen gas is burned in air.

10. Sulfuric acid solution is reacted with sodium hydroxide solution.

EXERCISE 9

Graphical Representation of Data

A. From the figure at the right, read values for the following:

1. The vapor pressure of ethyl ether at 20°C.

   _____

2. The temperature at which ethyl chloride has a vapor pressure of 620 torr.

   _____

3. The temperature at which ethyl alcohol has the pressure that ethyl chloride has at 2°C.

   _____

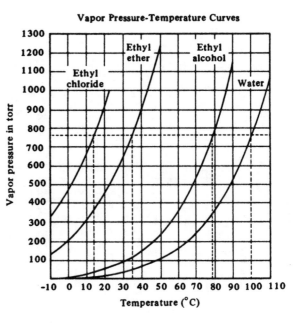

Vapor Pressure-Temperature Curves

B. Plotting Graphs

1. Plot the following pressure-temperature data for a gas on the graph below. Draw the best possible straight line through the data. Provide temperature and pressure scales.

Temperature, °C	0	20	40	60	80	100
Pressure, torr	586	628	655	720	757	800

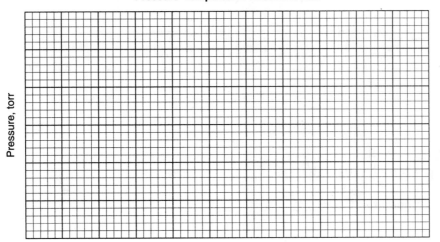

Pressure-Temperature Data for a Gas

2. (a) Study the data given below; (b) determine suitable scales for pressure and for volume and mark these scales on the graph; (c) plot eight points on the graph; (d) draw the best possible line through these points; (e) place a suitable title at the top of the graph.

Pressure-volume data for a gas

Volume, mL	10.70	7.64	5.57	4.56	3.52	2.97	2.43	2.01
Pressure, torr	250	350	480	600	760	900	1100	1330

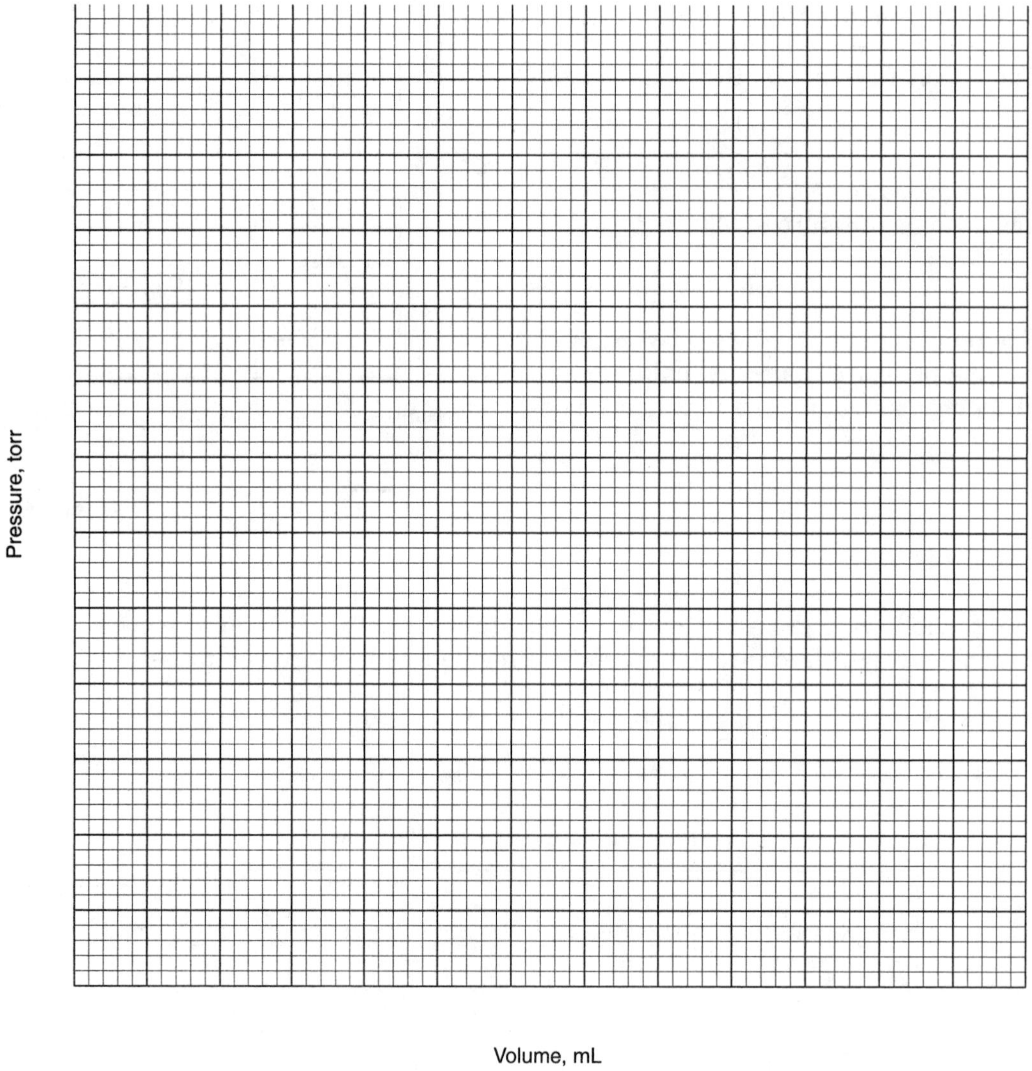

Pressure, torr

Volume, mL

Read from your graph:

(a) The pressure at 10.0 mL _____

(b) The volume at 700 torr _____

NAME ______________________

SECTION ______ DATE ____________

EXERCISE 10

INSTRUCTOR ______________________

Moles

Show calculation setups and answers for all problems.

1. Find the molar mass of (a) nitric acid, HNO_3; (b) potassium bicarbonate, $KHCO_3$; and (c) Nickel(II) nitrate, $Ni(NO_3)_2$.

(a) ______________

(b) ______________

(c) ______________

2. A sample of mercury(II) bromide, $HgBr_2$, weighs 8.65 g. How many moles are in this sample?

____

3. What is the mass of 0.45 mol of ammonium sulfate, $(NH_4)_2SO_4$?

____

4. How many molecules are contained in 6.53 mol of nitrogen gas, N_2?

____

5. Calculate the percent composition by mass of calcium sulfite, $CaSO_3$.

Ca ______________

S ______________

O ______________

6. An organic compound is analyzed and found to be carbon 51.90%, hydrogen 9.80%, and chlorine 38.30%. What is the empirical formula of this compound?

_____

7. A sample of oxygen gas, O_2, weighs 28.4 g. How many molecules of O_2 and how many atoms of O are present in this sample?

_____ molecules of O_2

_____ atoms of O

8. A mixture of sand and salt is found to be 48 percent NaCl by mass. How many moles of NaCl are in 74 g of this mixture?

_____

9. What is the mass of 2.6×10^{23} molecules of ammonia, NH_3?

_____

10. A water solution of sulfuric acid has a density of 1.67 g/mL and is 75 percent H_2SO_4 by mass. How many moles of H_2SO_4 are contained in 400. mL of this solution?

_____

NAME ____________________

SECTION ______ DATE ____________

INSTRUCTOR ____________________

EXERCISE 11

Stoichiometry I

Show calculation setups and answers for all problems.

1. Use the equation given to solve the following problems:

   $Na_3PO_4 + 3\ AgNO_3 \longrightarrow Ag_3PO_4 + 3\ NaNO_3$

   (a) How many moles of Na_3PO_4 would be required to react with 1.0 mol of $AgNO_3$?

   ____

   (b) How many moles of $NaNO_3$ can be produced from 0.50 mol of Na_3PO_4?

   ____

   (c) How many grams of Ag_3PO_4 can be produced from 5.00 g of Na_3PO_4?

   ____

   (d) If you have 9.44 g of Na_3PO_4, how many grams of $AgNO_3$ will be needed for complete reaction?

   ____

   (e) When 25.0 g of $AgNO_3$ are reacted with excess Na_3PO_4, 18.7 g of Ag_3PO_4 are produced. What is the percentage yield of Ag_3PO_4?

   ____

2. Use the equation given to solve the following problems:

$$2 \text{ KMnO}_4 + 16 \text{ HCl} \longrightarrow 5 \text{ Cl}_2 + 2 \text{ KCl} + 2 \text{ MnCl}_2 + 8 \text{ H}_2\text{O}$$

(a) How many moles of HCl are required to react with 35 g of KMnO_4?

_____

(b) How many Cl_2 molecules will be produced using 3.0 mol KMnO_4?

_____

(c) To produce 35.0 g of MnCl_2, what mass of HCl will need to react?

_____

(d) How many moles of water will be produced when 8.0 mol of KMnO_4 are consumed?

_____

(e) What is the maximum mass of Cl_2 that can be produced by reacting 70.0 g of KMnO_4 with 15.0 g of HCl?

_____

NAME ____________________

SECTION ______ DATE ____________

INSTRUCTOR ________________

EXERCISE 12

Gas Laws

Show calculation setups and answers for all problems.

1. A sample of nitrogen gas, N_2, occupies 3.0 L at a pressure of 3.0 atm. What volume will it occupy when the pressure is changed to 0.50 atm and the temperature remains constant?

____

2. A sample of methane gas, CH_4, occupies 4.50 L at a temperature of 20.0°C. If the pressure is held constant, what will be the volume of the gas at 100.°C?

____

3. The pressure of hydrogen gas in a constant-volume cylinder is 4.25 atm at 0°C. What will the pressure be if the temperature is raised to 80°C?

____

4. A 325 mL sample of air is at 720. torr and 30.°C. What volume will this gas occupy at 800. torr and 75.°C?

____

5. A sample of gas occupies 500. mL at STP. What volume will the gas occupy at 85°C and 525 torr?

____

6. A quantity of oxygen occupies a volume of 19.2 L at STP. How many moles of oxygen are present?

_____

7. A 425 mL volume of hydrogen chloride gas, HCl, is collected at 25°C and 720. torr. What volume will it occupy at STP?

_____

8. What volume would 10.5 g of nitrogen gas, N_2, occupy at 200. K and 2.02 atm?

_____

9. Calculate the density of sulfur dioxide, SO_2, at STP.

_____

10. In a laboratory experiment, 133 mL of gas was collected over water at 24°C and 742 torr. Calculate the volume that the dry gas would occupy at STP.

_____

11. A volume of 122 mL of argon, Ar, is collected at 50°C and 758 torr. What does this sample weigh?

_____

NAME ____________________

SECTION ______ DATE __________

EXERCISE 13

INSTRUCTOR ______________

Solution Concentrations

Show calculation setups and answers for all problems.

1. What will be the percent composition by mass of a solution made by dissolving 15.0 g of barium nitrate, $Ba(NO_3)_2$, in 45.0 g of water?

$Ba(NO_3)_2$ ______________

H_2O ______________

2. How many moles of potassium hydroxide, KOH, are required to prepare 2.00 L of 0.250 M solution?

____

3. What will be the molarity of a solution if 3.50 g of sodium hydroxide, NaOH, are dissolved in water to make 150. mL of solution?

____

4. How many milliliters of 0.400 M solution can be prepared by dissolving 5.00 g of NaBr in water?

____

5. How many grams of potassium bromide, KBr, could be recovered by evaporating 650. mL of 15.0 percent KBr solution to dryness (d = 1.11 g/mL)?

____

6. How many milliliters of 12.0 M HCl is needed to prepare 300. mL of 0.250 M HCl solution?

_____

7. A sample of potassium hydrogen oxalate, KHC_2O_4, weighing 0.717 g, was dissolved in water and titrated with 23.47 mL of an NaOH solution. Calculate the molarity of the NaOH solution.

_____

8. How many grams of hydrogen chloride are in 50. mL of concentrated (12 M) HCl solution?

_____

9. A sulfuric acid solution has a density of 1.49 g/mL and contains 59 percent H_2SO_4 by mass. What is the molarity of this solution?

_____

10. Sulfuric acid reacts with sodium hydroxide according to this equation:

$$H_2SO_4 + 2\ NaOH \longrightarrow Na_2SO_4 + 2\ H_2O$$

A 10.00 mL sample of the H_2SO_4 solution required 18.71 mL of 0.248 M NaOH for neutralization. Calculate the molarity of the acid.

_____

NAME ____________

SECTION ______ DATE ________

INSTRUCTOR ____________

EXERCISE 14

Stoichiometry II

Show calculation setups and answers for all problems.

1. Use the equation to solve the following problems:

$$6\ KI + 8\ HNO_3 \longrightarrow 6\ KNO_3 + 2\ NO + 3\ I_2 + 4\ H_2O$$

(a) If 38 g of KI are reacted, how many grams of I_2 will be formed?

____

(b) What volume of NO gas, measured at STP, will be produced if 47.0 g of HNO_3 are reacted?

____

(c) How many milliliters of 6.00 M HNO_3 will react with 1.00 mole of KI?

____

(d) When the reaction produces 8.0 mol of NO, how many molecules of I_2 will be produced?

____

(e) How many grams of iodine can be obtained by reacting 35.0 mL of 0.250 M KI solution?

____

2. Use the equation given to solve the following problems. All substances are in the gas phase.

$$N_2(g) + 3 H_2(g) \longrightarrow 2 NH_3(g)$$

(a) If 2.0 mol of H_2 react, how many moles of NH_3 will be formed?

_____

(b) When 5.50 mol of N_2 react, what volume of NH_3, measured at STP, will be formed?

_____

(c) What volume of NH_3 will be formed when 12.0 L of H_2 are reacted? All volumes are measured at STP.

_____

(d) How many molecules of NH_3 will be formed when 30.0 L of N_2 at STP react?

_____

(e) What volume of NH_3, measured at 25°C and 710. torr, will be produced from 18.0 g of H_2?

_____

(f) If a mixture of 9.00 L of N_2 and 30.0 L of H_2 are reacted, what volume of NH_3 can be produced? Assume STP conditions.

_____

NAME ____________

SECTION ______ DATE ________

INSTRUCTOR ____________

EXERCISE 15

Chemical Equilibrium

1. Consider the following system at equilibrium:

$$2\ CO_2(g) + 135.2\ \text{kcal} \rightleftharpoons 2\ CO(g) + O_2(g)$$

Complete the following table. Indicate changes in moles and concentrations by entering I, D, N, or ? in the table (I = increase, D = decrease, N = no change, ? = insufficient information to determine).

Change or stress imposed on the system at equilibrium	Direction of shift, left or right, to reestablish equilibrium	Change in number of moles			Change in molar concentrations		
		CO_2	CO	O_2	CO_2	CO	O_2
a. Add CO							
b. Remove CO_2							
c. Decrease volume of reaction vessel							
d. Increase temperature							
e. Add catalyst							
f. Add both CO_2 and O_2							

2. Consider the reaction $PCl_5(g) \rightleftharpoons PCl_3(g) + Cl_2(g)$.

At 250°C, PCl_5 is 45% decomposed.

(a) If 0.110 mol of $PCl_5(g)$ is introduced into a 1.00 L container at 250°C, what will be the equilibrium concentrations of PCl_5, PCl_3, and Cl_2?

PCl_5 ____________

PCl_3 ____________

Cl_2 ____________

(b) What is the value of K_{eq} at 250°C?

K_{eq} ____________

3. For the reaction $H_2(g) + I_2(g) \rightleftharpoons 2\,HI(g)$, $K_{eq} = 0.17$ at 500 K. What concentration of I_2 (g) will be in equilibrium with $H_2 = 0.040$ M, $HI = 0.015$ M?

_____

4. $CaCO_3$ has a solubility in water of 6.9×10^{-5} mol/L. Calculate the solubility product constant.

_____

5. A 0.40 M HClO solution was found to have an H^+ concentration of 1.1×10^{-4} M. Calculate the value of the ionization constant. The ionization equation is $HClO \rightleftharpoons H^+ + ClO^-$.

_____

6. Calculate (a) the H^+ ion concentration, (b) the pH, and (c) the percent ionization of a 0.40 M solution of $HC_2H_3O_2$ ($K_a = 1.8 \times 10^{-5}$).

(a) _____

(b) _____

(c) _____

EXERCISE 16

Oxidation-Reduction Equations I

Balance the following oxidation-reduction equations:

1. $\quad$ P + $\quad$ HNO_3 + $\quad$ $H_2O \longrightarrow$ $\quad$ H_3PO_4 + $\quad$ NO

2. $\quad$ H_2SO_4 + $\quad$ HI $\longrightarrow$ $\quad$ H_2S + $\quad$ I_2 + $\quad$ H_2O

3. $\quad$ $KBrO_2$ + $\quad$ KI + $\quad$ HBr $\longrightarrow$ $\quad$ KBr + $\quad$ I_2 + $\quad$ H_2O

4. $\quad$ Sb + $\quad$ $HNO_3 \longrightarrow$ $\quad$ Sb_2O_5 + $\quad$ NO + $\quad$ H_2O

5. $\quad$ NO_2 + $\quad$ $H_2O \longrightarrow$ $\quad$ HNO_3 + $\quad$ NO

6. Br_2 + NH_3 $\longrightarrow$ NH_4Br + N_2

7. KI + HNO_3 $\longrightarrow$ KNO_3 + NO + I_2 + H_2O

8. H_2SO_3 + $KMnO_4$ $\longrightarrow$ $MnSO_4$ + H_2SO_4 + K_2SO_4 + H_2O

9. $K_2Cr_2O_7$ + H_2O + S $\longrightarrow$ SO_2 + KOH + Cr_2O_3

10. $KMnO_4$ + HCl $\longrightarrow$ Cl_2 + KCl + $MnCl_2$ + H_2O

EXERCISE 17

Oxidation-Reduction Equations II

Balance the following oxidation-reduction equations using the ion-electron method.

1. $MnO_4^- +$ $Cl^- +$ $H^+ \longrightarrow$ $Mn^{2+} +$ $Cl_2 +$ H_2O

2. $Ag_2S +$ $NO_3^- +$ $H^+ \longrightarrow$ $S +$ $NO +$ $Ag^+ +$ H_2O

3. $ClO_4^- +$ $I^- +$ $H^+ \longrightarrow$ $I_2 +$ $Cl^- +$ H_2O

4. $Br_2 +$ $H_2O \longrightarrow$ $BrO_3^- +$ $Br^- +$ H^+

5. $MnO_4^- +$ $HS^- +$ $H_2O \longrightarrow$ $S +$ $MnO_2 +$ OH^-

6. $H_2O_2 +$ IO_3^- $\longrightarrow$ $I^- +$ O_2 (acid solution)

7. $Cl_2 +$ SO_2 $\longrightarrow$ $SO_4^{2-} +$ Cl^- (acid solution)

8. $U^{4+} +$ MnO_4^- $\longrightarrow$ $Mn^{2+} +$ UO_2^{2+} (acid solution)

9. $Fe(CN)_6^{3-} +$ Cr_2O_3 $\longrightarrow$ $Fe(CN)_6^{4-} +$ CrO_4^{2-} (basic solution)

10. $Cr(OH)_3 +$ O_2^{2-} $\longrightarrow$ CrO_4^{2-} (basic solution)

EXERCISE 18

Hydrocarbons

1. Write condensed structural formulas for the five isomers of hexane, all having the molecular formula C_6H_{14}.

2. Write condensed structural formulas for the six isomers of pentene, all having the formula C_5H_{10}.

3. Write condensed structural formulas for (a) acetylene, (b) 2-methylbutane, (c) benzene, (d) octane.

   (a)   (b)

   (c)   (d)

4. Write the formulas for the products of the following reactions:

(a) 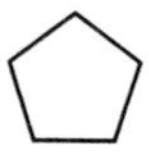 ⬠ + Br_2 $\xrightarrow{\text{UV light}}$

(b) $CH_2{=}CH{-}CH_3$ + Cl_2 $\xrightarrow{\text{dark}}$

(c)  ⬡ + H_2 $\xrightarrow[\text{and pressure}]{\text{catalyst, heat}}$

(d) $CH_3CH_2CH{=}CH_2$ + HCl $\longrightarrow$

(e) $CH_3CH_2CH_3$ + Cl_2 $\xrightarrow{\text{UV light}}$

5. Complete and balance the following combustion reactions (assume complete combustion).

(a) $CH_3CH_2CH_3$ + O_2 $\xrightarrow{\Delta}$

(b) $CH_3{-}C{\equiv}C{-}CH_3$ + O_2 $\xrightarrow{\Delta}$

(c)    ⬡ + O_2 $\xrightarrow{\Delta}$

EXERCISE 19

Alcohols, Esters, Aldehydes, and Ketones

1. Write condensed structural formulas of eight isomers of pentanol, all having the formula $C_5H_{12}O$.

2. Write condensed structural formulas for the ketones having the formula $C_6H_{12}O$.

3. Write condensed structural formulas for the aldehydes having the formula $C_6H_{12}O$.

4. Write the formulas for the products of the following reactions:

(a) $CH_3CH_2CH_2OH$ + CuO $\xrightarrow{\Delta}$

(b) 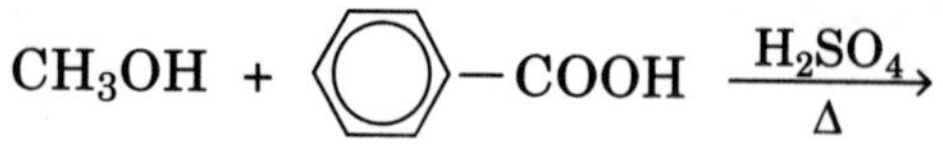 CH_3OH + ⬡$-COOH$ $\xrightarrow[\Delta]{H_2SO_4}$

(c) $2CH_3CH_2COOH$ + $HOCH_2CH_2CH_2OH$ $\xrightarrow[\Delta]{H_2SO_4}$

(d) 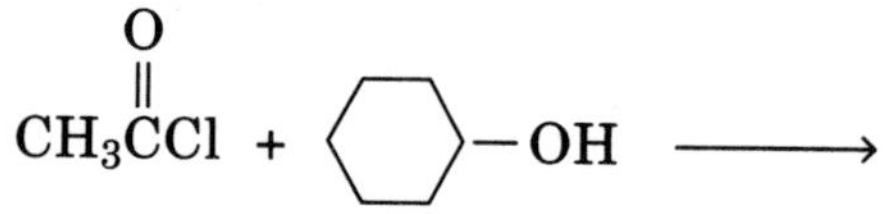 $CH_3\overset{\overset{\displaystyle O}{\|}}{C}Cl$ + ⬡$-OH$ $\longrightarrow$

(e) 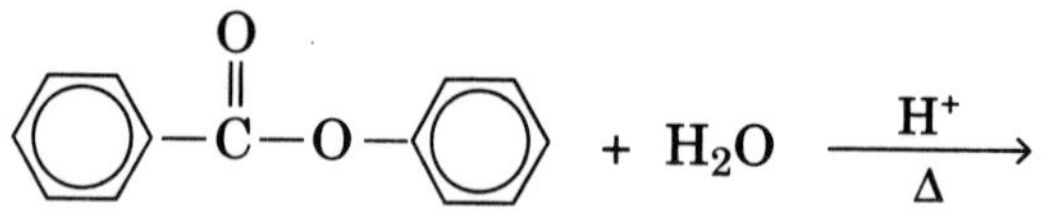 ⬡$-\overset{\overset{\displaystyle O}{\|}}{C}-O-$⬡ + H_2O $\xrightarrow[\Delta]{H^+}$

(f) $CH_3\overset{\overset{\displaystyle O}{\|}}{C}-O-\underset{\underset{\displaystyle CH_3}{|}}{C}HCH_3$ + NaOH $\xrightarrow[\Delta]{H_2O}$

5. Complete and balance the following combustion reactions (assume complete combustion).

(a) $CH_3-CH(OH)-CH_3$ + O_2 $\longrightarrow$

(b) 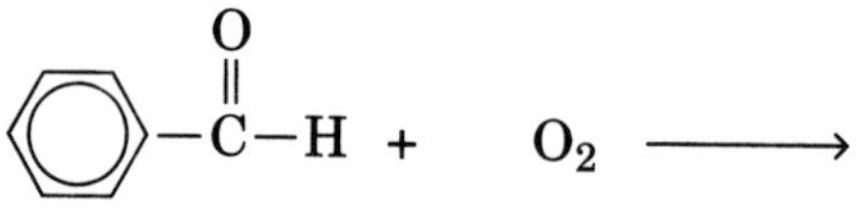   ⬡$-\overset{\overset{\displaystyle O}{\|}}{C}-H$ + O_2 $\longrightarrow$

EXERCISE 20

Functional Groups

1. Identify each functional group by name and circle the functional group on the formula.

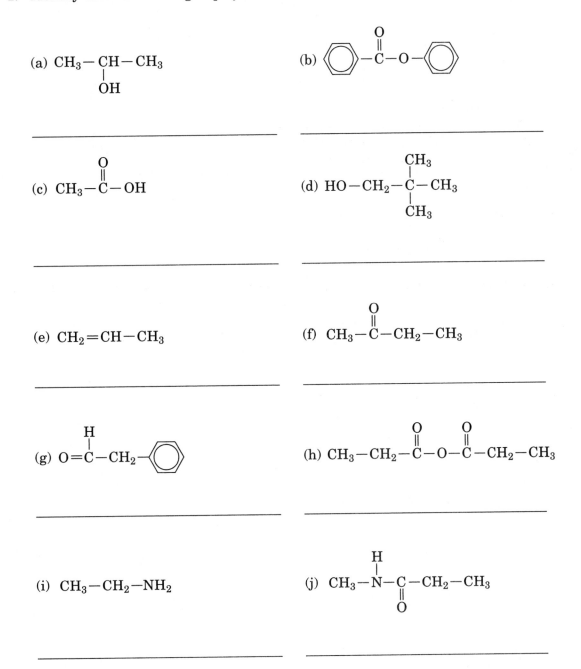

(a) $CH_3-CH-CH_3$
         $|$
         OH

_____

(b)

(c) $CH_3-\overset{\overset{O}{\|}}{C}-OH$

_____

(d) $HO-CH_2-\overset{\overset{CH_3}{|}}{\underset{\underset{CH_3}{|}}{C}}-CH_3$

_____

(e) $CH_2=CH-CH_3$

_____

(f) $CH_3-\overset{\overset{O}{\|}}{C}-CH_2-CH_3$

_____

(g) $O=\overset{\overset{H}{|}}{C}-CH_2-$

_____

(h) $CH_3-CH_2-\overset{\overset{O}{\|}}{C}-O-\overset{\overset{O}{\|}}{C}-CH_2-CH_3$

_____

(i) $CH_3-CH_2-NH_2$

_____

(j) $CH_3-\overset{\overset{H}{|}}{N}-\underset{\underset{O}{\|}}{C}-CH_2-CH_3$

_____

2. Name each of the compounds listed in Question 1.

(a) _____

(b) _____

(c) _____

(d) _____

(e) _____

(f) _____

(g) _____

(h) _____

(i) _____

(j) _____

EXERCISE 21

Synthetic Polymers

1. Complete the following reactions showing two units of the polymers formed.

   (a) $CH_3CH{=}CH_2$ $\xrightarrow{\text{Organic}}_{\text{peroxide}}$

   (b) $CCl_2{=}CH_2$ $\xrightarrow{\text{Organic}}_{\text{peroxide}}$

   (c) 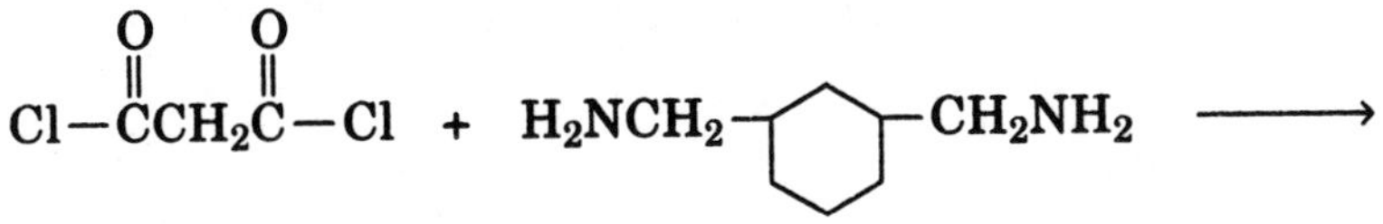 $Cl-\overset{\overset{O}{\|}}{C}CH_2\overset{\overset{O}{\|}}{C}-Cl$ + $H_2NCH_2{-}\bigcirc{-}CH_2NH_2$ $\longrightarrow$

   (d) 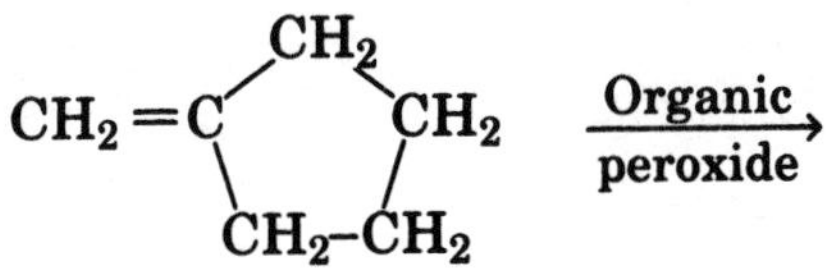 $CH_2{=}C\overset{\displaystyle CH_2}{\underset{\displaystyle CH_2-CH_2}{\diagdown}}CH_2$ $\xrightarrow{\text{Organic}}_{\text{peroxide}}$

   (e) 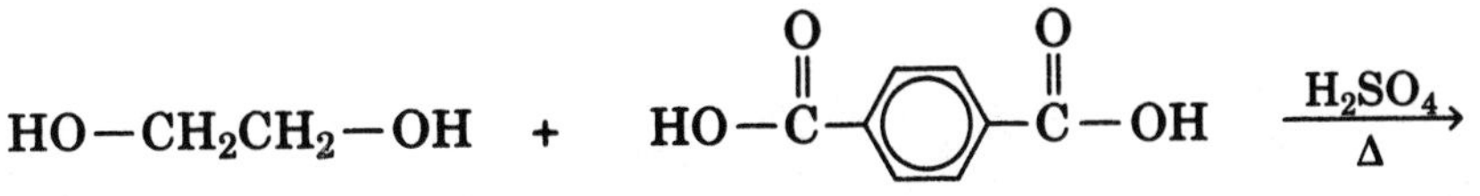 $HO-CH_2CH_2-OH$ + $HO-\overset{\overset{O}{\|}}{C}-\bigcirc-\overset{\overset{O}{\|}}{C}-OH$ $\xrightarrow[\Delta]{H_2SO_4}$

   (f) Why is a polymer usually a solid?

– 439 –

2. Write the structural formulas of the monomers from which the following polymers were obtained.

(a) $+CH_2-CCl_2-CHCl-CH_2-CH_2-CCl_2-CH_2-CHCl+_n$

(b) 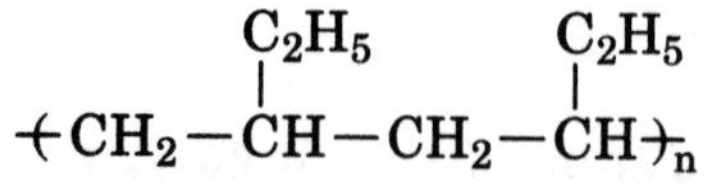 $+CH_2-\overset{\overset{\displaystyle C_2H_5}{|}}{CH}-CH_2-\overset{\overset{\displaystyle C_2H_5}{|}}{CH}+_n$

(c) $+\overset{\overset{\displaystyle CH_3}{|}}{CH}-\overset{\overset{\displaystyle CH_3}{|}}{CH}-\overset{\overset{\displaystyle CH_3}{|}}{CH}-\overset{\overset{\displaystyle CH_3}{|}}{CH}+_n$

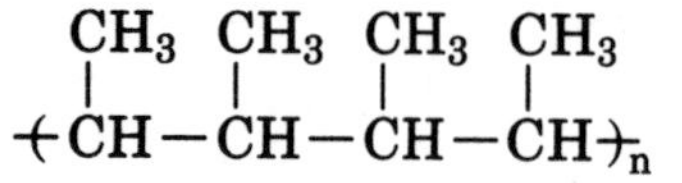

(d) 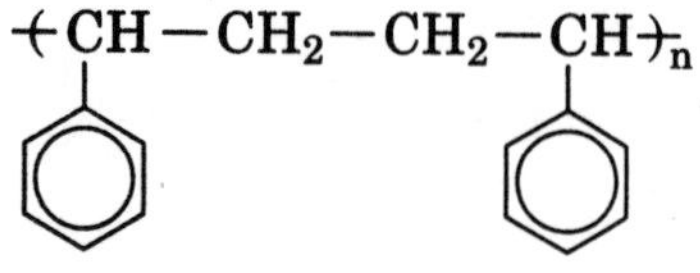 $+CH-CH_2-CH_2-CH+_n$

(e) 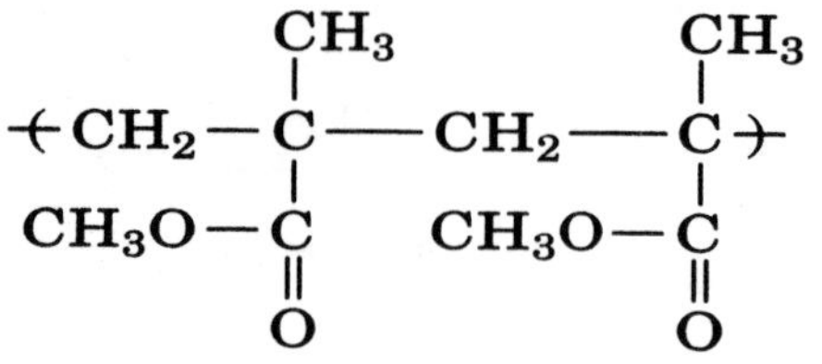 $+CH_2-\overset{\overset{\displaystyle CH_3}{|}}{\underset{\underset{\displaystyle O}{\parallel}}{\underset{CH_3O-C}{C}}}-CH_2-\overset{\overset{\displaystyle CH_3}{|}}{\underset{\underset{\displaystyle O}{\parallel}}{\underset{CH_3O-C}{C}}}+$

(f) 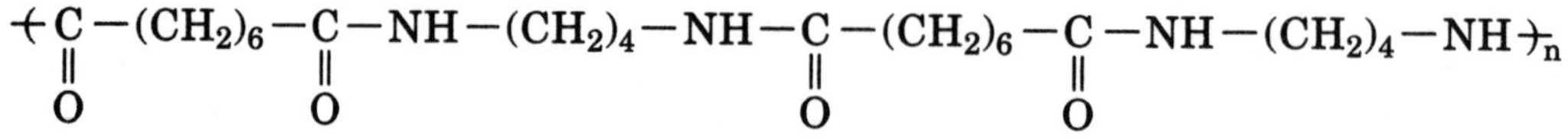 $+\overset{}{\underset{\underset{\displaystyle O}{\parallel}}{C}}-(CH_2)_6-\overset{}{\underset{\underset{\displaystyle O}{\parallel}}{C}}-NH-(CH_2)_4-NH-\overset{}{\underset{\underset{\displaystyle O}{\parallel}}{C}}-(CH_2)_6-\overset{}{\underset{\underset{\displaystyle O}{\parallel}}{C}}-NH-(CH_2)_4-NH+_n$

(g) $+CH_2-O-CH_2-O+_n$

EXERCISE 22

Carbohydrates

1. From the projection formulas given, write the Haworth perspective structures for α-D-glucopyranose and β-D-mannopyranose.

```
   H—C=O                      H—C=O
    |                          |
  H—C—OH                   HO—C—H
    |                          |
 HO—C—H                    HO—C—H
    |                          |
  H—C—OH                    H—C—OH
    |                          |
  H—C—OH                    H—C—OH
    |                          |
   CH₂OH                      CH₂OH

 α-d-glucose               β-d-mannose
```

2. Complete and balance:

```
   H—C=O
    |
 HO—C—H
    |
  H—C—OH   +   Cu²⁺   +   OH⁻  ⟶
    |
  H—C—OH
    |
   CH₂OH
```

3. Write Fischer projection formulas for (a) a d-aldotetrose, (b) a d-ketopentose, and (c) an l-aldohexose.

4. What is the monosaccharide compostion of

    (a) Maltose

    (b) Sucrose

    (c) Lactose

    (d) Cellulose

    (e) Starch

    (f) Glycogen

5. Circle the formulas of the reducing sugars:

(a)
```
      CH₂OH
        |
       C=O
        |
  HO — C — H
        |
   H — C — OH
        |
   H — C — OH
        |
      CH₂OH
```

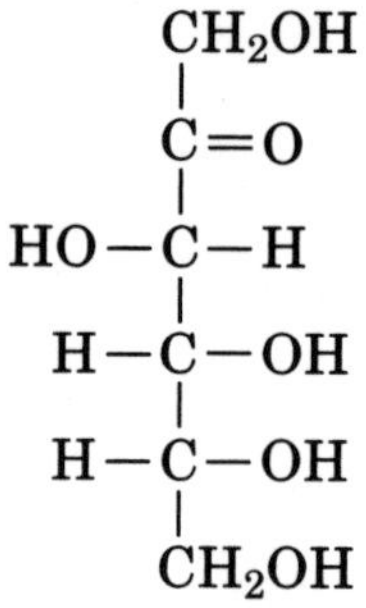

(b)
```
  H — C = O
        |
   H — C — OH
        |
   H — C — OH
        |
   H — C — OH
        |
      CH₂OH
```

(c)
```
  H — C = O
        |
   H — C — H
        |
   H — C — H
        |
   H — C — H
        |
       CH₃
```

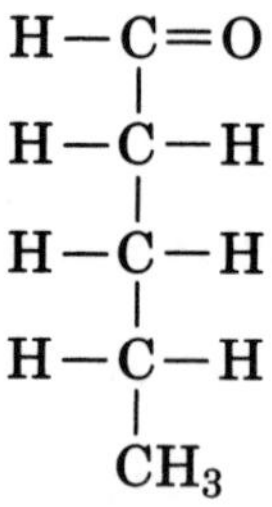

(d)
```
      CH₂OH
        |
   H — C — OH
        |
       C = O
        |
  HO — C — H
        |
   H — C — OH
        |
      CH₂OH
```

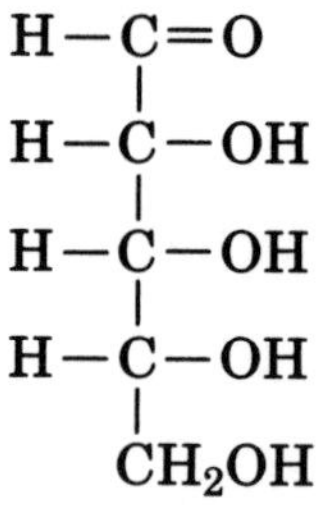

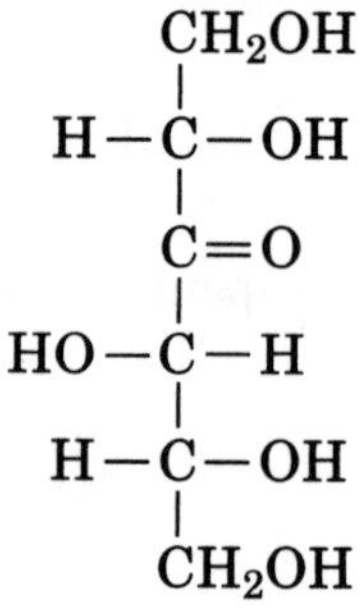

(e)
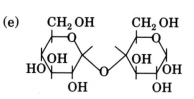

EXERCISE 23

Amino Acids and Polypeptides

1. Write structural formulas for:

   (a) Lysyltyrosine (Lys-Tyr)

   (b) Glycylvalylarginine (Gly-Val-Arg)

2. Nutrasweet artificial sweetener is a methyl ester of a dipeptide. The dipeptide is aspartylphenylalanine methyl ester. The methyl ester is on the carboxyl group of phenyl-alanine. Draw its structure.

3. Draw and name the hydrolysis products produced from the compound shown:

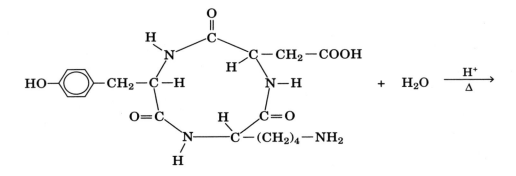

4. Indicate a positive result as (+), negative result as (−) for each of the following tests on
    polypeptides A and B.
    (A) contains 3 alanine, 2 aspartic acid, 2 cysteine, 4 glycine, 2 histidine, 1 methionine,
    and 3 proline residues.
    (B) contains 4 leucine, 5 glycine, 3 tryptophan, 3 valine, 2 serine, and 1 tyrosine residues.

Test	Results on Polypeptide A	Results on Polypeptide B
Biuret		
Ninhydrin		
Xanthoproteic		
Sulfur detection		

5. What is the sequence of amino acids in an octapeptide that contains one residue each
    of Glu, Arg, Thr, and Tyr, and two residues each of Phe and Gly, and hydrolyzes to the
    following fragments: (a) Arg-Gly-Phe, (b) Tyr-Thr, (c) Gly-Glu-Arg, and (d) Phe-Phe-Tyr?

EXERCISE 24

Lipids

1. Complete the following reactions:

(a)

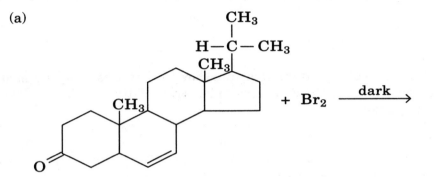

(b)

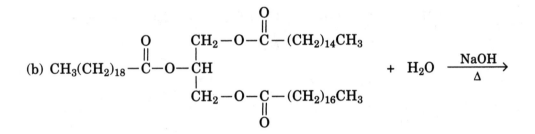

(c)

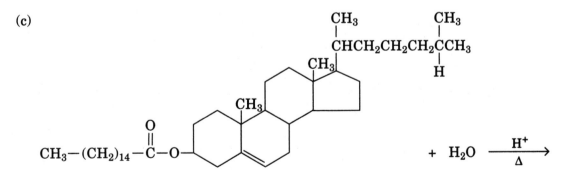

(d)

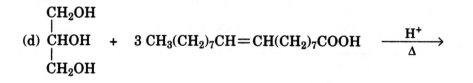

$$\begin{array}{l} CH_2OH \\ | \\ CHOH \\ | \\ CH_2OH \end{array} \quad + \quad 3\ CH_3(CH_2)_7CH=CH(CH_2)_7COOH \quad \xrightarrow[\Delta]{H^+}$$

(e) Phosphatidylserine, at pH 7, has the structure shown below. It is a major component of most membranes in cells. The R groups can be various fatty acid chains. Complete the hydrolysis equation.

$$R_2-\overset{\overset{\displaystyle O}{\|}}{C}-O-\overset{\overset{\displaystyle CH_2-O-\overset{\overset{\displaystyle O}{\|}}{C}-R_1}{|}}{CH}$$

$$CH_2-O-\overset{\overset{\displaystyle O}{\|}}{\underset{\underset{\displaystyle O^-}{|}}{P}}-O-CH_2-\overset{\overset{\displaystyle H}{|}}{\underset{\underset{\displaystyle NH_3^+}{|}}{C}}-COO^- \quad + \quad H_2O \quad \xrightarrow[\Delta]{H^+}$$

EXERCISE 25

Molecular Models and Isomerism

The objective of this exercise is to become familiar with the spatial arrangement of organic molecules and to observe the phenomenon of isomerism.

Bonding rules: Carbon must have 4 bonds to it; hydrogen 1 bond; oxygen 2 bonds; and chlorine 1 bond.

1. Construct four methane (CH_4) molecules using 1 C, 4 H, and 4 bonds for each molecule. Draw the three-dimensional structure for this CH_4 molecule.

2. Remove an H atom from each of two CH_4 molecules. You now have two methyl groups, CH_3—. Bond these two methyl groups together using a bond to give a model of ethane, CH_3—$CH_3(C_2H_6)$. Draw the three-dimensional structure of this ethane molecule.

3. Remove an H atom from the ethane molecule and an H atom from another methane molecule. You now have an ethyl, CH_3CH_2—, and a methyl, CH_3—, group. Bond these groups together with a bond to form a molecule of propane, $CH_3CH_2CH_3(C_3H_8)$. Draw the three-dimensional structure of propane.

4. Construct two ethyl groups and bond them together. You now have a model of butane, $CH_3CH_2CH_2CH_3(C_4H_{10})$. Draw the three-dimensional structure of this molecule.

5. Rearrange the butane model (from #4) to form isobutane, C_4H_{10}: (a) remove one methyl group from the butane model; (b) now remove one H from the central carbon atom of the remaining propyl, $CH_3CH_2CH_2$— group. Bond the methyl group to the central carbon atom and the hydrogen to the end carbon atom. The compound you now have is isobutane (2-methylpropane). Draw its structure.

6. Redraw the two butanes and compare their structures. This is an example of isomerism. **Isomerism** is the phenomenon of two or more compounds having the same molecular formula but different structural formulas.

7. Construct two molecules of propane, $CH_3CH_2CH_3$. On one molecule replace an H atom from an end carbon with a Cl atom. On the other molecule replace an H atom from the central carbon atom by a Cl atom. **This is another example of isomerism.** Both molecules have the formula C_3H_7Cl. The first molecule is named propyl chloride or 1-chloropropane; the second molecule is named isopropyl chloride or 2-chloropropane. Draw and label both structures.

8. Construct two different molecules, each having the molecular formula C_2H_6O. Follow the rules for bonding given earlier. One compound is called dimethyl ether and the other one is called ethyl alcohol or ethanol. Draw their three-dimensional structural formulas and give each its proper name.

9. If an H atom is replaced by a Cl atom in a molecule of butane, how many structural isomers of chlorobutane, C_4H_9Cl, can be formed? Draw the three-dimensional structures for these isomers.

10. (a) Suppose that two H atoms are replaced by two Cl atoms in a molecule of ethane. How many isomers can be formed? Draw their three-dimensional structural formulas. The molecular formula is $C_2H_4Cl_2$.

    (b) Suppose that two Cl atoms replaced two H atoms in a molecule of propane. How many isomers can be formed? Draw their dimensional structural formulas. The molecular formula is $C_3H_6Cl_2$. Construct models if necessary.

11. Construct and draw models of the following alkyl groups:
    methyl, ethyl, propyl, isopropyl, butyl, sec-butyl, isobutyl, and t-butyl

12. Construct and draw models of:
    pentane   2-methylbutane   cyclopentane
    C_5H_{12} C_5H_{12} C_5H_{10}

EXERCISE 26

Stereoisomerism–Optical Isomers

Stereoisomers are compounds that have the same structural formulas, but differ in their spatial arrangements. Two major types of stereoisomers are geometric isomers (cis-trans) and optical isomers. We will construct and study models of optical isomers in this exercise. The outstanding property of optical isomers is their optical activity; that is, the ability to rotate plane-polarized light. We will not study optical activity per se in this exercise, but we will observe the spatial structures and the criteria for compounds which have this property.

Definitions

Chiral (asymmetric) carbon atom: A carbon atom that is bonded to four different atoms or groups of atoms.

Superimposable molecules: When one molecule is laid on top of another molecule and all the atoms of both molecules coincide exactly, the molecules are identical and are said to be superimposable on one another.

Optical activity: Molecules that have the ability to rotate plane-polarized light are said to have optical activity.

Mirror image isomers: Non-superimposable molecules that appear as mirror images of one another (like the right and left hands).

Enantiomers: Non-superimposable molecules that appear as mirror images of one another are called enantiomers. Enantiomers have the property of optical activity.

Chiral molecule: A molecule that is not superimposable on its mirror image is said to be chiral or to have chirality.

Achiral molecule: A molecule that is superimposable on its mirror image is said to be achiral (it does not have chirality).

Projection formula: The three-dimensional and projection formulas for 1-bromo-1-chloroethane, $CH_3CHBrCl$, are illustrated as follows:

2   1

three-dimensional formula

projection formula

C-1 is a chiral carbon with four different groups attached to it: CH_3, H, Cl, and Br. In the three-dimensional formula the bonds from carbon to H and Cl are coming out of the plane of the paper toward the observer. The bonds from carbon to Br and CH_3 are projecting away from the observer. When we squash the molecule flat to illustrate it on a two-dimensional basis, this two-dimensional formula is known as a *projection formula*. In the projection formula it is understood that the H and Cl are coming out of the plane of the paper toward the observer and Br and CH_3 are going away from the observer.

Procedure

A. Construct two models of 1-bromo-1-chloroethane. Test to see if the two models are superimposable on one another. If they are superimposable, change the position of Br and Cl on one model. Now you have two models that are not superimposable. These two are enantiomers, non-superimposable mirror image isomers. (Turn, twist, rotate, do anything that you want to these models except interchange the positions of the groups; they remain non-superimposable.) Position these enantiomers so that they appear as mirror images. Arrange them so that the Br is on top and the CH_3 is on the bottom of the chiral carbon atom, and the H and Cl are coming out of the plane toward you. Now draw the projection formulas for these enantiomers.

Since these two molecules are not superimposable they must indeed be different compounds, for only superimposable molecules are identical.

As you observe these two projection structures (and the two models) you should note that they are mirror image isomers. It is as if you had one structure placed alongside a mirror which reflects its mirror image. With respect to the optical properties of these enantiomers, one isomer will rotate plane-polarized light to the right and is said to be *dextrorotatory;* the other isomer will rotate plane-polarized light to the left and is said to be *levorotatory.* The number of degrees that the light is rotated is not important to this exercise.

B. Construct two models of lactic acid. $\overset{3}{C}H_3 - \overset{2}{C}H - \overset{1}{C} = O$ C-2 is a chiral carbon atom.
$$\qquad\qquad\qquad\qquad\qquad\qquad\quad | \qquad |$$
$$\qquad\qquad\qquad\qquad\qquad\qquad\; OH \quad OH$$

Test to see if the two models are superimposable. If they are superimposable exchange the H and OH groups on C-2, which will make them non-superimposable. Once again the two non-superimposable isomers are enantiomers, and one compound will rotate plane-polarized light to the right and the other will rotate the light to the left.

Position the models so that they appear as mirror images. Arrange them so that the COOH group is on the top and the CH_3 is on the bottom of the chiral carbon atom, and the H and OH are coming toward you. Now draw projection structures of the two lactic acid enantiomers and observe that they are mirror images.

Question 1. One criterion for an organic compound to show optical activity is that the compound has at least one chiral carbon atom. Circle any of the following compounds that will show optical activity.

(a) CH_2Cl_2

(b) CH_2ClBr

(c) $CHClBrI$

(d) CH_3CHCl_2

(e) CH_2ClCH_2Cl

(f) $CH_3CHBrCl$

(g) CH_2ClCH_2Br

(h) $CH_3CH_2CH(CH_3)CH_2CH_3$

(i) $CH_3CH_2CH(CH_3)CH_2CH_2CH_3$

(j) $CH_3CBrClCH_2CH(CH_3)CH_3$

(k) $CH_2ClCHClCH_2Cl$

A compound that has only one chiral carbon atom can exist as two optically active isomers (enantiomers). In fact, the maximum number of optical isomers of a compound that may exist is related to the number of chiral carbon atoms and can be calculated from the formula 2^n, where n is the number of chiral carbon atoms in the compound. For example, when n = 1, two isomers exist; when n = 2, four isomers may exist; when n = 3, eight isomers may exist, etc. The formula 2^n only gives you the maximum number of optical isomers that may exist. Other structural features of the compound may preclude optical isomers from existing.

Question 2. What is the maximum number of stereoisomers for the following compounds:

(a) CH_2ClCH_2Br _____

(b) $CHClBrCHClBr$ _____

(c) $CH_2ClCHBrCH_2Cl$ _____

(d) $CH_3CHClCHClCH_3$ _____

(e) $CH_2=CHCHClCH_3$ _____

(f) $CH_3(CHCl)_3CH_2CH_3$ _____

(g) $CH_2ClCHClCOOH$ _____

(h) $CH_2ClCHClCHBrCH(OH)CH_3$ _____

C. (a) Construct four models of the compound $\overset{1}{C}H_3\overset{2}{C}HCl\overset{3}{C}HBr\overset{4}{C}H_3$. Take two models and hold them with the two CH_3 groups away from you. Now arrange the Cl on C-2 and the Br on C-3 so that both groups are on your right on one model and both are on your left on the other model. Test to see if these models are superimposable or are enantiomers (mirror image isomers). Draw projection formulas for these models. If they are the same, only one structure needs to be drawn. If they are enantiomers, draw both structures as mirror image isomers.

(b) Take the other two models and arrange them as follows: with the two CH_3 groups away from you place the Cl on the right of C-2 and the Br on the left of C-3 in one model, and the Cl on the left of C-2 and the Br on the right of C-3 on the other model. Test to see if these two models are superimposable or enantiomers. Draw their projection formulas.

You now have two pairs of enantiomers in these four structures. This corresponds to the four stereoisomers for a compound with two chiral carbon atoms as indicated by the formula 2^n ($2^2 = 4$). All four of these isomers are optically active. If we label these isomers as E-I and E-II for one set of enantiomers and E-III and E-IV for the other enantiomers, E-I and E-II are said to be diastereoisomers or diastereomers of E-III and E-IV. Thus, for example, E-I and E-III or E-IV are diastereomers. *Diastereomers* are stereoisomers of the same compound, but are not enantiomers.

D. (a) Construct four models of the compound $\overset{1}{C}H_3\overset{2}{C}HCl\overset{3}{C}HCl\overset{4}{C}H_3$. Take two of these models and hold them with the two CH_3 groups away from you. Place one Cl on the right of C-2 and the other Cl on the left of C-3 in one model. In the other model place one Cl on the left of C-2 and the other Cl on the right of C-3. Now test to see if these models are superimposable or enantiomers. Draw projection formulas for these structures.

(b) Take the other two models and arrange them as follows: with the two CH_3 groups away from you place one Cl on the right of C-2 and the other Cl on the right of C-3 in one model. In the other model, place both Cl groups on the left of C-2 and C-3. Test these two models to see if they are superimposable or enantiomers. If they are superimposable, draw one projection structure; if they are enantiomers, draw both projection structures as mirror image isomers.

Question 3.

(a) How many chiral carbon atoms are in 2,3-dichlorobutane? _____

(b) How many stereoisomers of 2,3-dichlorobutane did you find? _____

(c) Calculate the maximum number of stereoisomers possible for 2,3-dichlorobutane.

(d) Why is the number of isomers in Q3(c) different from that in Q3(b)?

In procedures D.(a) and (b) you found three stereoisomers of 2,3-dichlorobutane. The formula 2^n states that the maximum number of stereoisomers for a compound with two chiral carbon atoms is four. Why, then, are there only three isomers? The models constructed in Procedure D.(b) are superimposable. Although these models contain chiral carbon atoms, the molecule is symmetrical about a horizontal plane passing through its center, between C-2 and C-3. Thus the molecule shows no optical activity. Stereoisomers that contain chiral carbon atoms and are superimposable on their own mirror images are called *meso compounds*.

Question 4. Draw projection formulas for (a) $CH_3CHBrCHBrCl$ and (b) $CH_2BrCHBrCHBrCH_2Br$. Label all pairs of enantiomers and meso compounds.

Question 5. Which of the structures shown are enantiomers of Compound **A** and which structures are the same as Compound **A**? For comparison purposes, projection formulas may be rotated 180°; may not be rotated 90°; and may not be lifted out of the plane of the paper.

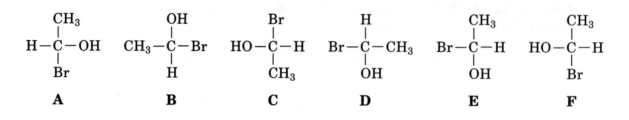

APPENDIX 1

Suggested List of Equipment

Equipment for Student Lockers

1. 5 Beakers: 50, 100, 150, 250, 400 mL
2. 1 Burner, tirrill (optional)
3. 1 Ceramfab pad
4. 1 Clay triangle
5. 2 Crucibles, size 0
6. 2 Crucible covers, size F
7. 1 Crucible tongs
8. 1 Evaporating dish, size 1
9. 1 File, triangular
10. 1 Filter paper (box)
11. 2 Flasks, Erlenmeyer, 125 mL
12. 2 Flasks, Erlenmeyer, 250 mL
13. 1 Flask, florence, 500 mL
14. 5 Glass plates, 3 × 3 in.
15. 1 Graduated cylinder, 10 mL
16. 1 Graduated cylinder, 50 mL
17. 2 Litmus paper (vials), red and blue
18. 2 Medicine droppers/disposable pipets
19. 1 Pipet, volumetric, 10 mL
20. 8 Rubber stoppers: 3 No. 1, solid; 1 No. 1, 1-hole; 1 No. 4, 1-hole; 1 No. 4, 2-hole; 1 No. 5, solid; 1 No. 6, 2-hole
21. 2 Rubber tubing (about 25 cm), 3/16 in. diameter
22. 1 Screw clamp
23. 1 Spatula
24. 1 Sponge
25. 12 Test tubes, 18 × 150 mm (or culture tubes)
26. 1 Test tube, ignition, 25 × 200 mm
27. 1 Test tube brush
28. 1 Test tube holder, wire
29. 1 Test tube rack
30. 1 Thermometer, 110°C
31. 1 Thistle top, plastic
32. 1 Utility clamp (single buret clamp)
33. 1 Wash bottle (plastic)
34. 2 Watch glasses, 4 in.
35. 5 Wide-mouth bottles, 8 oz.
36. 1 Wing top
37. 1 Wire gauze

Auxillary Equipment Not Supplied in Student Lockers

1. Aluminum foil (7 × 7 cm)
2. Balances
3. Beakers, 600 mL
4. Boyle's law apparatus
5. Buchner funnels, suction flasks, and suction rubber tubing
6. Burets, 25 or 50 mL
7. Buret clamps
8. Burner's tirrill (if not individually supplied)
9. Capillary tubes (sealed at one end)
10. Centrifuges
11. Centrifuge tubes
12. Chromatography columns (polypropylene from Kontes)
13. Deflagration spoons
14. Erlenmeyer flasks, 500 mL
15. Filter paper, Whatman #1 (14 × 14 cm)
16. Glass rod, 5 or 6 mm
17. Glass tubing, 6 mm
18. Glass wool (pyrex)
19. Glass writing markers
20. Hair dryers
21. Magnetic stirrers with bars
22. Metric rulers
23. Oil baths
24. pH meters
25. Pipets, graduated, 1 mL, 5 mL, and 10 mL
26. Pipets, micro
27. Pipets, Pasteur
28. Pneumatic troughs
29. Protective rubber gloves
30. Reflux and distillation equipment
    100 mL or 200 mL round-bottom distilling flasks
    Distillation take-off heads
    Condensers
    200° or 250° thermometers
    250 mL separatory funnels
31. Ring stands
32. Ring supports, 4 to 5 in. diameter
33. Rubber bands cut from 3/16 inch rubber tubing
34. Spectrophotometers
35. Spray applicators
36. Suction bulbs for pipets
37. Wire stirrers for oil and water baths

APPENDIX 2

List of Reagents Required and Preparation of Solutions

Solids

Acetamide, C_2H_5NO

Ammonium chloride, NH_4Cl

Barium chloride, $BaCl_2 \cdot 2\,H_2O$

Barium sulfate, $BaSO_4$

Benzoic acid, C_6H_5COOH

Benzophenone, $C_6H_5-CO-C_6H_5$

Benzoyl peroxide, $(C_6H_5COO)_2$

Boiling stones

Candles

Calcium carbide, CaC_2 (small lumps)

Calcium hydroxide, $Ca(OH)_2$

Calcium oxide, CaO

Cholesterol, $C_{27}H_{45}OH$

trans-Cinnamic acid, $C_9H_8O_2$

Cobalt chloride paper

Copper strips, Cu

Copper wire, #18, Cu

Copper(II) sulfate pentahydrate, $CuSO_4 \cdot 5\,H_2O$

Cotton

Diphenylacetic acid, $C_{14}H_{12}O_2$

Diphenylacetic acid-cholesterol, 50:50

Glass wool, pyrex

Glucose, $C_6H_{12}O_6$

Ice

Iron wire (20–24 gauge), Fe

Lead strips, Pb

Lead(II) chromate, $PbCrO_4$

Magnesium strips, Mg

Magnesium oxide, MgO

Magnesium sulfate, anhydrous, $MgSO_4$

Manganese dioxide, MnO_2

Marble chips, $CaCO_3$

Marbles, about 20 mm diameter

Menthol, $C_{10}H_{20}O$

1-Naphthol, $C_{10}H_8O$

4-Nitroaniline, $C_6H_6N_2O_2$

Potassium acid phthalate, $KHC_8H_4O_4$

Potassium bisulfate, $KHSO_4$

Potassium chlorate, C.P., $KClO_3$

Potassium chloride, C.P., KCl

Potassium chromate, C.P., K_2CrO_4

Potassium nitrate, KNO_3

Potato (fresh)

Sand paper or emery cloth

Salicylic acid, $C_6H_4(COOH)(OH)$

Sodium, Na

Sodium bicarbonate, $NaHCO_3$

Sodium chloride (coarse crystals), NaCl

Sodium chloride (fine crystals), NaCl

Sodium nitrite, $NaNO_2$

Sodium peroxide, Na_2O_2

Sodium sulfate, Na_2SO_4

Sodium sulfite, Na_2SO_3

Stearic acid, $CH_3(CH_2)_{16}COOH$

Steel wool, Fe (Grade 0 or 1)

Appendix 2 (continued)

Sucrose, $C_{12}H_{22}O_{11}$

Sulfur, S

Tyrosine

Urea, $(NH_2)_2CO$

Urea-trans-cinnamic acid, 50:50

Vegetable shortening

Wood splints

Zinc, mossy, Zn

Zinc strips, Zn (0.01 inch thick)

Pure Liquids/Commercial Mixtures

Acetic acid (glacial), CH_3COOH

Acetic anhydride, $(CH_3CO)_2O$

Acetone, CH_3COCH_3

Aniline, C_6H_7N

Benzylamine, C_7H_9N

n-Butyl alcohol (1-butanol), C_4H_9OH

Chloroform, $CHCl_3$

Decane, $C_{10}H_{22}$

Diethylamine, $C_4H_{11}N$

Ethyl alcohol (ethanol), 95%, C_2H_5OH

Glycerol, $C_3H_5(OH)_3$

Heptane (or low boiling petroleum ether), C_7H_{16}

Hexane, C_6H_{14}

n-Hexylamine, $C_6H_{13}NH_2$

Isoamyl alcohol (3-methyl-1-butanol), $C_5H_{11}OH$

Isopropyl alcohol (2-propanol) C_3H_7OH

Kerosene (alkene free)

Methyl alcohol (methanol), CH_3OH

Methyl methacrylate, $CH_2=CH(CH_3)COOCH_3$

Oleic acid, $CH_3(CH_2)_7CH=CH(CH_2)_7COOH$

Pentene (amylene), C_5H_{10}

Pyridine, C_5H_5N

Red wine

Sulfuric acid, conc., H_2SO_4

Toluene, C_7H_8

1,1,1-Trichloroethane, CCl_3CH_3

Vegetable oils (corn, cottonseed, peanut, soybean, etc.)

Solutions

All solutions, except where otherwise directed, are prepared by dissolving the designated quantity of solute in distilled water and diluting to 1 liter.

Acetic acid, concentrated (glacial), concentrated reagent $HC_2H_3O_2$

Acetic acid, dilute, 6 M; 355 mL concentrated $HC_2H_3O_2$/liter

Acetic acid-1-butanol-water (1:3:1 by volume); 200 mL CH_3COOH + 600 mL C_4H_9OH/liter

Adipoyl chloride, 0.4 M in cyclohexane; 18.3 g adipoyl chloride/250 mL cyclohexane

Alanine, 0.2 M; 1.78 g alanine/100 mL

Alanine · HCl, 0.1 M; 1.26 g alanine · HCl/100 mL

Alanine-aspartic acid-leucine-lysine solution (each 0.2 M); 1.78 g alanine + 2.66 g aspartic acid + 2.62 g leucine + 2.92 g lysine/100 mL

Albumin, 2%; 20 g albumin/liter (Make slurry with about 50 mL water, then add additional water slowly while stirring.)

Aluminum chloride, 0.10 M; 24.1 g $AlCl_3 \cdot 6\ H_2O$/liter

Ammonium chloride, 0.1 M; 5.4 g NH_4Cl/liter

Ammonium chloride, saturated; 60 g NH_4Cl/liter

Ammonium hydroxide, concentrated; concentrated reagent

Ammonium hydroxide, dilute, 6 M; 400 mL concentrated/liter

Arabinose, 1%; 10 g arabinose/liter

Arginine-tyrosine solution (each 0.1%); 100 mg arginine + 100 mg tyrosine/100 mL pH 6.0 phosphate buffer

Arsenomolybdate reagent; Nelson's arsenomolybdate reagent is commercially available from Sigma Chemical Co., St. Louis, Missouri

Aspartic acid, 0.2 M; 2.66 g aspartic acid/100 mL

Barfoed reagent; dissolve 13.3 g $Cu(C_2H_3O_2)_2 \cdot H_2O$ in 200 mL H_2O. (Filter if necessary), add 1.8 mL $HC_2H_3O_2$ (glacial). Copper(II) acetate is slow to dissolve.

Barium chloride, 0.10 M; 24.4 g $BaCl_2 \cdot 2\ H_2O$/liter

Barium hydroxide, saturated; 10 g $Ba(OH)_2 \cdot 8\ H_2O$/100 mL

Barium hydroxide, 0.2 M; 15.8 g $Ba(OH)_2 \cdot 8\ H_2O$/250 mL

Benedict reagent; dissolve 86.5 g sodium citrate and 50 g anhydrous Na_2CO_3 in 400 mL water with heating. Dissolve 8.65 g $CuSO_4 \cdot 5\ H_2O$ in 50 mL water. Mix these two solutions slowly and add water to produce 500 mL of solution.

Bial reagent; dissolve 1.5 g orcinol (5-methyl resorcinol) in 500 mL conc. HCl and add 1.5 mL of 10% aqueous $FeCl_3$

Blood, simulated blood or aseptic blood (Ward's Scientific or Carolina Biological Supply, respectively)

Bromine in 1,1,1-trichloroethane, 5% solution; 2.5 mL Br_2 plus 100 mL CCl_3CH_3

Buffer solution, standard pH 7.0; commercially available

Calcium chloride, 0.1 M; 14.7 g $CaCl_2 \cdot 2\ H_2O$/liter

Chlorine water; dilute 150 mL of 5.25% NaOCl (household bleach) to 1 liter. Add 15 mL concentrated HCl and mix gently.

Cholesterol standard solution; $C_{27}H_{45}OH$ (dissolve 120 mg cholesterol in 60 mL isopropyl alcohol)

Cobalt(II) chloride, 0.1 M; 23.8 g $CoCl_2 \cdot 6\ H_2O$/liter

Copper(II) nitrate, 0.1 M; 24.2 g $Cu(NO_3)_2 \cdot 3\ H_2O$/liter

Copper reagent; dissolve 24 g of anhydrous Na_2CO_3, 16 g of sodium potassium tartrate, 4 g of $CuSO_4 \cdot 5\ H_2O$, and 180 g of anhydrous $NaSO_4$ in water and dilute to 1 liter.

Copper(II) sulfate, 0.1 M; 25.0 g $CuSO_4 \cdot 5\ H_2O$/liter

Copper(II) sulfate, 0.2 M; 50.0 g $CuSO_4 \cdot 5\ H_2O$/liter

Cysteine $\cdot$ HCl, 0.10 M; 1.56 g cysteine $\cdot$ HCl/100 mL

1,6-Diaminohexane (Hexamethylenediamine), 0.40 M in 0.40 M NaOH; dissolve 11.6 g $(CH_2)_6(NH_2)_2$/250 mL 0.4 M NaOH

Dichlorofluorescein; dissolve 0.25 g of 2,7-dichlorofluorescein in 250 ml of 70% ethanol

Digitonin solution; dissolve 1.0 g digitonin in 50 mL ethanol

Dowex-50 slurry; The resin must be in the H^+ form. Newly purchased resin or used resin is washed sequentially with deionized water, acetone (use only for the new resin), 1 M NaOH, deionized water, 1 M HCl, and deionized water. The washed resin is suspended in the pH 6 phosphate buffer two or three times or until the slurry has a pH of 6. Store the slurry in the cold.

Ethanol-Acetone (1:1 by volume); 500 mL CH_3CH_2OH + 500 mL CH_3COCH_3

Ethanol, denatured, CH_3CH_2OH

Ethanol/water, 500 mL CH_3CH_2OH + 500 mL H_2O

Food colors, blue, green, and yellow; commercially available products

Formaldehyde, 10% solution; 25 mL formalin (40%) plus 75 mL H_2O

Fructose, 1% solution; 1.0 g fructose/100 mL

Fruit juices; orange, lemon, lime, grapefruit, apple, etc. (fresh if available)

Gelatin, 2% solution; 10 g gelatin/500 mL (Dissolves slowly)

Glucose, 1% solution; 5 g $C_6H_{12}O_6$/500 mL H_2O

Glucose, 10% solution; 10 g $C_6H_{12}O_6$ plus 90 mL H_2O

Glucose, standard solutions; made up to contain 2.0, 5.0, 8.0, 12.0, 15.0, and 18 mg glucose/100 mL solution

Glutamic acid $\cdot$ HCl, 0.10 M; 1.84 g glutamic acid $\cdot$ HCl/100 mL

Glycine $\cdot$ HCl, 0.10 M; 1.40 g glycine $\cdot$ HCl/100 mL

Glycine, 1% solution; 1.0 g glycine/100 mL

Histidine $\cdot$ HCl, 0.10 M; 2.10 g histidine $\cdot$ HCl/100 mL

Hydrochloric acid, concentrated, concentrated reagent, HCl

Hydrochloric acid, dilute, 6 M; 500 mL concentrated HCl/liter

Hydrochloric acid, 0.1 M; 8.33 mL concentrated acid HCl/liter; (or dilute 10 mL of 6 M HCl to 600 mL)

Hydrochloric acid, 0.01 M; 0.83 mL concentrated acid HCl/liter; (or dilute 10 mL of 0.1 M HCl to 100 mL)

Hydrochloric acid, 0.001 M; 0.083 mL concentrated acid HCl/liter; (or dilute 10 mL 0.1 M to 1000 mL)

Hydrogen peroxide, 3%; reagent solution or 100 mL 30% H_2O_2/liter

Hydrogen peroxide, 9%; 300 mL 30% H_2O_2/liter; store cold

Hydrogen peroxide, 30% (for dilution); store cold

Iodine in potassium iodide, 1%; 10 g I_2 + 20 g KI/liter (Dissolve I_2 and KI in about 50 mL H_2O, then dilute to 1 liter)

Iodine water, saturated; 5 g I_2/liter

Iron(III) chloride, 0. 1 M; 27.1 g $FeCl_3 \cdot 6\ H_2O$ + 5 mL concentrated HCl/liter

Iron reagent; 2.5 g $FeCl_3 \cdot 6\ H_2O$/100 mL phosphoric acid

Isopropyl alcohol-water (2:1 by volume); 667 mL $CH_3CH(OH)CH_3$/liter

Lead(II) acetate, 0.1 M; 3.25 g $Pb(C_2H_3O_2)_2$/100 mL

Lead(II) nitrate, 0.1 M; 33.1 g $Pb(NO_3)_2$/liter

Lead(II) nitrate, 0.50 M; 165.6 g $Pb(NO_3)_2$/liter (for Experiment 15 only)

Leucine, 0.2 M; 2.62 g leucine/100 mL

Lysine, 0.2 M; 2.92 g lysine/100 mL

Lysine $\cdot$ HCl, 0.1 M; 1.83 g lysine $\cdot$ HCl/100 mL

Magnesium sulfate, 0.1 M; 24.6 g $MgSO_4 \cdot 7\ H_2O$/liter

Maltose, 1% solution; 5.0 g maltose/500 mL

Mercury(I) nitrate, 0.1 M; dissolve 26.1 g $HgNO_3 \cdot H_2O$ in 50 mL of concentrated HNO_3 and slowly dilute with water to 1 liter (Prepare in a well-ventilated hood.)

Milk, fat free (skim)

Molisch reagent; dissolve 2.5 g α-naphthol in 50 mL 95% C_2H_5OH

Nickel nitrate, 0.1 M; 29.1 g $Ni(NO_3)_2 \cdot 6\ H_2O$/liter

Ninhydrin, 0.3%; 1.5 g ninhydrin/500 mL acetone

Ninhydrin, 0.2%; dissolve 0.2 g of ninhydrin in 100 mL of 1-butanol which is saturated with water

Nitric acid, concentrated; concentrated reagent, HNO_3

Nitric acid, dilute, 6 M; 375 mL concentrated HNO_3/liter

1-Nitroso-2-naphthol, 0.1% in acetone, $C_{10}H_7NO_2$; dissolve 0.17 g in 100 mL acetone

Phenol, 1% solution; 5.0 g C_6H_5OH/500 mL

Phenolphthalein, 0.2% solution; dissolve 2 g phenolphthalein in 600 mL ethanol (95%) and dilute with water to 1 liter

Phosphoric acid, 85% reagent, H_3PO_4

Phosphoric acid, dilute, 3 M; 201 mL 85% H_3PO_4 solution/liter

Phosphate buffer, 0.2 M; 34.8 g K_2HPO_4/liter and adjust pH to 6.0 using a pH meter and dilute H_3PO_4 or KOH solution

Potassium chlorate, 0.1 M; 12.2 g $KClO_3$/liter

Potassium chloride, 0.1 M; 7.46 g KCl/liter

Potassium chloride, saturated; 380 g KCl/liter

Potassium chromate, 0.1 M; 19.4 g K_2CrO_4/liter

Potassium nitrate, 0.1 M; 10.1 g KNO_3/liter

Potassium permanganate, 0.1 M; 15.8 g $KMnO_4$/liter

Potassium permanganate, 0.002 M; 0.16 g $KMnO_4$/500 mL H_2O

Potassium thiocyanate, 0.1 M; 9.7 g KSCN/liter

Seliwanoff reagent; dissolve 0.50 g resorcinol in 1000 mL 4 M HCl (333 mL conc. HCl diluted to 1000 mL)

Silver nitrate, 0.10 M; 17.0 g $AgNO_3$/1liter

Sodium arsenate, 0.1 M; 31.2 g $Na_2HAsO_4 \cdot 7 H_2O$ + 4.0 g NaOH/liter

Sodium bicarbonate, 5% solution; 50 g $NaHCO_3$/liter

Sodium bicarbonate, saturated solution; 125 g $NaHCO_3$/liter

Sodium bromide, 0.1 M; 10.3 g NaBr/liter

Sodium carbonate, 0.1 M; 10.6 g Na_2CO_3/liter

Sodium chloride, 0.1 M; 5.85 g NaCl/liter

Sodium chloride, saturated; 60 g NaCl/liter

Sodium hydroxide, 10% solution; 111 g NaOH/liter

Sodium hydroxide, 1% solution; 11.1 g NaOH/liter

Sodium hydroxide, 0.1 M; 4 g NaOH/liter

Sodium iodide, 0.1 M; 15.0 g NaI/liter

Sodium nitrite, 0.1 M; 6.9 g $NaNO_2$/liter

Sodium phosphate, 0.1 M; 38.0 g $Na_3PO_4 \cdot 12 H_2O$/liter

Sodium sulfate, 0.1 M; 14.2 g Na_2SO_4/liter

Starch, 1% solution; 5 g/500 mL (Make slurry and disperse in hot water.)

Sucrose, 1% solution; 10 g sucrose ($C_{12}H_{22}O_{11}$)/liter (freshly prepared)

Sulfuric acid, concentrated; concentrated reagent H_2SO_4

Sulfuric acid, dilute, 9 M; carefully, with stirring, slowly add 500 mL concentrated H_2SO_4 to 400 mL H_2O, cool and dilute to 1 liter

Sulfuric acid, dilute, 3 M; 167 mL concentrated H_2SO_4/liter

Vinegar, commercial (colorless)

Xylose, 1% solution; 1.0 g xylose/100 mL

Zinc nitrate, 0.1 M; 29.8 g $Zn(NO_3)_2 \cdot 6 H_2O$/liter

Zinc sulfate, 0.2 M; 14.4 g $ZnSO_4 \cdot 7 H_2O$/250 mL

APPENDIX 3

Special Equipment or Preparations Needed

Experiment 1.   Laboratory Techniques

A small sample of solid lead(II) chromate and potassium nitrate are needed for comparison purposes only.

Experiment 2.   Measurements

An assortment of metal slugs or other solid objects are needed as unknowns for density determination. The diameter of the slugs should be such that they will fit into the 50 mL graduated cylinder. Suggested materials are aluminum, brass, magnesium, steel, etc. Slugs should be numbered for identification.

Experiment 3.   Preparation and Properties of Oxygen

Three demonstrations are suggested (see experiment for details).

Experiment 4.   Preparation and Properties of Hydrogen

For safety: Instructor should dispense sodium metal (size of pieces should be no larger than a 4 mm cube).

Experiment 5:   Calorimetry and Specific Heat

An assortment of metal objects like those used for the density determinations in Exp. 2 are needed. They must be small enough to fit into the test tube with id=22 mm. Styrofoam cups and cardboard cut into 4" squares with a small thermometer hole in the middle should also be available.

Experiment 6.   Freezing Points—Graphing of Data

Slotted corks or stoppers, crushed ice

Experiment 7.   Water in Hydrates

An assortment of samples for unknowns for determination of percent water is needed. Samples can be issued in small coin envelopes or plastic vials. See the Instructor's Manual for the suggested list of samples.

Appendix 3 (continued)

Experiment 12.   Ionization—Acids, Bases, and Salts

Conductivity apparatus is needed for the demonstration. The procedure is based on the apparatus described in the experiment but other types may be used without detracting from the results of the demonstration. A magnetic stirrer greatly facilitates the last part of the demonstrations. Two or three pH meters are recommended for student use, set up at stations with the solutions described in the experiment.

Experiment 13.   Identification of Selective Anions

Two unknown solutions (in test tubes) are to be issued to each student. Stock reagents used in the experiment are satisfactory for unknowns. See Instructor's Manual for details.

Experiment 14.   Properties of Lead(II), Silver, and Mercury(I) Ions

An unknown solution containing one or more of the silver group cations is to be issued to each student. See Instructor's Manual for details.

Experiment 15.   Quantitative Precipitation of Chromate Ion

Lead(II) nitrate solution is 0.50 M for this experiment.

Experiment 16.   Electromagnetic Energy and Spectroscopy

Hand-held spectroscopes, 1.75 m springs for simulating wave motion (1 per 5 students), vapor lamps with power supplies (2 hydrogen and 2 neon); spectrum chart, incandescent and fluorescent lights, spectrophotometers with range from 350–700 nm, colored pencils, meter sticks, stopwatches (recommended).

Experiment 17.   Lewis Structures and Molecular Models

Ball-and-stick molecular model sets. Two students can share one kit. The number of sets required depends on how many labs are run simultaneously. It is also possible to purchase a large class set of components and divide them into smaller custom kits.

Experiment 18.   Boyle's Law

Boyle's law apparatus is needed. The kits for this experiment can be purchased from several vendors as "Simple Form Boyle's Law Apparatus" or "Elasticity of Gases Kit." The kits include the silicone grease but not the applied weights and vernier calipers. Slotted masses of 0.5 kg and 1 kg allow the applied weights to lie flat on the platform. If not enough slotted masses are available, a combination of bricks and slotted masses works well. One balance per laboratory with the capacity for weighing the heaviest mass to three significant digits. All masses can be preweighed and labeled with tape displaying their mass.

Experiment 20. Molar Volume of a Gas

A 3.0 cc or 5.0 cc disposable syringe is needed for each setup. Needle–rubber stopper assemblies that contain a rubber stopper and syringe needle should be preassembled and checked out and in by students. An additional safety feature is to snip off the end of the needle with a wire cutter after it is in the stopper. The needles need to be heavy enough to push through a rubber stopper without bending. 2 L beakers or battery jars.

Experiment 21. Neutralization—Titration I

The following are needed by each student: A small vial or test tube containing about 4 grams of potassium hydrogen phthalate (KHP) (these vials are collected for reuse), one 25 or 50 mL buret, a buret clamp, and 250 mL of unknown NaOH solution. The NaOH solution is used in Experiments 21 and 22. See Instructor's Manual for details.

Experiment 22. Neutralization—Titration II

The following are needed by each student: A 10 mL volumetric pipet, one 25 or 50 mL buret, 50 mL of unknown acid solution, 50 mL of vinegar, and 125 mL of standard NaOH solution if Experiment 21 is not done. See Instructor's Manual for details.

Experiment 23. Chloride Content of Salts

An unknown solution for the determination of chloride molarity is to be issued to each student. A solution of dichlorofluorescein indicator is needed. See Instructor's Manual for details.

Experiment 25. Heat of Reaction

Styrofoam cups are needed.

Experiment 26. Distillation of Volatile Liquids

Distillation setup using a 125 mL or 250 mL flask (see Figure 26.1); hot plates or heating mantles (with rheostats). Red wine as the alcoholic beverage for distillation.

Experiment 27. Boiling Points and Melting Points

Boiling point and melting point apparatus (see experiment for details), 200°–250°C thermometers, wire stirrers, capillary melting point tubes, and unknown solids are required. See Instructor's Manual for details.

Experiment 28. Hydrocarbons

Lumps of calcium carbide are needed. Test kerosene to see if it is free of alkenes. Toluene is *not* reacted with bromine.

Experiment 29.   Alcohols, Esters, Aldehydes, and Ketones

Furnish No. 18 copper wire with five or six spiral turns at one end. Wire should be about 20 cm overall in length.

Experiment 30.   Esterification—Distillation: Synthesis of n-Butyl Acetate

This is a two laboratory period experiment. Reflux and distillation equipment (see experiment for details), 200°–250°C thermometer, and 250 mL separatory funnel are required.

Experiment 31.   Synthesis of Aspirin

Buchner funnel, suction flask, suction tubing, melting point apparatus, and capillary melting point tubes are needed.

Experiment 32.   Amines and Amides

An ice bath is needed to cool the dye reaction.

Experiment 33.   Polymers—Macromolecules

Benzoyl peroxide is a shock and heat sensitive material. It should be dispensed to each student by the instructor or by qualified stock room personnel.

Experiment 34.   Carbohydrates

Pure fresh of frozen fruit juices, such as orange, lemon, lime, grapefruit, and apple are needed.

Experiment 35.   Glucose Concentration in Aseptic or Simulated Blood

Spectrophotometer, 10 mL graduated pipets, 20 mm diameter marbles, protective gloves, and aseptic or simulated blood are needed.

Experiment 37.   Paper Chromatography

Five hundred mL Erlenmeyer flasks, 14 × 14 cm squares Whatman No. 1 filter paper, 7 × 7 cm squares of Al foil, micropipets and a spray applicator for ninhydrin are needed. An unknown amino acid or amino acid mixture is required for each student.

Experiment 38.   Ion-Exchange Chromatography

Chromatography columns (see experiment for details), Dowex-50 resin slurry, and 600 mL beakers must be provided.

Experiment 39.   Identification of an Unknown Amino Acid by Titration

pH meter, pH 7.0 buffer, and magnetic stirrer are needed. An unknown amino acid is issued to each student or student pair.

Experiment 40.   Enzymatic Catalysis—Catalase

A potato is needed for each student or student pair.

Experiment 42.   Cholesterol Levels in Aseptic or Simulated Blood

Aseptic or simulated blood, spectrophotometer, centrifuge, centrifuge tubes, 1 mL and 5 mL graduated pipets, and protective gloves are needed.

APPENDIX 4

Units of Measurements

Numerical Value of Prefixes with Units

Prefix	Symbol	Number	Power of 10
mega	M	1,000,000	1×10^6
kilo	k	1,000	1×10^3
hecto	h	100	1×10^2
deca	da	10	1×10^1
deci	d	0.1	1×10^{-1}
centi	c	0.01	1×10^{-2}
milli	m	0.001	1×10^{-3}
micro	μ	0.000001	1×10^{-6}
nano	n	0.000000001	1×10^{-9}

Conversion of Units

1 m	=	1000	mm
1 cm	=	10	mm
2.54 cm	=	1	in.
453.6 g	=	1	lb
1 kg	=	2.2	lb
1 g	=	1000	mg
1 L	=	1000	mL
1 mL	=	1	cm^3
0.946 L	=	1	qt
1 cal	=	4.184	J
1 Torr	=	1	mm Hg
760 torr	=	1	atm

Metric Abbreviations

meter	m
centimeter	cm
millimeter	mm
nanometer	nm
liter	L
milliliter	mL
kilogram	kg
gram	g
milligram	mg
mole	mol

Temperature Conversion Formulas

$$°C = \frac{(°F - 32)}{1.8}$$

$$°F = 1.8°C + 32$$

$$K = °C + 273$$

APPENDIX 5

Solubility Table

	$C_2H_3O_2^-$	AsO_4^{3-}	Br^-	CO_3^{2-}	Cl^-	CrO_4^{2-}	OH^-	I^-	NO_3^-	$C_2O_4^{2-}$	O^{2-}	PO_4^{3-}	SO_4^{2-}	S^{2-}	SO_3^{2-}
Al^{3+}	aq	I	aq	–	aq	–	I	aq	aq	–	I	I	aq	d	–
NH_4^+	aq	aq	aq	aq	aq	aq	aq	aq	aq	aq	–	aq	aq	aq	aq
Ba^{2+}	aq	I	aq	I	aq	I	sl.aq	aq	aq	I	sl.aq	I	I	d	I
Bi^{3+}	–	sl.aq	d	I	d	–	I	I	d	I	I	sl.aq	d	I	–
Ca^{2+}	aq	I	aq	I	aq	aq	I	aq	aq	I	I	I	I	d	I
Co^{2+}	aq	I	aq	I	aq	I	I	aq	aq	I	I	I	aq	I	I
Cu^{2+}	aq	I	aq	I	aq	I	I	–	aq	I	I	I	aq	I	–
Fe^{2+}	aq	I	aq	sl.aq	aq	–	I	aq	aq	I	I	I	aq	I	aq
Fe^{3+}	I	I	aq	I	aq	–	I	–	aq	aq	I	I	aq	I	–
Pb^{2+}	aq	I	I	I	I	I	I	I	aq	I	I	I	I	I	I
Mg^{2+}	aq	d	aq	I	aq	aq	I	aq	aq	I	I	I	aq	d	sl.aq
Hg^{2+}	aq	I	I	I	aq	sl.aq	I	I	aq	I	I	I	d	I	–
K^+	aq	aq	aq	aq	aq	aq	aq	aq	aq	aq	aq	aq	aq	aq	aq
Aq^+	sl.aq	I	I	I	I	I	–	I	aq	I	I	I	I	I	I
Na^+	aq	aq	aq	aq	aq	aq	aq	aq	aq	aq	aq	aq	aq	aq	aq
Zn^{2+}	aq	I	aq	I	aq	I	I	aq	aq	I	I	I	aq	I	I

Key: aq = Soluble in water   I = Insoluble in water (less than 1 g/100 g H_2O)

sl.aq = Slightly soluble in water   d = Decomposes in water

APPENDIX 6

Vapor Pressure of Water

Temperature (°C)	Vapor Pressure torr (or mm Hg)	Temperature (°C)	Vapor Pressure torr (or mm Hg)
0	4.6	26	25.2
5	6.5	27	26.7
10	9.2	28	28.3
15	12.8	29	30.0
16	13.6	30	31.8
17	14.5	40	55.3
18	15.5	50	92.5
19	16.5	60	149.4
20	17.5	70	233.7
21	18.6	80	355.1
22	19.8	90	525.8
23	21.2	100	760.0
24	22.4	110	1074.6
25	23.8		

APPENDIX 7

Boiling Points of Liquids

Liquid	Boiling Point °C
Acetone	56.5
Ethanol	78.4
Diethyl ether	34.6
Methanol	64.7
1-propanol	82.5
Water	100.0

APPENDIX 8

Waste Disposal Requirements for Each Experiment

Listed below are special waste containers specified in the experiments for student disposal of waste. Where students are instructed to dispose of wastes in the sink, or where the experiment does not generate waste, the requirements are listed as NONE.

We use the same Waste Heavy Metal bottle for many experiments by combining all the ions poured into it on the label. The same can be done for Organic Solvent Waste bottles.

Exp	Title	Waste Containers That Should Be Available to Students	
1	Laboratory Techniques	Waste Heavy Metals (Pb^+ and CrO_4^{2-}) Waste $PbCrO_4$ on filter paper	bottle jar
2	Measurements	None	
3	Prep. and Prop. of Oxygen	Recycled 9% H_2O_2, unreacted Unreacted metal strips	bottle jar
4	Prep. and Prop. of Hydrogen	Recycled Mossy Zinc, rinsed	jar
5	Calorimetery and Specific Heat	None	
6	Freezing Points	Waste Acetic/Benzoic Acid Mixture	bottle
7	Water in Hydrates	Waste Heavy Metal Residues (Cu^{2+}, Zn^{2+}, Sr^{2+}, Ba^{2+})	jar
8	Properties of Solutions	Waste Organic Solvent (decane) Waste Kerosene Mixtures Waste Heavy Metal Solutions (Ba^{2+})	bottle bottle bottle
9	Composition of Potassium Chlorate	Waste Heavy Metals (Ag^+) Waste $KClO_3$	bottle bottle
10	Double Displacement Reactions	Waste Heavy Metals (Ag^+, Ba^{2+}, Cu^{2+}, Zn^{2+})	bottle
11	Single Displacement Reactions	Waste Heavy Metals (Ag^+, Ba^{2+}, Cu^{2+}, Pb^{2+}, Zn^{2+})	bottle

Exp	Title	Waste Containers That Should Be Available to Students	
12	Ionization—Acids, Bases, and Salts	None (Students do not handle the heavy metal solutions in the demonstration.)	
13	Identification of Selected Anions	Waste Arsenic Compounds Waste Organic Solvents (decane) Waste Heavy Metals (Ag^+, Ba^{2+})	bottle bottle bottle
14	Properties of Pb^{2+}, Ag^+, Hg^+	Waste Heavy Metals (Pb^{2+}, Ag^+, Hg^+, CrO_4^{2-})	bottle
15	Quantitative Precipitation of CrO_4^{2-}	Waste Heavy Metals (Pb^{2+}, CrO_4^{2-}) Waste $PbCrO_4$ residue on filter paper	bottle jar
16	EM Energy and Spectroscopy	Waste Heavy Metals (Ni^{2+}, MnO_4^-)	bottle
17	Lewis Structures /Molecular Models	None	
18	Boyle's Law	None	
19	Charles' Law	None	
20	Molar Volume of a Gas	None	
21	Neutralization—Titration I	None	
22	Neutralization—Titration II	None	
23	Chloride Content of Salts	Waste Heavy Metals (Ag^+)	bottle
24	Chemical Equilibrium	Waste Heavy Metals (Ag^+, Co^{2+}, CrO_4^{2-})	bottle
25	Heat of Reaction	None	
26	Distillation of Volatile Liquids	Recycled Ethanol/Ethanol Distillate	bottle
27	Boiling Points and Melting Points	Waste Organic Solvents Used melting point tubes	bottle jar
28	Hydrocarbons	Waste Organic Solvents	bottle
29	Alcohols, Esters, Aldehydes, Ketones	Waste Organic Solvents Waste Heavy metals (MnO_4^-, Ag^+)	bottle bottle

Exp	Title	Waste Containers That Should Be Available to Students	
30	Esterification—Distillation	Waste Organic Solvents Solid wastes	bottle jar
31	Synthesis of Aspirin	none	
32	Amines and Amides	Waste Organic Solvents	bottle
33	Polymers—Macromolecules	Waste Organic solvents Solid waste (nylon and Lucite)	bottle jar
34	Carbohydrates	Molisch test Seliwanoff test Benedict and Barfoed tests Bial test Dehydration (carbon)	bottle bottle bottle bottle jar
35	Glucose Concentration in Blood	Filter paper waste Arsenic waste	jar bottle
36	Amino Acids and Proteins	Biuret test (Cu^{2+}) Tyrosine and Ninhydrin tests Waste heavy metals (Pb) Solid wastes	bottle bottle bottle jar
37	Paper Chromatography	Waste Organic Solvents Solid waste	bottle jar
38	Ion-Exchange Chromatography	Waste Organic Solvents Used resin (for recycling)	bottle jar
39	Unknown Amino Acid by Titration	none	
40	Enzyme Catalysis—Catalase	none	
41	Lipids	Waste Organic Solvents Solid wastes	bottle jar
42	Cholesterol Level in Blood	RBC solids	jar

Periodic Table of the Elements

Current ACS and IUPAC Preferred U.S.

Atomic masses are based on carbon-12. Elements marked with † have no stable isotopes. The atomic mass given is that of the isotope with the longest known half-life.

Key:
Atomic number → 11
Symbol → **Na**
Name → Sodium
Atomic mass → 22.99

Legend: Metals | Metalloids | Nonmetals

Group 1 IA	2 IIA	3 IIIB	4 IVB	5 VB	6 VIB	7 VIIB	8 VIII	9 VIII	10	11 IB	12 IIB	13 IIIA	14 IVA	15 VA	16 VIA	17 VIIA	18 0 (Noble Gases)
1 H Hydrogen 1.008																	**2 He** Helium 4.003
3 Li Lithium 6.941	**4 Be** Beryllium 9.012											**5 B** Boron 10.81	**6 C** Carbon 12.01	**7 N** Nitrogen 14.01	**8 O** Oxygen 16.00	**9 F** Fluorine 19.00	**10 Ne** Neon 20.18
11 Na Sodium 22.99	**12 Mg** Magnesium 24.31											**13 Al** Aluminum 26.98	**14 Si** Silicon 28.09	**15 P** Phosphorus 30.97	**16 S** Sulfur 32.07	**17 Cl** Chlorine 35.45	**18 Ar** Argon 39.95
19 K Potassium 39.10	**20 Ca** Calcium 40.08	**21 Sc** Scandium 44.96	**22 Ti** Titanium 47.87	**23 V** Vanadium 50.94	**24 Cr** Chromium 52.00	**25 Mn** Manganese 54.94	**26 Fe** Iron 55.85	**27 Co** Cobalt 58.93	**28 Ni** Nickel 58.69	**29 Cu** Copper 63.55	**30 Zn** Zinc 65.39	**31 Ga** Gallium 69.72	**32 Ge** Germanium 72.61	**33 As** Arsenic 74.92	**34 Se** Selenium 78.96	**35 Br** Bromine 79.90	**36 Kr** Krypton 83.80
37 Rb Rubidium 85.47	**38 Sr** Strontium 87.62	**39 Y** Yttrium 88.91	**40 Zr** Zirconium 91.22	**41 Nb** Niobium 92.91	**42 Mo** Molybdenum 95.94	**43 Tc** Technetium 97.91†	**44 Ru** Ruthenium 101.1	**45 Rh** Rhodium 102.9	**46 Pd** Palladium 106.4	**47 Ag** Silver 107.9	**48 Cd** Cadmium 112.4	**49 In** Indium 114.8	**50 Sn** Tin 118.7	**51 Sb** Antimony 121.8	**52 Te** Tellurium 127.6	**53 I** Iodine 126.9	**54 Xe** Xenon 131.3
55 Cs Cesium 132.9	**56 Ba** Barium 137.3	**57 La*** Lanthanum 138.9	**72 Hf** Hafnium 178.5	**73 Ta** Tantalum 180.9	**74 W** Tungsten 183.8	**75 Re** Rhenium 186.2	**76 Os** Osmium 190.2	**77 Ir** Iridium 192.2	**78 Pt** Platinum 195.1	**79 Au** Gold 197.0	**80 Hg** Mercury 200.6	**81 Tl** Thallium 204.4	**82 Pb** Lead 207.2	**83 Bi** Bismuth 209.0	**84 Po** Polonium 209.0†	**85 At** Astatine 210.0†	**86 Rn** Radon 222.0
87 Fr Francium 223.0†	**88 Ra** Radium 226.0†	**89 Ac**** Actinium 227.0†	**104 Rf** Rutherfordium 261.1†	**105 Db** Dubnium —	**106 Sg** Seaborgium —	**107 Bh** Bohrium —	**108 Hs** Hassium —	**109 Mt** Meitnerium —	**110 Uun** Ununnilium —	**111 Uuu** Unununium —	**112 Uub** Ununbium —						

Transition Elements

Inner Transition Elements

Lanthanide Series * (Period 6)

58 Ce Cerium 140.1	59 Pr Praseodymium 140.9	60 Nd Neodymium 144.2	61 Pm Promethium 144.9†	62 Sm Samarium 150.4	63 Eu Europium 152.0	64 Gd Gadolinium 157.3	65 Tb Terbium 158.9	66 Dy Dysprosium 162.5	67 Ho Holmium 164.9	68 Er Erbium 167.3	69 Tm Thulium 168.9	70 Yb Ytterbium 173.0	71 Lu Lutetium 175.0

Actinide Series ** (Period 7)

90 Th Thorium 232.0	91 Pa Protactinium 231.0	92 U Uranium 238.0	93 Np Neptunium 237.0	94 Pu Plutonium 244.1†	95 Am Americium 243.1†	96 Cm Curium 247.1†	97 Bk Berkelium 247.1†	98 Cf Californium 251.1†	99 Es Einsteinium 252.1†	100 Fm Fermium 257.1†	101 Md Mendelevium 258.1†	102 No Nobelium 259.1†	103 Lr Lawrencium 262.1†

Atomic Masses of the Elements
Based on the IUPAC Table of Atomic Masses

Name	Symbol	Atomic Number	Atomic Mass	Name	Symbol	Atomic Number	Atomic Mass
Actinium*	Ac	89	227.0277	Mercury	Hg	80	200.59
Aluminum	Al	13	26.981538	Molybdenum	Mo	42	95.94
Americium*	Am	95	243.0614	Neodymium	Nd	60	144.24
Antimony	Sb	51	121.760	Neon	Ne	10	20.1797
Argon	Ar	18	39.948	Neptunium*	Np	93	237.0482
Arsenic	As	33	74.92160	Nickel	Ni	28	58.6934
Astatine*	At	85	209.9871	Niobium	Nb	41	92.90638
Barium	Ba	56	137.327	Nitrogen	N	7	14.00674
Berkelium*	Bk	97	247.0703	Nobelium*	No	102	259.1011
Beryllium	Be	4	9.012182	Osmium	Os	76	190.23
Bismuth	Bi	83	208.98038	Oxygen	O	8	15.9994
Bohrium	Bh	107	—	Palladium	Pd	46	106.42
Boron	B	5	10.811	Phosphorus	P	15	30.973762
Bromine	Br	35	79.904	Platinum	Pt	78	195.078
Cadmium	Cd	48	112.411	Plutonium*	Pu	94	244.0642
Calcium	Ca	20	40.078	Polonium*	Po	84	208.9824
Californium*	Cf	98	251.0796	Potassium	K	19	39.0983
Carbon	C	6	12.0107	Praseodymium	Pr	59	140.90765
Cerium	Ce	58	140.116	Promethium*	Pm	61	144.9127
Cesium	Cs	55	132.90545	Protactinium*	Pa	91	231.03588
Chlorine	Cl	17	35.4527	Radium*	Ra	88	226.0254
Chromium	Cr	24	51.9961	Radon*	Rn	86	222.0176
Cobalt	Co	27	58.933200	Rhenium	Re	75	186.207
Copper	Cu	29	63.546	Rhodium	Rh	45	102.90550
Curium*	Cm	96	247.0703	Rubidium	Rb	37	85.4678
Dubnium	Db	105	—	Ruthenium	Ru	44	101.07
Dysprosium	Dy	66	162.50	Rutherfordium	Rf	104	261.1089
Einsteinium*	Es	99	252.0830	Samarium	Sm	62	150.36
Erbium	Er	68	167.26	Scandium	Sc	21	44.955910
Europium	Eu	63	151.964	Seaborgium	Sg	106	—
Fermium*	Fm	100	257.0951	Selenium	Se	34	78.96
Fluorine	F	9	18.9984032	Silicon	Si	14	28.0855
Francium*	Fr	87	233.0197	Silver	Ag	47	107.8682
Gadolinium	Gd	64	157.25	Sodium	Na	11	22.989770
Gallium	Ga	31	69.723	Strontium	Sr	38	87.62
Germanium	Ge	32	72.61	Sulfur	S	16	32.066
Gold	Au	79	196.96655	Tantalum	Ta	73	180.9479
Hafnium	Hf	72	178.49	Technetium*	Tc	43	97.9072
Hassium	Hs	108	—	Tellurium	Te	52	127.60
Helium	He	2	4.002602	Terbium	Tb	65	158.92534
Holmium	Ho	67	164.93032	Thallium	Tl	81	204.3833
Hydrogen	H	1	1.00794	Thorium*	Th	90	232.0381
Indium	In	49	114.818	Thulium	Tm	69	168.93421
Iodine	I	53	126.90447	Tin	Sn	50	118.710
Iridium	Ir	77	192.217	Titanium	Ti	22	47.867
Iron	Fe	26	55.845	Tungsten	W	74	183.84
Krypton	Kr	36	83.80	Ununnilium	Uun	110	—
Lanthanum	La	57	138.9055	Unununium	Uuu	111	—
Lawrencium*	Lr	103	262.110	Ununbium	Uub	112	—
Lead	Pb	82	207.2	Uranium*	U	92	238.0289
Lithium	Li	3	6.941	Vanadium	V	23	50.9415
Lutetium	Lu	71	174.967	Xenon	Xe	54	131.29
Magnesium	Mg	12	24.3050	Ytterbium	Yb	70	173.04
Manganese	Mn	25	54.938049	Yttrium	Y	39	88.90585
Meitnerium	Mt	109	—	Zinc	Zn	30	65.39
Mendelevium*	Md	101	258.0984	Zirconium	Zr	40	91.224

*This element has no stable isotopes. The atomic mass given is that of the isotope with the longest known half-life.

NAMES, FORMULAS AND CHARGES OF COMMON IONS

Positive Ions (Cations)			Negative Ions (Anions)		
1+	Ammonium	NH_4^+	**1−**	Acetate	$C_2H_3O_2^-$
	Copper(I)	Cu^+		Bromate	BrO_3^-
	(Cuprous)			Bromide	Br^-
	Hydrogen	H^+		Chlorate	ClO_3^-
	Potassium	K^+		Chloride	Cl^-
	Silver	Ag^+		Chlorite	ClO_2^-
	Sodium	Na^+		Cyanide	CN^-
2+	Barium	Ba^{2+}		Fluoride	F^-
	Cadmium	Cd^{2+}		Hydride	H^-
	Calcium	Ca^{2+}		Hydrogen carbonate	HCO_3^-
	Cobalt(II)	Co^{2+}		(Bicarbonate)	
	Copper(II)	Cu^{2+}		Hydrogen sulfate	HSO_4^-
	(Cupric)			(Bisulfate)	
	Iron(II)	Fe^{2+}		Hydrogen sulfite	HSO_3^-
	(Ferrous)			(Bisulfite)	
	Lead(II)	Pb^{2+}		Hydroxide	OH^-
	Magnesium	Mg^{2+}		Hypochlorite	ClO^-
	Manganese(II)	Mn^{2+}		Iodate	IO_3^-
	Mercury(II)	Hg^{2+}		Iodide	I^-
	(Mercuric)			Nitrate	NO_3^-
	Nickel(II)	Ni^{2+}		Nitrite	NO_2^-
	Tin(II)	Sn^{2+}		Perchlorate	ClO_4^-
	(Stannous)			Permanganate	MnO_4^-
	Zinc	Zn^{2+}		Thiocyanate	SCN^-
3+	Aluminum	Al^{3+}	**2−**	Carbonate	CO_3^{2-}
	Antimony(III)	Sb^{3+}		Chromate	CrO_4^{2-}
	Arsenic(III)	As^{3+}		Dichromate	$Cr_2O_7^{2-}$
	Bismuth(III)	Bi^{3+}		Oxalate	$C_2O_4^{2-}$
	Chromium(III)	Cr^{3+}		Oxide	O^{2-}
	Iron(III)	Fe^{3+}		Peroxide	O_2^{2-}
	(Ferric)			Silicate	SiO_3^{2-}
	Titanium(III)	Ti^{3+}		Sulfate	SO_4^{2-}
	(Titanous)			Sulfide	S^{2-}
				Sulfite	SO_3^{2-}
4+	Manganese(IV)	Mn^{4+}	**3−**	Arsenate	AsO_4^{3-}
	Tin(IV)	Sn^{4+}		Borate	BO_3^{3-}
	(Stannic)			Phosphate	PO_4^{3-}
	Titanium(IV)	Ti^{4+}		Phosphide	P^{3-}
	(Titanic)			Phosphite	PO_3^{3-}
5+	Antimony(V)	Sb^{5+}			
	Arsenic(V)	As^{5+}			